Avian Subspecies

Ornithological Monographs

Editor: Michael L. Morrison

Department of Wildlife and Fisheries Sciences
210 Nagle Hall, 2258 TAMU
Texas A&M University
College Station, Texas 77843-2258

Managing Editor: Mark C. Penrose
Copy Editor: Richard D. Earles

Editors of this issue: Kevin Winker and Susan M. Haig

Spanish translation of abstracts: Carlos Daniel Cadena,
Cintia Cornelius, and Lucio R. Malizia

The *Ornithological Monographs* series, published by the American Ornithologists' Union, has been established for major papers and presentations too long for inclusion in the Union's journal, *The Auk*.

Back issues of *Ornithological Monographs* from earlier than 2007 are available from Buteo Books at http://www.buteobooks.com. Issues from 2008 forward are available from University of California Press at http://www.ucpressjournals.com. For complete abstracting and indexing coverage for *Ornithological Monographs*, please visit http://www.ucpressjournals.com.

Library of Congress Control Number 2010903695
ISBN: 978-0-943610-86-3
Issued 30 April 2010
Ornithological Monographs, No. 67 viii + 200 pp.

Avian Subspecies

EDITED BY

KEVIN WINKER[1] AND SUSAN M. HAIG[2]

[1]*University of Alaska Museum, 907 Yukon Drive, Fairbanks, Alaska 99775; and*
[2]*U.S. Geological Survey Forest and Rangeland Ecosystem Science Center,*
3200 SW Jefferson Way, Corvallis, Oregon 97331, USA

ORNITHOLOGICAL MONOGRAPHS NO. 67

PUBLISHED BY
THE AMERICAN ORNITHOLOGISTS' UNION
WASHINGTON, D.C.
2010

Cover: Representative members of the *Rhipidura rufifrons* complex: (a) *elegantula*, eastern Lesser Sundas; (b) *dryas*, northern Australia; (c) *squamata*, West Papuan and Banda Is.; (d) *rufifrons*, northeastern Australia; (e) *brunnea*, Malaita, northeastern Solomon Is.; (f) *saipanensis*, Saipan and Tinian, Mariana Is.; (g) *utupuae*, Utupua, Santa Cruz Is.; (h) *ugiensis*, Uki, off San Cristobal, Solomon Is.; (i) *russata*, San Cristobal, Solomon Islands; (j) *kubaryi*, Pohnpei, Mironesia; (k) *semiruba*, Admiralty Islands. These forms have been variously regarded as forming one, two, three, or seven biological species (see Pratt, this volume). Composed from paintings by H. Douglas Pratt in *Handbook of the Birds of the World*, Vol. 11, Plate 14; used by permission from Lynx Edicions.

TABLE OF CONTENTS

From the Editor

Subspecies have been a focus of much of my research, albeit from an ecological rather than a taxonomic perspective. I have never much troubled myself about whether my species of interest was a good or bad subspecies (I try not to judge!); rather, the point was that it was thought to be different from other members of its species. My interest was in why this group of animals was different, how they used their environment compared with other groups of the species, and what these differences (if found) meant to the viability of the group under changing conditions. Because certain subspecies or populations of species are rare, there is a (legislated and regulated) overlap between being very rare and the availability of research funding. I have studied and continue to study many subspecies, both avian and mammalian. Studying what are usually small and isolated groups of animals is a challenge, regardless of their official scientific designation. Ecologists will continue to focus on the rare and isolated, regardless of what we are required to call them, because of the wealth of knowledge we gain from such a research focus.

Earlier in my career, I considered the question of subspecies and hybridization. What I gained from my brief, rather unsophisticated exercises was a knowledge of and keen respect for the work of earlier naturalists—without all of our fancy statistics and genetic analyses—on identifying differences and patterns in nature. Today, many scientists belittle the hands-on, "measure it with a caliper and color chart" approach to taxonomy and the descriptive nature of ecological investigations. But if you read the writings of these earlier workers, you will find a rich knowledge of nature and of the way things seemed to work. The history of the study of subspecies that is contained in this volume is tightly associated with the history of how we came to study and better understand birds.

In Chapter 1, you will read about the history of the designation of subspecies, changes in methods of identification and analysis, the Endangered Species Act in relation to subspecies and conservation, and also several case studies that identify subspecies. The authors who have contributed to this monograph are passionate about the issue of subspecific designation, and this passion comes through in all the papers (I prefer to call people "passionate" rather than argumentative—or related descriptive terms—when they strongly disagree with most of my suggested changes to their text). Several of the papers are more opinion than fact, and the authors were somewhat less than willing to consider any changes to their writing. Such papers would perhaps not have passed muster for publication in *The Auk*. But I thought that including them was appropriate as part of the overall volume because my philosophy is that it is the volume, and not individual papers, that tell this story about subspecies.

It was interesting to me that many authors focused on new and better statistical analyses to help decipher distinct segments of a species. Such a focus on statistics is the same issue that has confronted ecologists for some time, perhaps best stated by Douglas Johnson in his comments on how insignificant statistical significance can be. In ecology, we worry so much about a P value that we forget about biological relevance; that is, how different does something need to be to be ecologically meaningful? Does a male bird need to forage at a rate that is 10% or 20% (or another percent) different from a female for this to be related in an ecologically meaningful way to their survival and fitness? Likewise, does one group of birds need to be 10% or 20% different in some morphological characteristic to warrant a trinomial? Again, as an ecologist, I care most about what these differences mean for population viability; yet I need to know whether the differences in morphology are, indeed, biologically real. Thus, I ultimately need to know that we have the taxonomy correct. Although many have criticized the "75% rule" (you will read plenty about this herein), at least it is an attempt to put biological relevance into the discussion of subspecific designation.

Thus, ecologists really would like for those who work on taxonomy to get it correct, although I imagine what is correct will always (and I mean always) be a subject of debate. But it seems to me that ecologists and taxonomists could actually be working more closely together on specific groups to see whether there is, indeed, a relationship between what is different, say, morphologically, and how that relates to various population-level metrics. If, for example, we find that morphological differences do not appear to equate to differences in survival, then are those longer wings anything more than a result of chance? Perhaps the way to solve the subspecies dilemma is to look closely at what the differences in size, shape, and color mean for survival and fitness.

I am confident that *Ornithological Monographs No. 67* will challenge you to think about the designation of groups of animals. The point is not whether subspecies exist or whether you agree with the conclusions of individual papers. Rather, the point is that there is much to be learned in examining morphological and ecological differences in nature—and the links between them. Enjoy.

Michael L. Morrison

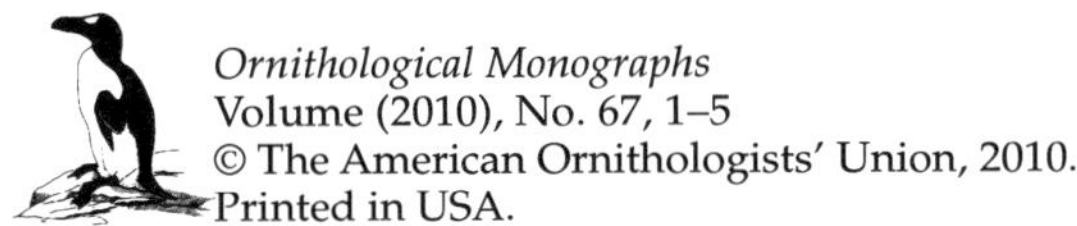
Ornithological Monographs
Volume (2010), No. 67, 1–5
© The American Ornithologists' Union, 2010.
Printed in USA.

AVIAN SUBSPECIES: INTRODUCTION

Frances C. James[1]

Department of Biological Science, Florida State University, Tallahassee, Florida 32306-4295, USA

Most of the 13 papers in this monograph were delivered at the meeting of the American Ornithologists' Union, the Cooper Ornithological Society, and the Society of Canadian Ornithologists in Portland, Oregon, during 4–9 August 2008, in a symposium session organized and chaired by Susan M. Haig and Kevin Winker. The purpose of the symposium was to review the history of subspecies in ornithology, the relevance of subspecies to studies of biodiversity and to conservation policy under the Endangered Species Act, and examples of recent research that involves subspecies of birds. The papers added after the symposium complement the others. As organized here, the first seven papers are general in nature and the last six are examples of recently analyzed case histories. The authors present various interpretations of the concept of subspecies and of methods for assigning individuals to subspecies, but they all favor the retention of subspecies as a taxonomic category. Arguments to the contrary are presented to provide perspective.

The term "subspecies" applies to either a taxonomic category in the International Code of Zoological Nomenclature (ICZN; Ride 1999) or to a particular trinomial example (e.g., *Turdus migratorius migratorius*, which was listed in the first AOU Check-list in 1886). In the classic summaries for North American birds (Peters et al. 1934–1987, Ridgway and Friedmann 1901–1950), subspecific designations were based on variation in measurements and plumage among specimens prepared as museum study skins. The most comprehensive recent list of birds of the world (Clements 2007) reports that ~57% of the more than 2,000 bird species from Canada to Panama have subspecies. The validity of such lists is questionable, however, because many continental subspecies were described from small or geographically isolated samples and may therefore reflect arbitrary breaks in clines (continuous patterns of character variation). I prefer to think of the early studies of subspecies as based on admittedly inadequate samples from major patterns of intraspecific variation. When the clinal variation is examined more closely, it is often found to be concordant across species and is itself of substantial evolutionary interest (James 1991). Nevertheless, the designation of subspecies on the basis of arbitrary divisions of clinal character variation is not warranted by today's standards. As our understanding of the patterns and processes of intraspecific differentiation has expanded, the impossibility of readily assigning complex patterns of variation to discrete categories has become ever clearer.

In modern ornithology, a subspecies is usually defined as a breeding population that occupies a distinct segment of the geographic range of its species and that is measurably distinct in phenotype, genotype, or both (for various definitions, see Mayr 1969; Avise 2004; Patten, this volume; Remsen, this volume). Although such designations have always been controversial, criticism of the subspecies concept intensified when researchers began to apply genetic tools such as haplotypes of mtDNA to studies of intraspecific variation (Ball and Avise 1992, Zink et al. 2000, Zink 2004). Others (e.g., Edwards and Beerli 2000) noted the wide confidence limits of such estimates and implored that any such conclusions be based on multilocus data. The extent to which gene trees are reliable predictors of population history, even when they are based on multiple loci, is actively debated in molecular evolutionary circles (Barrowclough and Zink 2009, Edwards 2009). Advocates of subspecies are interested in

---

[1]E-mail: james@bio.fsu.edu

Ornithological Monographs, Number 67, pages 1–5. ISBN: 978-0-943610-86-3. © 2010 by The American Ornithologists' Union. All rights reserved. Please direct all requests for permission to photocopy or reproduce article content through the University of California Press's Rights and Permissions website, http://www.ucpressjournals.com/reprintInfo.asp. DOI: 10.1525/om.2010.67.1.1.

both the phenotypic variation, which is interpreted as mostly genetically based and adaptive, and the phylogeographic patterns as revealed by neutral genetic variation in markers such as that in mtDNA and microsatellites (Winker 2009). The three-way choice seems to be whether to admit defeat and abandon subspecies as a taxonomic category, to restrict the diagnosis of subspecies to molecular criteria, or to devise criteria that combine molecular and morphological variation.

The official nomenclature of the ICZN, which includes subspecies as its lowest rank, is consistent with the "biological species concept" of Mayr (1942b, 1963), in which taxa below the species level are recognized as relevant to the process of speciation. Distinctive parapatric (adjacent) and allopatric (isolated) populations that are considered not reproductively isolated from other populations of the same species are judged to be subspecies. Note, however, that not all subspecies should be viewed as incipient species (Mayr and Ashlock 1991). Adherents to alternative views, such as the "phylogenetic species concept" (Cracraft 1983), consider that assessment of reproductive isolation should not enter into taxonomic decisions and that the designation of taxa should be based only on evolutionary history. The "evolutionary species concept" (Wiley 1981), which focuses on the establishment of lineage independence, incorporates elements of reproductive isolation, monophyly, and fixation of diagnostic characters. In a recent review of species concepts, Coyne and Orr (2004) concluded by supporting the biological species concept as a paradigm but with important caveats about hybridization. For examples of some of these alternative views, see Cracraft (1983, 1997, 2000), Zink (2004, 2006), Navarro-Sigüenza and Peterson (2004), and Peterson et al. (2006). Not surprisingly, members of each of various camps think that their paradigm is the better one for serving conservation (biological species concept: O'Brien and Mayr 1991, Haig et al. 2006; others: Peterson and Navarro-Sigüenza 1999, Cracraft 2000). Advocates of the phylogenetic species concept would elevate all diagnosable populations to full species status (thereby possibly doubling the number of species of birds).

Among the most recent and most stinging criticisms of subspecies of birds are those of Zink (2004, 2006), who claimed that the mismatch between subspecies of continentally distributed North American and Eurasian birds and their geographic patterns of mitochondrial haplotypes

is so great that the category of subspecies is actually misleading studies of biodiversity and should perhaps be abandoned. The symposium in Portland was organized partly as a response to such challenges. Haig and Winker agree with advocates of the phylogenetic species concept that many current subspecies may be based on arbitrary divisions of phenotypic clines and that others should probably be elevated to full species status. They think, however, that the deficiencies of subspecies can be accommodated by (1) more attention to definition, (2) more explicit methods of diagnosis, and (3) a modern examination of each case. Please see their summary and prospectus.

Molecular methods have become a powerful way to reveal intraspecific differentiation and even cryptic species (Bickford et al. 2007). One conservation-related example is the recommendation based on mtDNA and microsatellite data that the geographic ranges of the three morphologically defined subspecies of the Snowy Plover (*Charadrius alexandrinus*) be changed (Funk et al. 2007). Another conservation-related recommendation is that because the endangered Ivory-billed Woodpecker (*Campephilus principalis bairdii*) in Cuba is as genetically distinct in mtDNA from the U.S. population as it is from the Imperial Woodpecker (*C. imperialis*) in Mexico, the Cuban population merits a species name of its own (Fleischer et al. 2006).

If finding the upper limit of the subspecies category is partly a conceptual issue (what model to use for species), delimitation at the other end of the spectrum, finding the lower limit, involves an even grayer area. At this end, the position of Zink (2004, 2006) that sequences of portions of a single gene like mtDNA are a sufficient index for defining biodiversity and evolutionary differentiation below the species level seems to be an extremely conservative one (Greenberg et al. 1998, Bulgin et al. 2003, Phillimore and Owens 2006). Winker (2009, this volume) emphasizes that intraspecific phenotypic variation has a genetic basis (even though it may also have a nongenetic component). This variation, properly analyzed, is what most clearly reveals patterns of genetically based local adaptation and regional differentiation in the face of gene flow. This adaptive variation has not been well detected by studies of neutral elements of the genome, yet this aspect of genetically based variation can develop in a few generations and is essential to understanding evolution (Badyaev and Hill 2000, Badyaev 2005).

The difficulties of statistically characterizing subspecies once they have been defined by morphological criteria have not yet been fully confronted by adherents of the biological species concept. An important step was made by Amadon (1949), who proposed an admittedly arbitrary guideline, that 75% of a sample of specimens should be distinguishable from 99% of a sample from a reference population. Recently this idea has been extended by Patten and Unitt (2002), who developed their D statistic to test differences between localities determined *a priori*. Unfortunately, any such comparisons between two populations will be flawed unless their spatial relationship is considered (see Fig. 1). Skalski et al. (2008) contend that, unless the populations are allopatric, the proper null hypothesis is actually that no valid subspecies exist. Even when sample-size problems with *t*-tests are accommodated, *a posteriori* tests risk Type 1 error, the designation of subspecies when the geographic variation, at least in the character under consideration, is clinal. According to Skalski et al. (2008), neither univariate nor multivariate tests are appropriate for testing the presence of a cline, because the locations of the samples are decided *a priori*. The important question is whether an abrupt change, or a change in the rate of change, occurs with linear clines or with isopleth maps in two dimensions. The paper on diagnosis by Patten (this volume) suggests appropriate methods of analysis (step regression, spline regression, kriging) and some available software.

The tough statistical problem of distinguishing subspecies in the presence of clinal variation is partly illustrated in Figure 1, which plots the average value of a hypothetical quantitative character against its location on a geographic cline. The line with filled squares could be a subspecies distinction at location 5 on the cline because of the sharp discontinuity in the value of the character, and the line with filled circles is just clinal variation, but if the populations at location 1 on the cline and location 10 on the cline are compared, they are equally distinct. Any comparison between locations 1 and 10 based on population

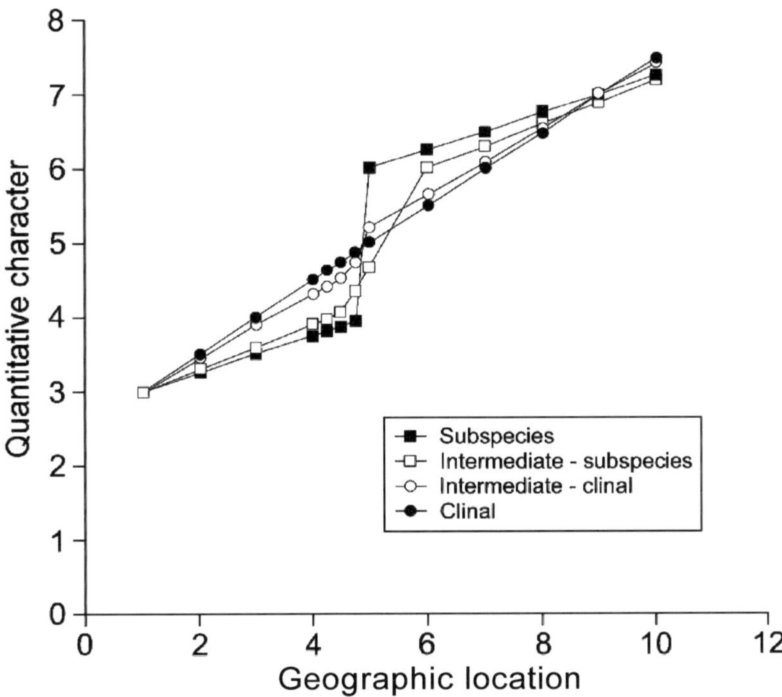

Fig. 1. Along a linear transect within the geographic range of a species, a morphometric character like size could change in a stepwise fashion as expected with a subspecies (closed squares) at location 5 or continuously as expected with a cline (closed circles), or the variation could be intermediate (open squares, open circles). Given the same pattern of variation, the scale of the analysis (locations 4–5) affects the interpretation.

distinctness (like the 75% rule) would not be able to distinguish the difference. And what about the intermediate cases, open squares and open circles? Those could be considered either smooth variation (clinal) or a distinct jump (subspecies), depending on how fine a scale was applied. At a fine scale, the changes around location 5 look like smooth changes (albeit faster than those on either side). At a coarser scale, they might look like abrupt changes, so even approaches that look for distinct changes will run into subjective issues in the choice of scale. This problem can be solved only by greater explicitness about the description and diagnosis of a subspecies than has been used in the past, and this case is just one character and one dimension. Covariation in multiple characters is reassuring (Barrowclough 1982), but despite birds' exceptionally determinate growth, various characters of their size, shape, and plumage often show different patterns of clinal variation across the two-dimensional map.

In Chapter 2 of this monograph and elsewhere (Haig et al. 2006), Haig and colleagues summarize the complexities of the administration of the Endangered Species Act by the U.S. Fish and Wildlife Service (USFWS). Acknowledging that protection should apply below the species level, at least for vertebrates, the act applies not only to full species but also to subspecies and distinct population segments (DPS) that are deemed to be at risk. Included among the criteria for a DPS are that it be a somewhat discrete unit and that it have some unique characteristics, traits that we expect would apply to subspecies by definition. Haig et al. (2006) proposed that the AOU take the USFWS criteria into account when considering the formulation of more explicit criteria for subspecies.

Because many hundreds of subspecies have not had recent review, we should not be surprised that the USFWS is asking professional societies like the AOU for an updated definition of subspecies and for clearly defined criteria for their diagnosis. Without this guidance, decisions about whether a subspecies is the unit for protection can fall to nonbiologists. For example, major lawsuits involving development rights have been brought in which final decisions were made by juries or judges. One well-publicized case concerned the endangered subspecies of the California Gnatcatcher (*Polioptila californica*; Atwood 1988) in the coastal sage scrub. It was deemed to be neither genetically nor phenotypically distinct (Zink et al. 2000, Skalski et al. 2008). The population

would have lost its protection except for its subsequent designation as a DPS, on the basis of its endangered status.

In Chapter 1 of this monograph, Winker reviews the history of avian subspecies in detail and emphasizes that even though all species probably go through a subspecies stage, not all subspecies are incipient species. He and several authors in this volume are optimistic that statistical diagnosis of allopatric populations will be possible, so that reliable subspecies can be defined. In Chapter 2, Haig and D'Elia explain that the onus of finding a reliable definition of subspecies is on the scientific societies and that the request from the USFWS for such a definition is reasonable. In Chapter 3, Patten insists that subspecies should be phenotypically diagnosable, and he points out that even the "75% rule" of Amadon (1949), which emphasizes effect size rather than simply statistically significant mean differences between populations, is not fully satisfactory. He has developed a new statistic, D (for diagnosability), that distinguishes among locations (presumably for allopatric populations), and he also gives several statistical approaches for estimation in cases with continuous distributions and in which no locations are hypothesized *a priori*. Patten contends that testing subspecies for monophyly is misdirected, because subspecies should not be expected to be monophyletic.

In Chapter 4, Phillimore explores the idea that subspecies nomenclature as a whole may capture the early stages of the speciation process, and he applies a birth–death model to species-age and subspecies-richness data from avian subspecies worldwide. In Chapter 5, Fitzpatrick acknowledges the lack of consensus on the value of subspecies and what they mean. He argues that subspecies status is given to a heterogeneous mix of evolutionary phenomena and that genuine standardization is probably impossible. He considers subspecies a useful tool of convenience but does not consider them capable of resolving policy issues such as endangered species listings without additional criteria. Remsen's contribution in Chapter 6, like Patten's, views subspecies as minimum diagnosable units and adheres to Patten's modification of Amadon's 75% rule. By excluding fringe localities and preferring 95% diagnosability, he seems to equate subspecies with some versions of phylogenetic species. In Chapter 7, Pratt summarizes his views about what he considers to have been overlumping of related populations on different Pacific islands. He advocates elevating many of these populations—many of those in Hawaii, in particular—to full species status while retaining the framework of the biological species concept. Neither Pratt nor any of the authors here recommends that decisions about species and subspecies status be made on the basis of conservation, but of course elevating taxa to the species level raises their visibility.

The remaining papers, Chapters 8–13, are case studies. Chapter 8, by Pérez-Emán, Mumme, and Jabłonski, examines phylogenetic structure and variation in plumage in the Slate-throated Redstart (*Myioborus miniatus*), which occurs from northern Mexico south to Argentina. A Central American clade includes four named subspecies that are well differentiated in plumage but homogeneous in mtDNA. The authors report past field experiments that show that subspecific variation in the extent of the white in the tail reflects evolutionary adaptation to regional prey or habitat characteristics that maximizes flush-pursuit foraging. This adaptive evolutionary divergence was not revealed by the mtDNA data. Cicero, in Chapter 9, reviews research on three subspecies of the Sage Sparrow (*Amphispiza belli*), a particularly interesting case because initially Grinnell (1898a) had considered two of the subspecies, which coexist after the breeding season, to be separate species. Chapter 10, by Oyler-McCance, St. John, and Quinn, reports that, solely on the basis of the criterion of reciprocal monophyly, they would have failed to recognize five species of lek-breeding grouse. They warn against making taxonomic revisions even at the species level based solely on mtDNA data. In Chapter 11, Marantz and Patten analyze morphometric variation in the woodcreeper genus *Dendrocolaptes*. They examined more than 3,000 specimens and found that the differentiation based on plumage was not concordant with differentiation in measurements. In Chapter 12, Wilson, Valqui, and McCracken discuss geographic variation in the

Cinnamon Teal (*Anas cyanoptera*), which occurs broadly throughout the Western Hemisphere in five discrete populations. In this case, their application of the 75% rule, based on assignment of individuals to the five populations by a discriminant analysis, seems justified. Rounding out the set of papers with Chapter 13, Pruett and Winker describe geographic variation in Alaskan Song Sparrows (*Melospiza melodia*). They include data for body mass, mtDNA sequences, and microsatellite loci. The mtDNA did not show reciprocal monophyly among subspecies, but subspecies differed in body mass and microsatellite allele frequencies. The authors emphasize that multiple lines of evidence, genetic and morphological, should be used in assessing subspecific status.

Overall, this set of papers is an important update to the literature on intraspecific geographic variation in birds. Without solving the problem of diagnosing (as contrasted with defining) species or subspecies, the collection airs the philosophical position of advocates of the biological species concept. Each contribution emphasizes the importance of subspecies to our recognition of intraspecific genetically based geographic variation in the phenotypes of birds. This variation should be studied simultaneously with the neutral or nearly neutral genetic variation being detected by today's molecular methods. Ideally, these multiple characters covary, but that is not always the case. These papers do not resolve the problems discussed above, but they do present the case for continuing research within the paradigm of our current nomenclatural system.

ACKNOWLEDGMENTS

Thanks to J. A. Pourtless for technical help, to C. E. McCulloch for discussion and for Figure 1, and to R. M. Zink, A. T. Peterson, S. Haig, K. Winker, R. T. Chesser, and P. Beerli for helpful comments on the manuscript. A. Thistle transcribed notes from tapes and provided excellent editorial advice.

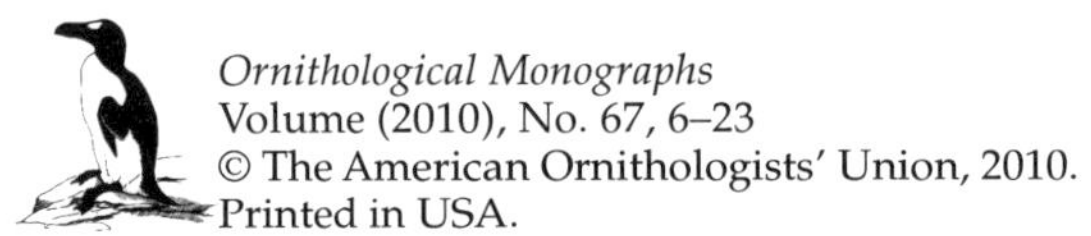

Ornithological Monographs
Volume (2010), No. 67, 6–23
© The American Ornithologists' Union, 2010.
Printed in USA.

CHAPTER 1

SUBSPECIES REPRESENT GEOGRAPHICALLY PARTITIONED VARIATION, A GOLD MINE OF EVOLUTIONARY BIOLOGY, AND A CHALLENGE FOR CONSERVATION

KEVIN WINKER[1]

University of Alaska Museum, 907 Yukon Drive, Fairbanks, Alaska 99775, USA

ABSTRACT.—In this review I summarize the history of the subspecies concept and the major debates and issues surrounding its use, with an emphasis on ornithology, in which the concept originated. The study of subspecific variation in birds has been an important driving force in the development of evolutionary biology. Subspecific study has also been essential in the description and preservation of biodiversity. Although controversy has surrounded the concept of subspecies since its inception, it continues to play an important role in both basic and applied science. I cover 10 relevant issues that have been largely resolved during this 150-year controversy, although not all are widely appreciated or universally accepted. These include nomenclature, sampling theory, evolutionary biology, and the heterogeneity of named subspecies. I also address three big unresolved questions and some of the philosophy of science related to them: What are subspecies, how do we diagnose them, and what does subspecific variation mean? Discordance between genotypic and phenotypic data at these shallow evolutionary levels should be expected. The process of diagnosing states that exist along a continuum of differentiation can be difficult and contentious and necessarily has some arbitrariness; professional standards can be developed so that such diagnoses are objective. Taxonomies will change as standards do and as more data accrue. Given present evidence, our null hypothesis should be that subspecific variation probably reflects local adaptation. In looking forward, it seems assured that geographically partitioned variation— and the convenient label "subspecies"—will continue to play an integral role in zoology.

Key words: adaptation, birds, diagnoses, evolution, history, philosophy of science, sampling error, speciation, taxonomy.

Las Subespecies Representan Variación Estructurada Geográficamente, una Mina de Oro de la Biología Evolutiva y un Desafío para la Conservación

RESUMEN.—En esta revisión hago un resumen sobre la historia del concepto de subespecie y los principales debates y asuntos que rodean su uso con énfasis en la ornitología, en donde el concepto se originó. El estudio de la variación subespecífica en las aves ha sido una fuerza importante que ha impulsado el desarrollo de la biología evolutiva. El estudio de las subespecies también ha sido esencial en la descripción y la preservación de la biodiversidad. Aunque la controversia ha rodeado el concepto de subespecie desde que fue acuñado, éste continúa jugando un papel importante tanto en la ciencia básica como en la aplicada. Abordo 10 asuntos relevantes que han sido resueltos en buena parte a lo largo de esta controversia de 150 años, aunque no todos son apreciados ampliamente ni aceptados de manera universal. Entre éstos se incluye la nomenclatura, la teoría sobre muestreos, la biología evolutiva y la heterogeneidad de las subespecies nombradas. También abordo tres preguntas grandes no resueltas y parte de la filosofía de la ciencia relacionada con ellas: ¿qué son las subespecies, cómo establecemos su diagnosis y qué significa la variación

---

[1]E-mail: kevin.winker@alaska.edu

Ornithological Monographs, Number 67, pages 6–23. ISBN: 978-0-943610-86-3. © 2010 by The American Ornithologists' Union. All rights reserved. Please direct all requests for permission to photocopy or reproduce article content through the University of California Press's Rights and Permissions website, http://www.ucpressjournals.com/reprintInfo.asp. DOI: 10.1525/om.2010.67.1.6.

subespecífica? La discordancia entre datos genotípicos y fenotípicos a estos niveles pandos de diferenciación evolutiva debería ser esperable. El proceso de establecer rasgos para diagnosis a partir de un continuo de diferenciación puede ser difícil y controvertido y necesariamente incluye algo de arbitrariedad; es posible establecer estándares profesionales para que el proceso de diagnosis se haga objetivo. Las taxonomías cambiarán conforme cambien los estándares y se acumulen más datos. Con base en la evidencia actual, nuestra hipótesis nula debería ser que la variación subespecífica probablemente refleja adaptación a nivel local. Mirando hacia el futuro, parece seguro que la variación estructurada geográficamente—y el nombre conveniente de "subespecie"—continuará jugando un papel integral en la zoología.

THE VERY FIRST volume of *The Auk* contained debates about subspecies, and it is noteworthy that 125 years later the subject still draws considerable interest. Unresolved issues in science remain so either because they are neglected or because they represent fundamentally difficult areas. Subspecies have by no means been neglected in zoological research, which leads one to conclude that this phenomenon of geographically partitioned morphological variation within species—to which we commonly apply the label "subspecies"—has some inherently interesting and difficult properties. Here, I first summarize the history of avian subspecies and consider the many relevant issues—from nomenclature, to sampling theory, to evolutionary biology—that have been largely resolved as the biological sciences have matured. I then examine several major questions that remain unresolved, including what subspecies are, how we should diagnose them, and the meaning of subspecific variation. This discussion requires consideration of the philosophy of science surrounding these issues. Finally, I consider likely aspects of the future use of subspecies.

HISTORY

The 19th century.—The concept of geographically partitioned variation below the species level was born of evidence and of need. The typological species concept of Linnaeus and taxonomists of his era did not explain geographic variation within putative species. An assortment of terms was used to label these variants: varieties, races, forms, subspecies, con-species, geographical races, incipient species, and other terms (Coues 1884, Cutright and Brodhead 1981, Mayr 1982a). Linnaeus's own term "variety" was applied to within-population variation and to variation among populations (e.g., breeds of dogs), and by the mid-1800s the term "subspecies" began to become established, taxonomically embodied by the trinomial: the addition of a third Latin name to the traditional binomial nomenclature

established by Linnaeus in the previous century (Mayr 1982a). Interestingly, this practice was perhaps first formally encoded in zoological nomenclature by the American Ornithologists' Union [AOU] in the first edition of the *Check-list of North American Birds* (AOU 1886).

The recognition of geographically partitioned variation (geographical varieties) was integrally important in stimulating scientific progress away from an essentialist, typological view of biological diversity and toward an evolutionary, populational perspective (Mayr 1982a). Thus, the variation now largely encompassed by the rubric "subspecies" played an important historical role in the development of evolutionary biology. Ornithologists were among the leaders in this scientific progress (Cutright and Brodhead 1981, Mayr 1982a), producing important work on the subject before and after Darwin's (1859) *On the Origin of Species*. Darwin used this type of variation in developing the theory of evolution (Darwin 1859, 1895), and its presence there is important:

Certainly no clear line of demarcation has as yet been drawn between species and sub-species— that is, the forms which in the opinions of some naturalists come very near to, but do not quite arrive at, the rank of species: or, again, between sub-species and well-marked varieties, or between lesser varieties and individual differences. These differences blend into each other by an insensible series; and a series impresses the mind with the idea of an actual passage.

Hence I look at individual differences, though of small interest to the systematist, as of the highest importance for us, as being the first steps towards such slight varieties as are barely thought worth recording in works on natural history. And I look at varieties which are in any degree more distinct and permanent, as steps towards more strongly-marked and permanent varieties; and at the latter, as leading to sub-species, and then to species. The passage from one stage of difference to another may, in many cases, be the simple result of the nature of the organism and of the different physical conditions to which it has long been exposed; but with respect to the more important and adaptive characters, the passage from one stage of difference to another, may be safely attributed to the cumulative action of natural selection. . . . (Darwin 1895:38–39)

Darwin (1859:47) emphasized a focus on subspecific variation that approached full species:

> Those forms which possess in some considerable degree the character of species, but which are so closely similar to some other forms, or are so closely linked to them by intermediate gradations, that naturalists do not like to rank them as distinct species, are in several respects the most important for us.

From these passages it is clear why subspecies were important very early in the field of evolutionary biology, and indeed they address topics that are at the forefront of the discipline today.

Stejneger (1884) considered that the Swedish ornithologist Carl Sundevall was the first to use trinomialism in a modern sense in ornithology when, in 1840, he treated poorly delimited species as geographic varieties, to which he gave a third name in addition to the binomial specific name. Although Stejneger (1884) stated that Sundevall's use of trinomials was closely followed by Herman Schlegel (1844), his own copy of his bound works at the U.S. National Museum in Washington, D.C., bears numerous annotations regarding his subsequent discoveries about the use of subspecific nomenclature in the 1840s and 1850s. In these annotations, Stejneger (loc. cit.) pointed to the early use of trinomials by Keyserling and Blasius (1840) and to the importance of Sélys Longchamps (1842). Mayr (1982a) omitted detailed historical discussion and considered that Schlegel (1844) was the first to routinely use trinomials. The use of trinomials grew rapidly during the remainder of the 19th century.

As Cutright and Brodhead (1981) summarized, the development and use of subspecies in the New World was particularly strong, likely for two reasons. First, North American ornithologists had a larger continent to understand, with more geographic and ecological variation than Europe, producing greater phenotypic variation in widespread taxa. Second, as Darwin (1859, quoted above) made clear, understanding this geographic variation provided insight into the process of natural selection and the origin of species. Developing an understanding of geographic variation was indeed rich ground, and a panoply

of the North American ornithologists of the day have been highlighted as practitioners, including Cassin (1856), Baird (1858), Lawrence (1864), Coues (1866, 1871, 1872, 1884), Allen (1871), and Ridgway (1881; Stejneger 1884, Cutright and Brodhead 1981).

Although subspecies were less accepted in Europe than in North America, important ornithologists there were also using trinomialism during this period; Bonaparte (Parzudaki 1856, Stejneger loc. cit.), Blasius (Newton 1862), Dubois (1871; Stejneger 1884), and Seebohm (1881) were among them. Curiously, although it may have been his irritation with the credit that Coues had been given for winning widespread acceptance of the use of subspecies that caused Stejneger to write his brief history (Cutright and Brodhead 1981; cf. Stresemann 1975), his 1884 paper (especially with his own annotations) remains one of the most important foundations for a detailed history of the development of the concept and use of subspecies in ornithology. The North American precedence in the widespread use of subspecies was largely one of promotion and acceptance only, however, for as Haffer (2001) observed, there were continental European workers who covered vast areas of Eurasia and adopted similar views; however, they were not associated with museums and thus were not leaders in taxonomy. The conservatism of the latter prolonged the practice of "essentialistic microtaxonomy" in Europe until well into the 20th century (Haffer 2001).

Elliott Coues's visit to Europe in 1884, where he was well received for his prominence in scientific ornithology (Cutright and Brodhead 1981), was an important event in the spread of the use of subspecies and trinomial nomenclature (Stresemann 1975). But Coues's visit alone was not unique in promoting subspecies in Europe; see, for example, Seebohm's (1881) introduction and his treatment of *Hypolais* [sic]. Nor was Coues's visit sufficient to overcome staunch opposition (Haffer 2001). Much of the debate over subspecies was related not to their existence, but to the adoption of trinomial nomenclature to denote these often minor variants. As R. Bowdler Sharpe wrote near the end of his life, "That races or subspecies of birds exist in nature, no one can deny, but, to my mind, a binomial title answers every purpose . . ." (Sharpe 1909:v; cf. Seebohm's [1881] treatment of subspecies as binomials, which he was forced to do by the editorial dictates of that publication). Philip L. Sclater, one of the world's most important and influential ornithologists during the second half of the 19th century and editor or co-editor of the journal *Ibis* from 1859 to 1864 and again from 1877 to 1912 (and thus an important gatekeeper for the publication of European ornithology in English), was also a major opponent of trinomialism (Elliott 1914, Stresemann 1975). The objections of Sharpe, Sclater, and others opposed to trinomialism stemmed largely from conservatism (the Linnaean binomial system was accepted tradition and sufficient) and from concern over "the danger of an outbreak of frivolous names given by scribblers who were only eager to publish and had no critical judgement . . ." (Stresemann 1975:252). Coues (1884:246) himself recognized this potential drawback, considering that trinomialism

> is so sharp a tool that without great care in handling, one is apt to cut his fingers with it. It is of such pliability and elasticity, and lends itself so readily to little things, that in naming forms one is tempted to push discrimination beyond reasonable and due bounds. . . . This is the real difficulty . . . its abuse in the hands of immature specialists.

The recognition and use of subspecies in the New World was standard by the late 1800s, and, despite such prominent opponents as Sclater and Sharpe, in Europe the use of subspecies by such careful and important workers as Ernst Hartert, Karl Jordan, Walter Rothschild, and others eventually prevailed (Stresemann 1975, Mayr 1982a, Rothschild 1983, Mallet 2007). But an important change was implemented in the process. Under the morphological species concept of the late 19th century, evidence of intergradation was key to the recognition of subspecies, whereas morphologically distinct isolates were usually treated as full species (Mayr 1982a).

In 1891, Hartert wrote

> I believe it is right to regard as subspecies forms that differ only in a small variation in size, lighter or darker coloring, or small variations in pattern, even though one does not have the intermediate forms at hand. This type of nomenclature shows the closeness of the relationship, whereas the simple specific name gives no indication whether the species are poles apart or very nearly related. (Stresemann 1975:259)

In 1892, Hartert met Walter Rothschild and was soon thereafter hired as a curator at the Tring Museum (Rothschild 1983). At Tring, the ornithologists and entomologists Rothschild, Hartert,

and Jordan produced a body of work that effectively expanded the definition of subspecies to include geographically isolated, closely related populations (Rothschild 1983, Mallet 2007). Mayr (1982a) considered this development to be based on the biological species concept, which had yet to be defined, but which decades later came to dominate taxonomy and systematics.

The 20th century.—The recognition of closely related but geographically isolated populations as subspecies was an important turning point in how the concept of subspecies was applied; it now encompassed both distinct but intergrading populations and distinct geographic isolates not sufficiently differentiated to warrant recognition as full species. Several more decades were required (until the 1940s) for modern terminology and concepts to be fully developed, but this concept of subspecies was an integral part of that process. Bernhard Rensch, Julian Huxley (who contributed, among other things, the term "polytypic species"), and Ernst Mayr were important contributors during this period (Mayr 1982a, Mallet 2007). Ultimately, the hierarchical levels of differentiation (populations, subspecies–polytypic species, superspecies) that could be seen within many taxa, among populations that were separated to greater or lesser degrees geographically, provided a framework for understanding biological diversity that contributed several key tenets to the Evolutionary Synthesis of the 1930s and 1940s (Futuyma 1998).

In taxonomy, the effects of adopting this expanded view of subspecies were also profound. Recognition of polytypic species (species that comprise two or more subspecies) and adoption of the biological species concept caused a major reduction in the number of species-level taxa, from >20,000 in the 1920s to ~9,000 in the 1980s (Mayr 1982a). But there was overlumping, with perfectly good allopatric species lumped into polytypic species and treated as subspecies. We are still rebounding from this process as an increasing number of allopatric subspecies are, with additional data, being recognized as full biological species.

Mayr (1982a) considered that birds were well suited to this new taxonomy that included polytypic species and that its application to Aves caused the group to be especially valuable for both evolutionary and ecological studies. But here he omitted mention of further disagreement over the utility of the subspecies concept, which has waxed and waned since its inception.

Wilson and Brown (1953) wrote an important and influential critique of subspecies, their main concerns being that (1) characters varied in different and nonconcordant ways, and that, depending on which character was used, different subspecific delineations could result; (2) there was a great deal of arbitrariness in defining subspecies, especially in delimiting the lower limits; and (3) subspecific taxonomy was getting in the way of real research (e.g., on geographic variation) by demanding an artificial formality and a system that lacked sufficient flexibility.

Mayr (in Inger 1961:283) observed that when Wilson and Brown (1953)

> recognized that these subspecies of the literature were not subspecies they acted just like the little boy who knocks himself against the corner of the chair and beats the chair for being so bad: they let out against the subspecies their anger at the specialist of ants for having mistreated and misused the subspecies concept.

Wilson (1994:208) later wrote that he and Brown had overstated their case. Concordant changes in multiple characters often occur; this is the geographically partitioned variation that is the hallmark of good subspecies. Wilson and Brown (1953) were correct that subspecific taxonomy is probably most difficult and arbitrary at the lower limits of the subspecies category and that undue focus on subspecific taxonomy had, in effect, created what they termed a "subspecies mill." Indeed, the fears of Sharpe came to pass, and even Mayr (1951:94) had to admonish ornithologists to put less effort into describing minor subspecies and more into studying the trends of geographic variation.

In 1982, an invited forum on avian subspecies by 11 authors was published in *The Auk* (Wiens 1982). As might have been expected, although this forum provided a series of strong essays on the subject, together they did not provide definitive resolution of the key issues debated in relation to subspecies since Darwin (Wiens 1982); such resolution remains evasive.

In sum, the use of trinomials in the literature of the time shows that, in referring to subspecies and subspecific variation, Darwin (1859, 1895) was communicating effectively with the leading scientists of the time on a subject important to them. Ornithologists were in the vanguard in adopting the use of subspecies, both because the concept helped researchers understand biological diversity and its generation (Cutright and Brodhead 1981, Mallet

2007) and because in many other taxa describing species-level diversity remained (or remains) a dominant pursuit (Mayr 1982a). The problem of an accepted definition, however, alluded to by the passages of Darwin (1859, 1895) quoted above, has persisted within and among taxa to the present (Mayr 1982a, Haig et al. 2006). Examples of how definitions have progressed are given below in my discussion of unresolved issues.

Sharpe's and others' fears that subspecies names would be inappropriately applied to many dubiously distinct forms (Stresemann 1975) were prophetic, and many unwarranted named subspecies remain to be eliminated. However, a certain naivete is evident among many authors in the second half of the 20th century, who seemed to desire that scientists forgo describing their discoveries and observations (in describing subspecies in this case) until enough specimens and data were in hand to understand true patterns of geographic variation. Or, conversely, they felt that such studies should be restricted to hypothesis-testing approaches rather than the descriptive approach most often used in subspecies-level, discovery science. These views do not sufficiently acknowledge the important stepwise progress of science in describing and understanding biodiversity at its finest levels. In addition to changing techniques and differences in data interpretation, specimen limitations alone have long precluded a shortcut to accurately summarizing biodiversity (e.g., might an apparent discontinuity instead represent clinal variation as yet insufficiently sampled?). Not only is the job not yet done (e.g., Zink and Remsen 1986), but the decline of taxonomy as a discipline in major universities during the second half of the 20th century has caused much of the unresolved mass of dubious names, questionable species limits, and phenotypically undescribed or unexamined populations to languish. And if new generations of students are not being trained in this presently less popular aspect of biodiversity science, we postpone a full understanding of biodiversity and hobble our abilities to document and manage this diversity during a period of major global changes. This is a huge challenge for 21st-century biology. How will we meet it?

Issues That Are Largely Resolved

Numerous issues have been largely resolved in the long debate over subspecies, even if all are not widely appreciated or universally accepted.

Nomenclature.—As Mallet (2007) pointed out, the International Code of Zoological Nomenclature (ICZN) regulates just that—nomenclature—and does not treat the difficult problem of how one determines what taxonomic rank a lineage or group of specimens represents. The subspecies concept thus comprises two important but now largely separate issues: trinomial nomenclature (that battle is over and is encoded in the Code; ICZN 1999) and the thornier question of what constitutes a subspecies—that is, in what cases do we apply trinomialism? The nomenclatural tool is here to stay; we as taxonomists and systematists simply need to do as Darwin (1859:47) suggested and come to some agreement on how and where to apply it.

Subspecies can be genuine biological units.—There are many examples of good subspecies, cases in which multiple species concepts would agree that populations that exhibit diagnosable (see below) differentiation from their nearest related population(s) in more than one phenotypic character do not achieve the threshold required of full species. Some of the best examples of subspecies occur among Song Sparrow (*Melospiza melodia*) populations in northwestern North America (Pruett and Winker, this volume). One reason I think that Song Sparrows are a good example is that there are no obvious species limits in this group under the phylogenetic species concept; probably because of the apparent recent colonization and ongoing gene flow (though frequently low), reciprocally monophyletic units (using any character) would be difficult to recover in this species when good samples (numerically and geographically) are examined. Yet the populations in Alaska recognized as legitimate subspecies exhibit increasingly concordant divergence in multiple characters as isolation increases (Pruett and Winker, this volume). Another good series of subspecies occurs in Rock Ptarmigan (*Lagopus muta*) across the Holarctic; some of the more pronounced occur among the Aleutian Islands (Fig. 1). Some of these subspecies might be considered phylogenetic species, but evidence of gene flow and the capacity for gene flow among populations in this species causes the taxonomic fit to be better with the polytypy of the biological species concept. Importantly, in both of these examples the recognized variation is discontinuous (i.e., it does not occur in a smoothly clinal way). Populations that exhibit clinal variation in one or more characters are likely to be legitimately subdivided

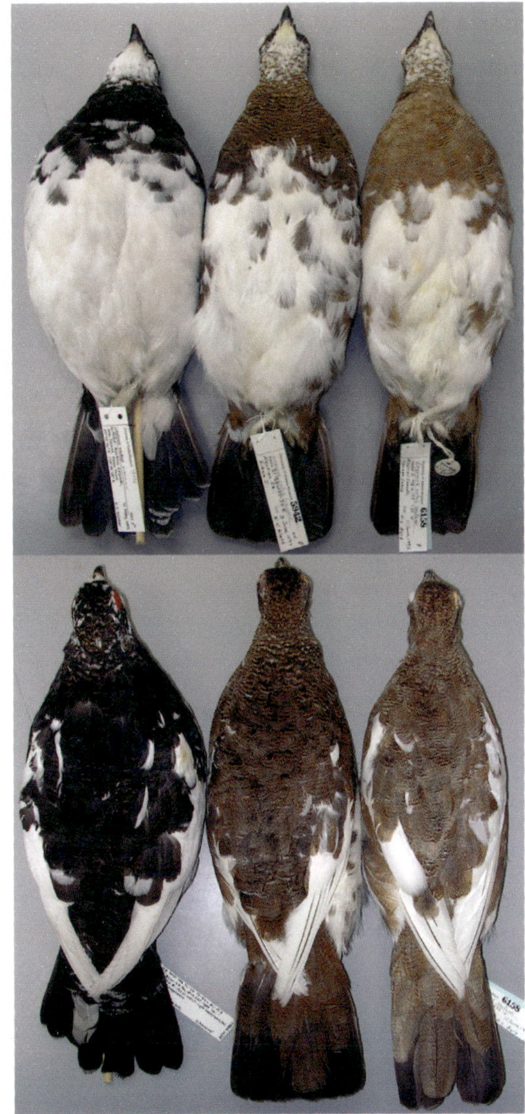

Fig. 1. Summer adult male Rock Ptarmigan (*Lagopus muta*) from the Aleutian Islands, Alaska. From left to right are ventral and dorsal images of *L. m. evermanni* (from Attu Island), *L. m. townsendi* (from Kiska Island), and *L. m. atkhensis* (from Tanaga Island).

into subspecies only when discontinuities occur (classically termed "step clines"; Mayr 1963, Mayr and Ashlock 1991).

Sampling error.—The problem of sampling error—that summaries of diversity change with increased sampling—was recognized quite specifically by the first AOU Committee on Classification and Nomenclature in formally adopting trinomialism (AOU 1886:31). What has yet to be fully appreciated (though see Funk and Omland 2003, Brumfield 2005, DeSalle et al. 2005) by researchers using rather small data sets of DNA sequence in phylogeographic and barcoding studies is that this same problem of sampling error exists here, too. The parallels between the historic naming and lumping of subspecies and the recognition and subsequent (or future) loss of monophyletic mitochondrial DNA (mtDNA) clades are obvious but are nevertheless overlooked by many researchers today.

In the description of phenotypic diversity, initial specimens or data sets might indicate sufficient geographic variation to name subspecies (Fig. 2A). Increased geographic sampling might show clinal variation, necessitating the elimination of recognized subspecies (Fig. 2B), or sufficient discontinuous variation to warrant continued recognition of the subspecies initially described (Fig. 2C). Increased numeric sampling might show that individual variation within populations is too high to enable credible diagnoses of the individuals in each population and, thus, that they do not warrant subspecific labels (Fig. 2D), or that such variation is sufficiently low to enable a high degree of diagnosability between populations for which subspecies designation is appropriate (Fig. 2E).

Similarly, in the description of the distribution of genetic variation, initial data sets may indicate highly partitioned variation in the form of monophyletic clades (Fig. 3). It may be appropriate to label these clades with names (although I disagree that species or subspecies limits should be diagnosed using a locus or two of DNA sequence data alone; Winker et al. 2007). Such trees can be reconstructed from genetic or phenotypic data, but the same sampling-error issues remain. Increased geographic sampling might uncover a population that has characteristics of more than one described population (Fig. 3B) or lend further support to the initial summary by finding continued geographic partitioning of population-defining characters (Fig. 3C). Increased numeric sampling might reveal that one or more populations comprise an admixture of what were thought to be population-defining characters (Fig. 3D), or that these clade-defining characters are segregated, warranting continued recognition of these units (Fig. 3E). Power analyses addressing this problem are considered below.

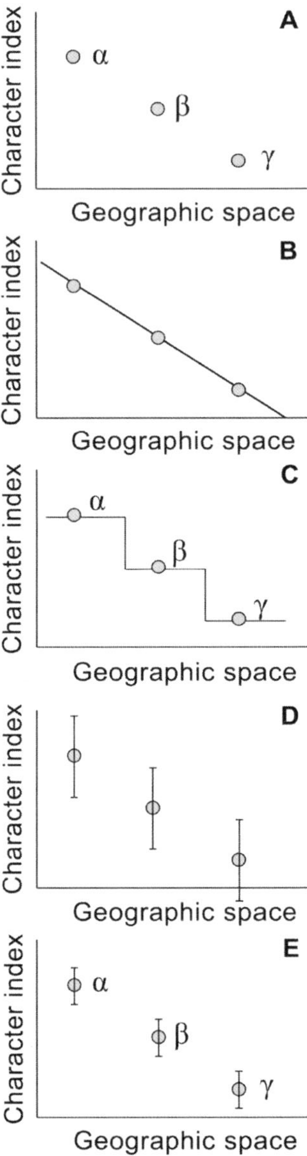

Fig. 2. A simple diagram of how initial sampling of biological diversity across geographic space might indicate (A) the existence (and cause the description) of multiple subspecies. Subsequently increased geographic sampling might show (B) that variation is clinal and the recognition of subspecies is unwarranted or (C) that variation occurs in a discontinuous (here, stepwise) manner and that continued recognition of subspecies is warranted. Similarly, increased numeric sampling might reveal (D) such degrees of intrapopulation variation that reliable diagnoses of previously described subspecies are not possible or (E) that such variation is sufficiently low to enable reliable subspecific diagnoses.

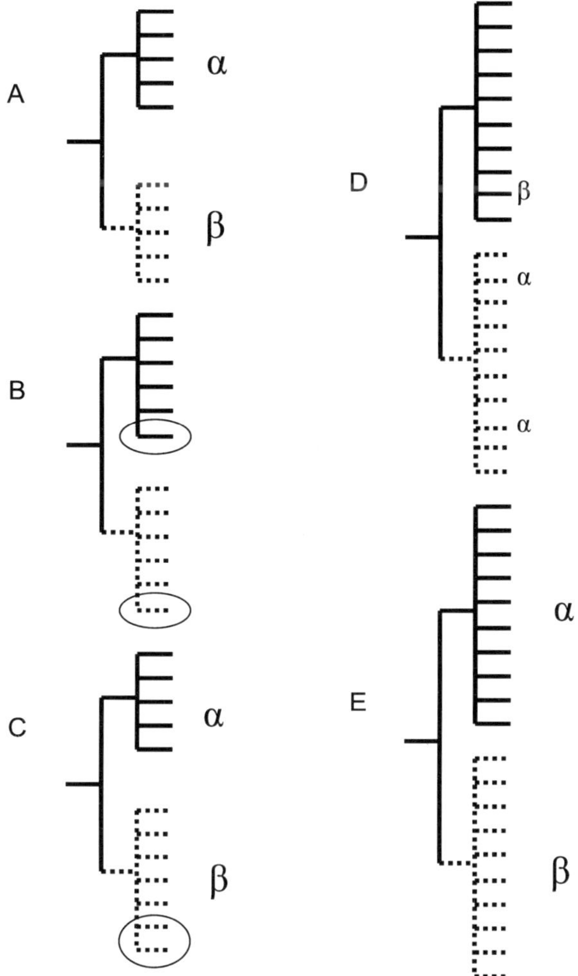

FIG. 3. A simple diagram of how initial sampling of biological diversity across geographic space might reveal two monophyletic clades, indicating (A) the existence (and possibly causing the naming) of different (sub)specific units. Subsequently increased geographic sampling might show (B) that a newly sampled population (here, $n = 2$, circled individuals) has individuals from each clade, suggesting that named units are not warranted, or (C) that this newly sampled population retains the initially described geographic partitioning of variation and that continued recognition of those units is warranted. Increased numeric sampling might reveal (D) that one or more populations comprise a mix of individuals that possess what initially seemed to be population-defining characters (e.g., wherein an individual from the β population is found with α population characters and two individuals from the α population possess β characters). Conversely, increased numeric sampling might continue to show (E) strong separation of these clade-defining characters, warranting continued recognition of these monophyletic units.

Importantly, both increased geographic and numeric sampling are usually needed to ascertain the veracity of initial diagnoses. For example, isolation-by-distance effects might cause increased numeric sampling to support unit recognition, but a lack of increased geographic sampling might cause zones of intergradation to be overlooked. Incomplete data will affect diagnoses and ensure continued modifications of taxonomies as understanding improves.

Diagnosable units.—It is important to emphasize that I use the term "diagnosable" in a probabilistic sense and not in a cladistic sense. Consider the now classic "75% rule" (Amadon 1949, Patten and

Unitt 2002) as our probabilistic threshold in recognizing subspecies (discussed more below). This level of diagnosability falls considerably short of the 100% diagnosability criterion of cladistics. In other words, reciprocal monophyly (two groups monophyletic with respect to each other), no matter what characters one chooses to use, provides a level of unit diagnosability in cladistics that is not appropriate as a lower threshold for evaluation of subspecies. Why? Because a probabilistic framework of diagnosability is more concordant with the processes of divergence between populations: gene flow can and often does still occur between populations, leading to an evolutionarily important relationship between diverging forms (in this case a lack of independence) that must be accounted for in assessing and managing biodiversity. The working hypothesis of the biological species concept (which allows some hybridization; Johnson et al. 1999, Winker et al. 2007, Price 2008) accounts for limited gene flow between species; subspecies, by definition, must also allow some gene flow to occur.

Typological thinking has no place in defining and recognizing subspecies. Further, in recognizing the populational processes of divergence, simple statistically significant differences between populations are expected and are decidedly not sufficient grounds for the recognition of subspecies (Mayr et al. 1953, Patten and Unitt 2002).

Gene flow.—It is widely accepted that gene flow inhibits divergence between populations. What is not often remembered, however, is that to the best of our knowledge the effects of gene flow on the process of divergence are highly nonlinear (Fig. 4). Even at low rates, gene flow can prevent speciation and even local adaptation under strong selection (Rice and Hostert 1993, Hostert 1997, Postma and van Noordwijk 2005). Another consideration is that populations may evolve collectively when alleles of high selective advantage spread among them at even very low levels of gene flow (Morjan and Rieseberg 2004). Because low rates of gene flow can act as a sort of evolutionary glue that prevents populations from diverging, sampling error becomes particularly important in assessing data on the distribution of genetic variation. Using sequence data from a genetic locus such as mtDNA and the coalescent, 10 individuals from each of two populations provide a good sample size for analyses of gene lineage histories and lineage sorting (Harding 1996, Rosenberg 2007). But gene flow is not adequately surveyed with such samples, because even

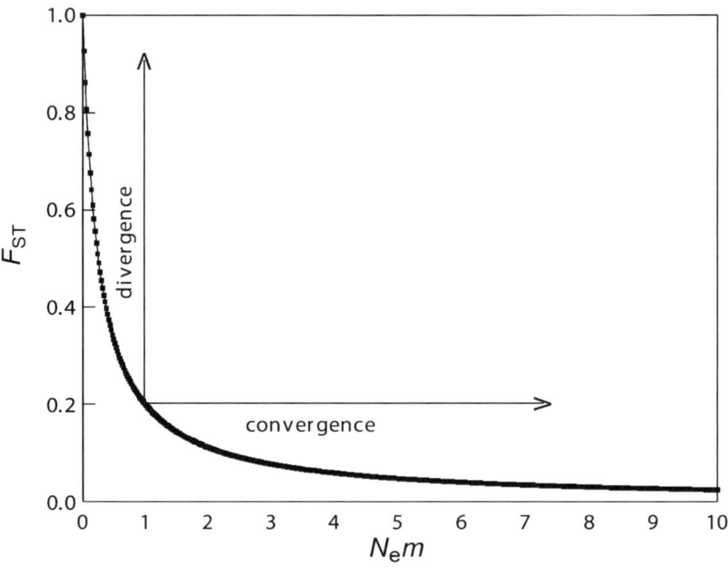

FIG. 4. The relationship of Wright's (1943) F_{ST}, an index of interpopulation genetic differentiation, to the product of effective population size (N_e) and the rate of gene flow (migration, or rate of effective gene flow), m. Note the highly nonlinear relationship and that the inflection point, at $N_e m = 1.0$ (one migrant per generation), marks a transition under neutral conditions between populations that are diverging and populations that are effectively fused. After Cabe and Alstad (1994).

moderate levels of gene flow could escape detection. Using Gregorius (1980) for power analysis, a sample size of 11 gives a 95% probability of detecting all alleles (or haplotypes) in a population that occur at ≥30% frequency. From Figure 4, it can be seen that this level of power does not come anywhere near the inflection point of divergence under neutral conditions (nor, likely, under strong divergent selection; Rice and Hostert 1993, Hostert 1997). This observation of the limited power of most phylogeographic studies is not meant to deter them, for more are urgently needed. Rather, it is to focus interpretation of these studies on the populational processes of divergence (for a good example in birds, see Brumfield 2005). Gene flow and phenomena such as clinal variation, isolation-by-distance, reticulation, and polyphyly and paraphyly are to be expected when populations are in the process of diverging (cf. Funk and Omland 2003). Knowing when these phenomena can be ruled out in explaining the distribution of genetic variation is very important.

Subspecies are interesting and have value.—Insofar as subspecific labels for geographic variation have not only been deemed necessary by historical leaders in biological research but have also been used as guides for both historic and modern research, the value of the concept has been proven despite a lack of agreement over what exactly a subspecies is. The continuing value of the concept can be seen in both applied and basic research. Two examples will demonstrate this.

Outside of game species, the classic study of museum specimens (subspecies) has provided most of our knowledge of the nonbreeding distributions of the birds of North America. Identification of population movements, especially of migratory forms during the nonbreeding season (except for game species, in which hunters effectively provide large band-recovery efforts), has been accomplished mostly on the basis of subspecific plumage characters. Classic examples are summarized throughout the fifth edition of the AOU Check-list (AOU 1957), exhibiting this practical utility of subspecies very well.

In basic research, one of the best examples of the interest and value of subspecies is a research program that ends up questioning the utility of the concept itself. In reviewing 18 species studied by R. M. Zink and colleagues (Table 1), it can be seen that the species they chose to work with had an average of 13.5 recognized subspecies each. The congeneric species not chosen

for these studies averaged about 3.1 recognized subspecies (Peters et al. 1931–1987). In 10 of these 18 studies, the species with the most recognized subspecies of any in the genus was the focal taxon (Peters et al. 1934–1987). The probability of this occurring randomly is quite small ($P \approx 0.008$). Regardless of why these species were chosen as the focal taxon of their respective genera, if subspecies were not often used as a guide for this body of research (and in some studies they clearly were not), then they were at least an important component.

I am not using this analysis to condemn the approach of Zink and his colleagues (Table 1). Whether they chose to work with species that have above-average subspecific diversity (in contrast with congeners) or whether this was merely an accidental (though highly significant) correlation, their research nevertheless tells us something about avian subspecies. Where I differ with Zink's (2004) conclusion is in our perspective on what the evidence tells us, a difference that is not simply a glass-half-empty versus glass-half-full situation. Zink's (2004) message was that the vessel itself (subspecies) is faulty, a condemnation reminiscent of that of Wilson and Brown (1953; cf. Mayr in Inger 1961:283). As Zink (2004) showed, much of this body of work could be focused around the concept of subspecies, and therefore that concept had scientific interest and some merit. In the process of speciation, discordant distributions of largely neutral genetic variation and phenotypic variation (which for many if not most characters will have a substantial non-neutral, adaptive component) are to be expected, as Price (2008), Winker (2009), and others have pointed out. Phillimore and Owens (2006), using a larger and geographically more diverse data set than Zink (2004), found a greater level of mtDNA and subspecies concordance. In fact, a large body of research has used subspecific designations to better understand diversity and its distribution and to ask important questions of microevolutionary processes (e.g., Remsen 1984).

Subspecies represent a heterogeneous taxonomic category.—Since the end of the 19th century, when allopatric forms that showed relatively small differences were included with intergrading forms as subspecies, the category has been one of heterogeneous named populations (e.g., fig. 1 in Fitzpatrick, this volume). This fact remains an irritant to many of those concerned with subspecies. The basis for subspecific names is not and never has

TABLE 1. Eighteen species chosen for research by R. M. Zink and colleagues, showing the study, the number of subspecies in the focal taxon, total species in the genus, average number of subspecies among species in that genus (minus the focal taxon), and whether the focal taxon had the highest number of recognized subspecies in the genus. Taxonomy is an amalgam of American Ornithologists' Union (1998) and Peters et al. (1934–1987).

Source	Species	Subspecies (n)	Species in genus (n)	Subspecies in genus[a]	Study species highest
Zink et al. 1987	*Callipepla californica*	4	5	2.25	
Karl et al. 1987	*Larus californicus*	0	35	0.91	
Zink et al. 2002a	*Dendrocopos major*	27	33	5.44	X
Zink et al. 2002b	*Picoides tridactylus*	11	2	—	X
Johnson and Zink 1985	*Vireo olivaceus* (complex)	14	25	2.58	
Zink et al. 2000	*Polioptila melanura* (or *californica*)	6	8	4.14	
Drovetski et al. 2004	*Troglodytes troglodytes*	35	5	12.75	X
Zink et al. 2001	*Campylorhynchus brunneicapillus*	7	11	3	X
Zink et al. 2003	*Luscinia svecica*	7	25	1.83	
Zink and Blackwell-Rago 2000	*Toxostoma curvirostre*	7	10	1.89	X
Pavlova et al. 2003	*Motacilla flava*	18	9	3.13	X
Zink and Klicka 1990	*Geothlypis trichas*	13	13	1.75	X
Zink 1988	*Pipilo fuscus*	18	7	5.5	
Zink and Dittmann 1993a	*Spizella passerina*	6	6	2.4	X
Zink et al. 2005	*Passerculus sandwichensis*	21	9	4.25	X
Zink 1994	*Passerella iliaca*	18	9	9.25	
Zink and Dittmann 1993b	*Melospiza melodia*	39	9	6.63	X
Zink et al. 1991	*Quiscalus quiscula*	3	8	4	

[a] Average excluding the focal taxon.

been reserved exclusively for evolutionary units (Mayr 1969:41). The existence of or potential for gene flow between subspecies precludes this exclusivity. In this, subspecies differ from higher taxonomic ranks, but as Smith and White (1956) pointed out, this is fully consistent with nomenclatural theory, which is based on denoting differences. And because most species have probably passed through some sort of subspecies stage of differentiation, Darwin's (1959, 1895) insight that this passage is important remains a stimulus for studying subspecies. As eloquently put by Smith and White (1956:190), "The gap from non-species to recognizable species is necessarily bridged by stages of major evolutionary import." Also, our recognition of a nomenclatural category in which some units may very well become full species and others will not gives us a conceptual tool of inestimable value in understanding the process of speciation (Smith and White 1956). Subspecies and their heterogeneity therefore do not represent a problem to basic science so much as they represent an opportunity.

We should not eliminate subspecies altogether.— Smith and White (1956) presented one of the best rebuttals to the condemnation of subspecies by Wilson and Brown (1953), and most of their observations remain applicable today. For those who study speciation, recognition of units only at the species level and above ignores stages of differentiation that are clearly important in the speciation process. Subspecies have a demonstrated utility if only for this research purpose alone, making their use largely unstoppable. This does not mean that they must be used by all. However, it must be recognized that the work that goes into subspecific taxonomic revision has become rather unpopular, leaving something of a mess of unresolved subspecific names, species limits, and specimens and data that remain to be critically reexamined in this context.

Subspecies are a challenge to conservation.— Subspecies represent a challenge to conservation that might be summed up in a relatively simple way: homogeneity is more easily managed than heterogeneity. Recognizing the conservation

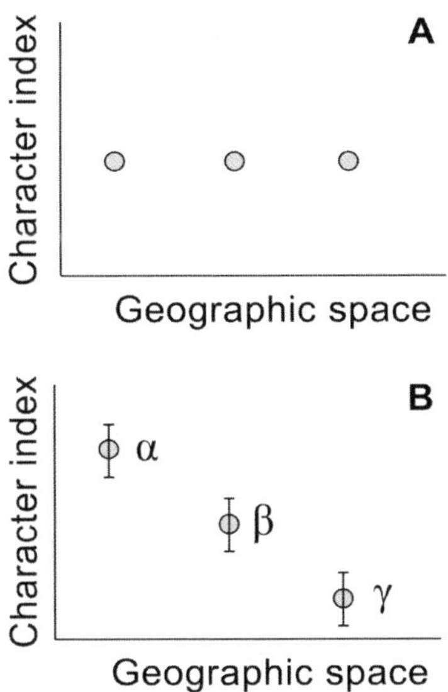

FIG. 5. Subspecies represent a challenge for conservation. The conservation implications of a taxon that (A) exhibits no structured geographic variation (the no-trinomial view) is a much simpler situation than that implied by (B) the more complex adaptive space likely occupied when trinomials are applicable.

implications of a taxon that exhibits no structured geographic variation (the no-trinomial view) is much simpler than responding effectively to the complexity that exists when different adaptive space is likely occupied by subsets of populations (Fig. 5). Use of subspecies in this context includes an important assumption: that subspecies labels are applied to populations that possess unique properties in an evolutionary, adaptive sense. Although many populations that possess very minor differences (i.e., invalidly considered diagnosable) have been named as subspecies and remain to be eliminated, this is probably a useful null hypothesis (see below).

Human imperfections play a considerable role.— Our own human tendencies have had predictable influences on the subspecies phenomenon. Sharpe's concern about the dangers posed by authors who lack critical judgment yet are eager to publish accurately anticipated what Smith and White (1956) termed the *"mihi"* itch of taxonomists to describe new taxa, and what Wilson and Brown (1953) called the "subspecies mill" generated a surfeit of names to describe, at times, even the most minor differences among museum specimens (i.e., differences that are not useful for

diagnosing populations at levels even less stringent than that of the 75% rule). But the opposite side of that proverbial coin—the bracketing human condition to writing up trivial phenomena— is laziness among those not wanting to recognize subspecies because of the work necessary to accurately determine and describe geographic variation (Smith and White 1956). Detailed study and description of biological diversity at and below the level of species is fundamentally hard work, whether one is performing a pioneering study or reevaluating the work of others with new data. Knowing that we scientists have both of these tendencies might help temper action at the extremes.

UNRESOLVED ISSUES AND PHILOSOPHY
OF SCIENCE

What is a subspecies?—In summarizing some of the key views on subspecies between Darwin (1859) and now, we might conclude that we have not come remarkably far from Darwin's (1859) view that in determining the taxonomic rank of a form that is near the species level we will have to be guided by scientists with judgment and experience.

The debate now is not over binomialism or trinomialism (taxonomy) but over how to diagnose and recognize geographically partitioned variation that may or may not exhibit intergradation. Allopatric populations that show comparatively minor differences but with no evidence of intergradation are common, especially among islands. Parapatric, intergrading populations (e.g., step clines) also occur. No species concept can decisively solve this problem; some subjectivity will be involved (e.g., Winker et al. 2007).

Allen (1871), in focusing on the fact that intergrading forms should not be considered full species, hastened adoption of the acceptance that "subspecies are distinguishable forms which intergrade, while species do not intergrade" (Stejneger 1884:75; cf. Coues 1871:371–372). In Allen (1871) and Coues (1871), one can see the beginning stages of what eventually would comprise the AOU's early views in adopting formal use of subspecies (Allen and Coues were members of the committee), which were summarized pithily: "In a word, *intergradation* is the touchstone of trinomialism" (AOU 1886:31). This school of subspecies was soon enhanced by including discontinuous forms that did not intergrade but that differed to a degree similar to that found among parapatric subspecies. As Hartert defined subspecies in 1903:

> We describe as subspecies the geographically separated forms of one and the same type, which taken together make up a species. Therefore not just a small number of differences, but differences combined with geographic separation, permit us to determine a form as a subspecies, naturally when there is general agreement of the main characters. (Stresemann 1975:262)

Amadon (1949) felt that if subspecies names were applied conservatively they would call attention to geographic variation and incipient speciation, which recalls Darwin's (1859, 1895) points quoted above. To help achieve that conservatism, he formalized the 75% rule, a popular lower limit for the delineation of subspecies. He defined this rule

> to mean that 75 per cent of a population must be separable from all (99+ per cent) of the members of overlapping populations to qualify as a subspecies. An equivalent statement is that 97 percent of one of two overlapping populations must be separable from 97 per cent of the other. (Amadon 1949:258)

Rand and Traylor (1950) recognized that subspecies were based on natural phenomena but that their delimitation was subjective, and they suggested that a less conservative quantitative rule be applied: to be recognizable, just 80–90% of one subspecies should be separable from the same amount of another. Amadon's (1949) more conservative rule seems to be the most widely adopted quantitative criterion (e.g., Mayr 1969, Patten and Unitt 2002).

Mayr (1982a:289) stated that

> A subspecies is now defined as "an aggregate of phenotypically similar populations of a species, inhabiting a geographic subdivision of the range of a species, and differing taxonomically from other populations of the species (Mayr, 1969:41)."

He also considered subspecies to be a category of convenience for taxonomists (Mayr 1982a), reflecting precisely the views of Darwin (1895:39) on the terms applied to what he called this "insensible series" of differentiation. In Mayr's (1969, 1982) usage, "differing taxonomically" meant differing "by sufficient diagnostic morphological

characters." He felt that taxonomic differences have to be observable in museum specimens (Mayr 1951:93) and that "what is taxonomically different can be determined only by agreement among taxonomists" (Mayr et al. 1953:31). Importantly, "no nonarbitrary criterion is available to define the category subspecies. Nor is the subspecies a unit of evolution, except where it happens to coincide with a geographical or other genetic isolate" (Mayr 1969:41).

Principles of Systematic Zoology (Mayr 1969, Mayr and Ashlock 1991) and its precursor, *Methods and Principles of Systematic Zoology* (Mayr et al. 1953), represented one of the most important touchstones of animal taxonomy for the second half of the 20th century, and they are still heavily used (although out of print). These works gave important common ground to the taxonomic practices among researchers working on various animal taxa. Between Mayr (1969) and Mayr and Ashlock (1991), however, there was a decrease in the detail with which subspecific diagnoses were treated, reflecting both a greater conservatism in the use and recognition of the category and the fact that standards for recognizing subspecies had become more rigorous (Mayr and Ashlock 1991:98). In ornithology (and vertebrate taxonomy in general), that rigor may be best expressed in the assertion of the 75% rule (Mayr et al. 1953, Simpson 1961, Patten and Unitt 2002); critically, however, determinations still depend on the material available and how it is used (e.g., Cicero and Johnson 2006).

Simpson (1961) corrected two misapprehensions about subspecies. First, he stated that they are not what he termed "little species" and usually are not incipient species: "They are taxa of a markedly different kind from species, and relatively few of them will ever become species although some are, to be sure, approaching that status" (Simpson 1961:175). Second, he observed that

> subspecies do not express the geographic variation of the characters of a species and are only partially descriptive of that variation. They are formal taxonomic population units, usually arbitrary, and cannot express or fully describe the variation in those populations any more than classification in general can express or fully describe phylogeny. They are not, for all that, any less useful in discussing variation. (Simpson 1961:175)

Simpson also observed that

> When there are semiarbitrary groups in a species, their designation as subspecies should

hardly raise a question if the data are adequate and the taxonomist finds the subspecies valuable for his purposes. It is no argument against such usage that recognizable subgroups do not occur in all species, because the subspecies is a nonobligate category and need not, even in principle, be used throughout the whole of a classification. (Simpson 1961:173)

Avise (2004:362–363) suggested a phylogenetic approach to intraspecific taxonomy that united strengths of the biological and phylogenetic species concepts, and he proposed that subspecies are

> groups of actually or potentially interbreeding populations (normally mostly allopatric) that are genealogically highly distinctive from, but reproductively compatible with, other such groups. Importantly, the empirical evidence for genealogical distinction must come, in principle, from concordant genetic partitions across multiple, independent, genetically based molecular (or phenotypic; Wilson and Brown 1953) traits.

The strongest disagreements over the taxonomic category of subspecies have tended to occur at both the lower and upper bounds of the unit: when does geographic variation become sufficiently partitioned (and in which and how many characters) to warrant a named subspecies; and when do allopatric populations (recognized as subspecies) achieve sufficient differentiation to be considered full species? The 75% rule is a widely accepted quantitative definition for the lower limits (Amadon 1949, Mayr et al. 1953, Simpson 1961, Mayr 1969, Mayr and Ashlock 1991, Patten and Unitt 2002, Haig et al. 2006). For the upper limits there are a series of now classic comparative criteria in which the phenomena that occur at contact zones (e.g., degrees of intergradation or hybridization) and/or in similar cases in closely related taxa are used to infer subspecies or species status (Stresemann 1921 [as translated by Haffer 2007:46], Mayr 1969, Mayr and Ashlock 1991). Advances in genetics and the allure of molecular taxonomy can provide insights (e.g., Petit and Excoffier 2009), but these tools still do not alleviate the difficulty of diagnosing allopatric taxa and will likely just be added to the taxonomist's toolbox and used in traditional ways. These classic comparative methods are not an exact methodology, nor are they purely objective, but they have proved highly effective in zoology for describing and categorizing diversity (e.g., Remsen 2005).

Biodiversity diagnoses.—In many respects, the diagnosis of subspecies is similar to determinations in other fields that are, of necessity, based on arbitrary thresholds. Importantly, objective criteria can be applied to diagnoses of conditions that occur on continua and bear arbitrarily determined thresholds. Hey et al. (2003), in discussing uncertainties in species determinations, made the important observation of parallels in other disciplines, but in their limited treatment emphasizing the social (or philosophical) over the scientific aspects they did not fully develop those parallels.

In biology we work daily with a hierarchical scale of differentiation: populations, subspecies, species, genera, families, and a succession of higher taxonomic categories. Parallels abound in everyday phenomena and in our language to describe them. There is always some water in the air; at what point does it become fog? Or when does a heavy fog or a drizzle become rain, or light rain become moderate or heavy? When do physical symptoms become full-blown pneumonia? When does a breeze become a wind, and when do strong winds become storms, and when do storms such as tornadoes and hurricanes reach certain force strengths? When is a celestial body a planet? In each of these examples we can point to genuine cases of rain, pneumonia, wind, and planets and recognize that there are precursors that do not qualify. As we have seen with subspecies and planets (Margot 2006), arguing over where to place the threshold (and what gets excluded) can be contentious.

These debates over how to categorize phenomena cannot effectively call into question the existence of real states that exist along a continuum (in the case of subspecies, a biological continuum of differentiation). Among cases, discontinuities usually exist that help us understand the nature of these phenomena. It is left to practitioners to work out the finer details of generally agreed-upon thresholds and diagnoses. Those who do not find these details useful are free to ignore them and simply use (or not) the products of this work (cf. Inger 1961 and the comments therein). But for those who study these phenomena, or affect (or are affected by) them, attention is warranted, and the finer details of diagnoses become important. And, because biodiversity is not an ephemeral phenomenon and is threatened by anthropogenic forces, when geographically partitioned variation is present it should be properly assessed and, if warranted, managed for retention. This latter is in fact legally mandated at critical levels in some countries (e.g., U.S. Department of the Interior and U.S. Department of Commerce 1996, Committee on the Status of Endangered Wildlife in Canada 2005), and subspecies designations can be objectively assessed within multiple legal and biological frameworks (e.g., Topp and Winker 2008).

Yes, there are many differences among meteorologists, astronomers, physicians, and zoologists in subject matter and in methodologies, but there are commonalities as well in our efforts to diagnose conditions that often exist between agreed-upon states. It should not be a surprise that meteorologists, physicians, and astronomers have professional criteria that they use to make their diagnoses. As zoologists we can point to clear discontinuities that might tempt us to consider that we are different from these other fields, but there are definite similarities when we come to the cases that do not fall clearly into one category or another, when thresholds and criteria are often fervently debated. Data, existing and new, and reexaminations (second opinions) effectively enable us to treat diagnoses as hypotheses. Placing thresholds at points along continuous processes will always have an arbitrary component, and we can expect difficulties when knowledge is incomplete or when there are not clear boundaries (Hey et al. 2003). But we can agree as professionals that exemplary, real states exist along such continua and that there are professional guidelines to help us determine where particular cases fall.

One of the beauties of science is the way in which we scientists contribute to determining the details of such evolving guidelines. The history of the subspecies concept exhibits utilitarian changes made through time that enhanced both the concept's usefulness and its acceptability among professionals. Interestingly, and not surprisingly, a similar process occurred in the definition and widespread adoption of the Beaufort Scale, which is used to categorize air movements when reporting the weather (Huler 2004). If the past 150 years are any indication, we can look forward to a continued spirited discussion of subspecies as a long-recognized state in the processes of adaptation and speciation and in the state of biodiversity.

What does subspecific variation mean?—There are two aspects to the question of what subspecific variation means: the conceptual tool itself and the phenomena that cause the variation that it attempts to categorize. With regard to the labels, subspecies represent a heterogeneous assemblage of units, reflecting the nomenclatural application of denoting differences; this has been true since the inception of the concept (see above). Although the application of subspecific names needs ongoing reevaluation, even an ideal taxonomy will not change this unresolvable aspect of subspecies; for example, it is not possible to make them all what Simpson (1961) termed "little species." Also, as long as data remain incomplete, the hypothetical ideal taxonomy of one moment will continue to undergo change as our data and science improve.

With respect to the variation that these labels attempt to denote, a deceptively simple answer is that such variation represents evolutionary and developmental responses to geographically heterogeneous phenomena (biotic and abiotic). The phenomena involved in creating geographically partitioned variation are the very building blocks of biological evolution, the details of which remain the basis for research programs throughout evolutionary biology. As we know, differentiation can be affected by geography (e.g., isolation-by-distance, parapatry, allopatry, and their effects on gene flow), by selection (natural and sexual), and by neutral processes (e.g., mutation and lineage sorting or drift). Added to this are uncertain influences from environmental effects and developmental plasticity. At present it is safe to say that, except perhaps in a very few well-studied instances, we cannot accurately predict how these phenomena are responsible for any specific case of partitioned phenotypic variation in wild populations. Not knowing the answer at this level of detail, however, is not particularly problematic. Although we can answer this question only in the broadest terms, the fact that we can only rarely provide a highly detailed answer is more an indication of the exciting research to be done at the subspecific level than a problem of the concept itself.

It seems that, to a considerable degree, the debate over the subspecies concept comes from not being satisfied with this broad answer and yet not having, or being able to develop, a detailed one in a particular system or organismal group. This philosophical discord has often stemmed from discord in data sets, for example among those studying some attribute of subspecific variation and finding that their focus does not accord with the named subspecies of a previous worker. This has perhaps never been as pronounced as with the advent of genetic data. However, at shallow levels of divergence, small genetic data sets in which variation is largely dominated by neutral or near-neutral processes (Kimura 1983, Ohta 2002) should be expected to be discordant with phenotypic data sets, in which variation is likely dominated by processes that are not selectively neutral. Thus, evidence of decoupled genotypic and phenotypic marker systems may not be particularly informative when using one data type to ask questions of the other at species and subspecies levels.

However, as genomic data sets have improved, the genetic bases (protein coding and regulatory) for local, intraspecific adaptation are being revealed in many taxa (Mitchell-Olds et al. 2007). Further, phenotypic attributes of the sort used to describe subspecies are being shown to be adaptive (Mumme et al. 2006) and genetically based (Hoekstra et al. 2006). Genomic studies suggest that changes in color intensity (often an important attribute of described subspecies) can be genetically simple, but pattern changes are likely to involve *cis*-regulatory elements and mosaically pleiotropic loci rather than coding-sequence changes (Carroll 2008). Further, interspecific morphological differences appear to involve *cis*-regulatory element changes more than intraspecific ones do, which led Stern and Orgogozo (2008) to conclude that speciation causes fixation of a certain subset of the genetic variation that causes phenotypic variation within species. This not only echoes Simpson's (1961), Mayr's (1969), and others' views that only some subspecies will become species, but it provides us with important tools and hypotheses for examining the genetic bases of partitioned geographic variation and understanding them in a better collective sense in relation to speciation. And we should not lose sight of what a rich harvest subspecies and their characteristics can continue to provide for these frontiers of research in evolutionary biology.

How frequently might neutral phenotypic variation occur among populations? We do not know. Remsen (1984) demonstrated that among subspecies of many birds in the Andes of South America, there is a pronounced stochasticity in the among-species distributions of a leapfrog pattern of subspecific variation in plumage color and pattern. This led him to conclude that phenotypic differentiation at this scale had a strongly random component (Remsen 1984). Even though leapfrog patterns of subspecific variation are rare, the stochastic component to geographic variation

that Remsen's study revealed confirms that evolution within populations can proceed at different rates and in different directions; it does not tell us how (or whether) selection (natural and sexual) within the subspecies involved has affected these processes or their outcomes. If multiple characters are involved, the role for selection is likely enhanced.

The question of neutral phenotypic evolution at the population and subspecies levels remains open, but even a putatively exemplary case for neutral phenotypic evolution has proved with greater study to be driven instead by selection (Schemske and Bierzychudek 2007). We cannot blindly attribute all partitioned phenotypic variation to local adaptation (which historically occurred to excess; Gould and Lewontin 1979, Mayr 1983), because factors such as environment and developmental plasticity can cause divergence between populations. The influences of these factors are not yet well understood, but progress is being made (Price et al. 2003, West-Eberhard 2003, Suzuki and Nijhout 2007). Two things are important to realize. First, phenotypic variation caused by environmental factors and the developmental plasticity underlying it can be under selection and be adaptive (James 1991, West-Eberhard 2003). Second, the processes of divergence are dominated more by selection than by neutral changes (Coyne and Orr 2004). Partitioned phenotypic variation in multiple characters is likely to reflect partitioned genetic variation and thus extend into unmeasured attributes of populations; thus, even if the characters used to define a particular subspecies should prove not to be under local adaptive selection, they may well be concordant with such.

LOOKING FORWARD

Recognition of the reality of geographically partitioned phenotypic variation within species is so ubiquitous that biologists confront it daily, from the legal and ethical promotion of protecting and enhancing diversity (even within humans) to taxonomic nomenclature (ICZN 1999). That this phenotypic reality is not predictably variable using simple tools (e.g., mtDNA data or other single-character attributes) is the problem of the tools and the practitioners, not of the concept of partitioned intraspecific variation.

Subspecies represent a rather crude depiction of among-population patterns without regard to the processes underlying the geographic variation they attempt to describe. Two convenient ways to consider subspecies are as hypotheses and as taxonomic bookmarks. Both views allow one to proceed with the understanding that a presumed expert found sufficient variation that "something more should be done than merely to lock together into one heterogeneous fold forms so different" (T. Gill, in Coues 1884). As with all scientific hypotheses, such determinations are subject to further study and testing. And as with bookmarks, subspecific labels highlight items of note below the species level.

Perhaps the most useful academic purpose to which subspecific variation has been put is as a frame of reference for posing questions about animals and evolution. The most practical purpose is undoubtedly to describe biodiversity and, since the mid-20th century, to focus management and conservation efforts upon intraspecific units above the level of populations but below the level of species. In both cases, an often unspoken assumption is that the partitioned morphological variation upon which subspecies are based reflects an adaptive component—that these populations, although not fully species, possess unique attributes with local adaptive value. Whether the variation that subspecies denote reflects important local adaptation is an unresolved issue, but increasing evidence suggests that we would be safer in both basic and applied biology not to dismiss such an assumption outright. Indeed, our null hypothesis should be that subspecific differentiation is indicative of local adaptation, not the converse.

If we think of partitioned geographic variation as a species' fit to a variable environmental (and adaptive) landscape (a natural extension of the null hypothesis that such variation reflects local adaptation), then we can gain insights both into the processes of biological divergence and, from the applied perspective, into the likely retention of biodiversity during periods of global change. On a global scale, populations are currently so challenged by anthropogenically forced environmental changes that we would be foolish indeed if we failed to recognize that trying to maintain a fit to a variety of geo-environmental facets is a better approach than one-size-fits-all. It would be a grave mistake to think that subspecific variation that does not lead to speciation is not of evolutionary importance; such variation might be critical for resilience and persistence, preventing extinction of existing species during episodes of environmental

change. It would also be a mistake to think that, if lost, such variation could be regenerated under similar environmental conditions.

Although for over half a century we have had wide acceptance of a quantified lower limit for subspecific recognition (the 75% rule; Amadon 1949, Mayr et al. 1953, Simpson 1961, Mayr 1969, Mayr and Ashlock 1991, Patten and Unitt 2002, Haig et al. 2006), this limit became widely accepted after most avian subspecies had been described (Remsen, this volume). Further, acceptance of this quantified approach grew during a period of waning traditional taxonomy and a decline in the collection of specimens upon which this research is based (e.g., Winker 1996). Inasmuch as the simple description of biodiversity is no longer considered cutting-edge research, some might be tempted to dismiss the entire subject as a historical fad that has little place in modern science. That would be an error. In fact, the underlying reasons for studying such subjects have never been stronger. In the past few decades, we have experienced not only a resurgence in systematics, but also robust growth in evolutionary and conservation biology and ecology as well as increased interdisciplinarity in wildlife biology. In addition, entirely new disciplines such as phylogeography (Avise 2000) and landscape genetics (Manel et al. 2003, Holdregger and Wagner 2006) continue to provide a modern cachet to the fundamental aspects of geographic variation that caused the subspecies concept to be developed in the first place. Subspecific variation in each of these areas is core material; it may only be taxonomic housekeeping that has not been kept up.

Using subspecies to help us understand fundamental aspects of biological differentiation, a longstanding tradition (Remsen 1984), continues in fields as diverse as ecology, evolutionary biology, wildlife management, and genetics, demonstrating both that subspecies retain heuristic value in research and that they are indeed often correlated with phenomena of longstanding interest. For example, Belliure et al. (2000) found that dispersal distances in birds were correlated with subspecies richness. Irwin et al. (2001, 2005), focusing on the classic case of the ring species in *Phylloscopus trochiloides* (Mayr 1942b), showed how song (a trait used in mate choice) exhibited parallel changes concordant with subspecific divergence as the ring progressed north into Siberia and that genetic evidence supported a cessation of gene flow between these most-diverged populations. Sol et al. (2005)

found that among Holarctic passerines, species with relatively larger brains tended to have more subspecies; these authors used subspecies as an index of intraspecific phenotypic diversification and relative brain size as a correlate of a propensity for behavioral changes, and their results strongly implicate behavioral change as a driving force in evolutionary diversification. In a global survey of avian subspecies, Phillimore and Owens (2006) found that ~36% were phylogenetically distinct (monophyletic with respect to individuals of different subspecies when using mtDNA sequence data), and further study revealed that biogeographic factors such as breeding range and habitat heterogeneity explained ~30% of subspecies richness (Phillimore et al. 2007). Also using subspecies, Martin and Tewksbury (2008) found that phenotypic differentiation within bird species is greater at lower latitudes, which fits evolutionary hypotheses for the latitudinal gradient in species diversity. Darwin (1859) was right in pointing out the importance of subspecies, and these and other recent efforts (e.g., see Price 2008) continue to show this.

Martin and Tewksbury (2008) summarized three lines of evidence that support using subspecies as an index of divergence in speciation research. The first, that subspecific variation ranges from subtle character differences to that approaching allopatric species and therefore reflects intermediate stages in the process, has been a mainstay of the use of subspecies in this context since Darwin (1859). The second, that ring species show the temporal progression of speciation with subspecific differentiation accruing to the point of reproductive isolation at the tips, has been another important focus in evolutionary biology (Mayr 1942b, 1963; Irwin et al. 2001, 2005; Coyne and Orr 2004). Their third point, which they demonstrated, is that the maximum number of subspecies in a species within a genus covaries positively with the number of species in that genus, which likely indicates that a lineage's propensity to differentiate is an attribute that it holds for a substantial period of evolutionary time (well through the speciation process and observable at the subspecies level).

Because in using subspecies we are trying to label and understand something that occurs at an interval along an inherently continuous process (divergence and differentiation), incomplete data and differing views, not to mention disagreement on species concepts, combine to produce ongoing debates about what subspecies are, what they reflect, and whether a particular example is one or not. But this debate can overthrow neither the biological reality that the concept embodies nor the utility that it has demonstrated for over 150 years.

Both hypothesis testing and descriptive (discovery science) approaches to our study of biodiversity continue to make subspecies useful. Consequently, the use of subspecies in ornithology remains strong. For example, major works such as del Hoyo et al. (1992–2008), Dickinson (2003), Poole and Gill (1992–2002), Cramp et al. (1977–1994), and Marchant et al. (1990–2006) all used the biological species concept and treated polytypic species at the subspecific level. Although the AOU has not given detailed treatment to the subspecies of North American birds since 1957 (AOU 1957), its commitment to that level of diversity remains strong (AOU 1986, 1998; AOU Check-list Committee pers. comm.). Indeed, legally mandated management obligations are beginning to push back on this neglect, and not just in birds (e.g., Haig et al. 2006). Moreover, there are hundreds of papers published each year in peer-reviewed journals with subspecies as a topic (Haig and Winker, this volume).

Subspecies have enthralled and frustrated zoologists—and especially ornithologists—since the inception of the concept. To rail against this concept, however, is as ineffective as railing against the wind. Subspecies have been and remain an important conceptual tool—and a taxonomic label—that can help zoologists categorize, study, and conserve biodiversity. As eternally controversial as politics, subspecies will likely remain integral to research and management in zoology.

Acknowledgments

I thank S. Haig, C. Cicero, G. Graves, D. Gibson, R. Dickerman, J. Rappole, R. Banks, R. Browning, J. Klicka, J. Cracraft, T. Peterson, M. Patten, R. Zink, and J. V. Remsen, Jr., for discussions about variation within biological species, and Remsen, Haig, F. James, and M. Morrison for comments on an earlier draft.

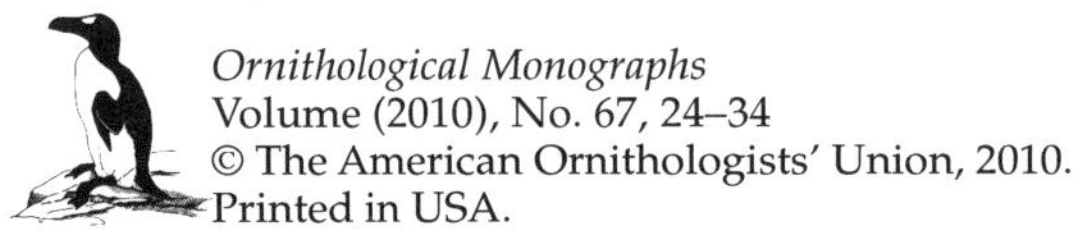

Ornithological Monographs
Volume (2010), No. 67, 24–34
© The American Ornithologists' Union, 2010.
Printed in USA.

CHAPTER 2

AVIAN SUBSPECIES AND THE U.S. ENDANGERED SPECIES ACT

Susan M. Haig[1,4] and Jesse D'Elia[2,3]

[1]*U.S. Geological Survey Forest and Rangeland Ecosystem Science Center, 3200 SW Jefferson Way,
Corvallis, Oregon 97331, USA;*
[2]*U.S. Fish and Wildlife Service, 911 NE 11th Avenue, Portland, Oregon 97232, USA; and*
[3]*Department of Fisheries and Wildlife, Oregon State University, Corvallis, Oregon 97331, USA*

Abstract.—Scientific debate over identification of taxa below the species level has persisted for centuries. This issue can be especially problematic for avian species, because dispersal is often orders of magnitude greater than in other vertebrates, leaving genetic differences among groups proportionately smaller. While the debate lingers, management decisions, often with millions of dollars and potential extinctions resting on the outcome, are regularly made by agencies tasked with maintaining lists of threatened and endangered taxa. With outdated taxonomic treatments and no formal policy or guidelines for defining species or subspecies, agencies have no authority to cite in determining limits to species or subspecies ranges. Lack of guidance from professional organizations regarding taxonomic criteria and lists does not benefit these species of concern. Here, we describe how subspecies designations are evaluated under the Endangered Species Act, tradeoffs between maintaining the biological species concept in avian taxonomy versus adopting a phylogenetic species approach, and why it is imperative for scientific organizations to maintain updated taxonomic treatments regardless of the species concept they use.

Key words: biological species concept, distinct population segment, Endangered Species Act, phylogenetic species concept, species, subspecies.

Subespecies de Aves y el Acta de Especies Amenazadas de los Estados Unidos

Resumen.—El debate científico sobre la identificación de taxones por debajo del nivel de especie ha persistido por siglos. Este asunto puede ser especialmente problemático para las especies de aves, debido a que la dispersión en éstas con frecuencia es órdenes de magnitud mayor que en otros vertebrados, lo que conduce a que las diferencias entre grupos sean proporcionalmente más pequeñas. Mientras el debate continúa sin resolverse, las agencias que mantienen listas de taxones amenazados regularmente toman decisiones de manejo que ponen en juego millones de dólares y extinciones potenciales. Al contar con tratamientos taxonómicos desactualizados y al carecer de políticas o lineamientos formales para definir especies y subespecies, las agencias no tienen autoridades a las cuales citar al determinar los límites de las distribuciones de las especies y subespecies. La falta de guianza por parte de organizaciones profesionales con respecto a los criterios taxonómicos y a las listas no beneficia a las especies de interés en la conservación. En este trabajo, describimos cómo las designaciones de subespecies son evaluadas bajo el Acta de Especies Amenazadas, los compromisos entre mantener el concepto biológico de especie en la taxonomía de las aves versus adoptar un enfoque de especies filogenéticas, y por qué es imperativo que las organizaciones científicas mantengan tratamientos taxonómicos actualizados, independientemente del concepto de especie que utilicen.

---

[4]E-mail: susan_haig@usgs.gov

Ornithological Monographs, Number 67, pages 24–34. ISBN: 978-0-943610-86-3. © 2010 by The American Ornithologists' Union. All rights reserved. Please direct all requests for permission to photocopy or reproduce article content through the University of California Press's Rights and Permissions website, http://www.ucpressjournals.com/reprintInfo.asp. DOI: 10.1525/om.2010.67.1.24.

THE SUBSPECIES CONCEPT may be "the most critical and disorderly area of modern systematic theory" (Wilson and Brown 1953:100), and debate over the existence and definition of subspecies will likely continue for years to come because of fundamental differences of opinion regarding species concepts and inherent difficulties in objectively determining intraspecific units. In the meantime, bird species worldwide face an extinction crisis of epic proportions (e.g., 2–3 times the prehuman rates of extinction; Brooke et al. 2008). This crisis results from factors such as anthropogenic habitat destruction, climate change, and introduction of invasive species. As a result, many government and nongovernmental conservation agencies around the globe seek to efficiently and effectively identify and prioritize taxonomic and conservation units eligible for protection (Table 1; Phillimore and Owens 2006). Garnett and Christidis (2007) summarized international endangered-species legislation and definitions of what taxonomic groups are eligible for listing under each system and found that it tends to be in the poorer countries, many of which are highly subspeciose, that subspecies assessments have not been undertaken. Some of those poorer countries rely exclusively or heavily on assessments by the International Union for the Conservation of Nature, which does not currently assess avian subspecies because of a lack of resources. However, in countries where these assessments are undertaken, subspecies taxonomy can have considerable influence on the allocation of limited conservation resources. To provide some perspective on this situation, we discuss below why subspecies designations matter for avian species listed under the U.S. Endangered Species Act. We consider the legal and conservation implications of avian taxonomists officially adopting either the biological species concept or a phylogenetic species concept.

WHY SUBSPECIES MATTER UNDER THE ENDANGERED SPECIES ACT

Subspecies have been eligible for protection since the inception of endangered species laws in the United States. In 1966, 13 of the 36 avian taxa listed under the 1966 Endangered Species Preservation Act were subspecies. Today, under the Endangered Species Act of 1973 (as amended, 16 U.S.C. 1531 et seq.), 40 of 90 avian taxa listed are subspecies, which represents one of the highest percentages (44%) of avian subspecies listings

among national and international classification systems of imperiled species (Table 1). Yet most taxonomists are not aware of how subspecies designations affect the listing and recovery of populations under the Endangered Species Act, despite the fact that taxonomic descriptions often have real conservation consequences. To begin to understand the complexities of the issue, one must first appreciate what is eligible for listing under the Endangered Species Act below the species level and how those listings are affected by trinomial nomenclature.

The Endangered Species Act allows listing of species, subspecies, and "distinct population segments" of vertebrates. The Endangered Species Act and its implementing regulations do not include "evolutionarily significant units" in their definition of units eligible for protection. The concept of using evolutionarily significant units for Endangered Species Act listings was first introduced in a National Marine Fisheries Service (NMFS) Policy on the Definition of Species under the Endangered Species Act (NMFS 1991). The policy on distinct population segments considers evolutionarily significant units to be equivalent to distinct population segments, but in listing species under the Endangered Species Act, evolutionarily significant units have been applied only to Pacific salmon and therefore are not applicable to avifauna.

It has also been argued that "significant portions of a species' range" are eligible for Endangered Species Act protections, because this phrase is included in the act's definitions of "threatened" and "endangered." However, there is no consensus on what a significant portion of a species' range is or on whether the entire species gets listed if only a significant portion of the range is at risk (see Vucetich et al. 2006; Waples et al. 2007a, b; Nelson et al. 2007; Office of the Solicitor 2007; D'Elia et al. 2008). Litigation on this point is ongoing.

Defining the significance of a population is also a key factor in determining eligibility for status as a distinct population segment (see below). Although distinct population segments can be population segments of either species or subspecies (U.S. Fish and Wildlife Service [USFWS] and NMFS 1996; Center for Biological Diversity v. USFWS, 9th Cir. 2008), because significance is evaluated against the taxon to which it belongs (per the policy on distinct population segments), subspecies designations play a critical role in the legitimacy of some distinct-population-segment designations. Without subspecific status, some distinct population segments simply would not meet the significance

TABLE 1. Classification systems for imperiled species and the number of avian species and subspecies listings under each system.

Classification system	Number of avian taxa listed	Allows subspecific listing?	Number of avian subspecies listed	Percentage of avian taxa listed as subspecies	Categories[a]
International					
CITES Appendices	1,455	Yes	17	1	Appendix I, II, and III
IUCN Red List	1,217	Yes	0	0	CE, E, V
NatureServe Conservation Status Assessments	225	Yes	128	57	CI, I, V
U.S. Endangered Species Act List of Endangered and Threatened Wildlife (International)	186	Yes	37	20	T, E
Australia					
Australia's Environment Protection and Biodiversity Conservation Act List of Threatened Fauna	108	Yes	55	51	CE, E, V
Brazil					
Lista Nacional das Espécies da Fauna Brasileira Ameaçadas de Extinção Grupo	160	Yes	44	28	CE, E, V
Canada					
Federal List of Widlife Species at Risk	53	Yes	19	36	E, T, SC
Costa Rica					
Lista de Especies de Aves con Poblaciones Reducidas y en Peligro de Extincion para Costa Rica (2005)	17	No	—	—	E
Europe					
European Union Birds Directive Species	193	Yes	18	9	Annex I species
Mexico					
Especies enlistadas en la NOM-059-SEMARNAT-2001 y de especies prioritarias	361	Yes	85	24	E, T, SSP
New Zealand					
New Zealand Threat Classification List	66	Yes	22	33	NC, E, V, SD
Panama					
Animales en Peligro de Extinción en Panamá	27	Yes	0	0	E
Peru					
Especies de Fauna Amenazada del Peru	108	No	—	—	CE, E, V
Russia					
Red Data Book for the Russian Federation (1997)	123	Yes	11	9	—
South Africa					
South Africa's Red List	32	Yes	0	0	CE, E, V, PS
United States					
U.S. Endangered Species Act List of Endangered and Threatened Wildlife (Candidates)	13	Yes	3	23	C
U.S. Endangered Species Act List of Endangered and Threatened Wildlife (Domestic)	90	Yes	40	44	T, E

[a] C = candidate, CE = critically endangered, CI = critically imperiled, E = endangered, I = imperiled, NC = nationally critical, PS = protected species, SC = special concern, SD = serious decline, SSP = subject to special protection, T = threatened, and V = vulnerable.

test of the policy (i.e., those populations that are significant to the subspecies but not to the more widely distributed species) and therefore would not merit the substantial protections provided under the Endangered Species Act.

The Endangered Species Act does not define distinct population segments. However, the USFWS and the NMFS (1996), the two agencies tasked with implementing the Endangered Species Act, established a joint formal policy interpreting what the Endangered Species Act means by distinct population segments. According to the policy, two tests must be satisfied for a population segment to qualify as a distinct population segment: discreteness of the population segment in relation to the remainder of the taxon and significance of the population segment to the taxon. If a population segment qualifies as a distinct population segment, the conservation status of that distinct population segment is evaluated to determine whether it is threatened or endangered.

A population segment of a vertebrate species may be considered discrete by the USFWS if it satisfies either of the following conditions: (1) it is markedly separated from other populations of the same taxon as a consequence of physical, physiological, ecological, or behavioral factors; or (2) it is delimited by international governmental boundaries between which there are differences in control of exploitation, management of habitat, conservation status, or significant differences in regulatory mechanisms.

If a population is found to be discrete, it is evaluated for significance under the policy on distinct population segments on the basis of its importance to the taxon to which it belongs. This consideration may include, but is not limited to, the following: (1) persistence of the discrete population segment in an ecological setting unusual or unique to the taxon, (2) evidence that loss of the discrete population segment would result in a significant gap in the range of a taxon, (3) evidence that the population represents the only surviving natural occurrence of a taxon that may be more abundant elsewhere as an introduced population outside its historical range, or (4) evidence that the population differs markedly from other populations of the species or subspecies in its genetic characteristics. Thus, the policy on distinct population segments clearly contemplated the potential to conserve threatened or endangered populations that have a distinct evolutionary history from other populations of the same species

or subspecies. This policy affords the USFWS substantial flexibility to ensure that distinct vertebrate populations are protected before the entire taxon is imperiled (Pennock and Dimmick 1997).

If a population segment is discrete and significant (i.e., it is a distinct population segment), its evaluation for endangered or threatened status is based on a thorough review of population numbers, trends, and threat factors that affect the population segment. Although the policy on distinct population segments has been upheld by several courts, individual listing decisions that rely on this policy remain heavily litigated and subject to policy interpretation regarding what qualifies as a discrete and significant population. Conversely, subspecific designations backed by taxonomic authorities are usually not subject to this kind of policy interpretation or litigation when they are kept current, because the courts generally defer to the scientific authority.

LISTING PRIORITIZATION

Subspecific designations also come into play in prioritizing candidate species for listing under the Endangered Species Act and in prioritizing recovery planning. Currently, 2 of 11 candidate avian taxa are subspecies—Streaked Horned Lark (*Eremophila alpestris strigata*) and Red Knot (*Calidris canutus rufa*)—and >18% of all candidate taxa (46 of 247) are subspecies or varieties (see http://ecos.fws.gov/tess). The Endangered and Threatened Species Listing and Recovery Priority Guidelines (USFWS 1983) indicate that species will be afforded priority over subspecies in listing and recovery actions. Later, in its policy on distinct population segments (USFWS and NMFS 1996), the USFWS established that distinct population segments would be afforded the same considerations as subspecies, but when a subspecies and distinct population segments have the same numerical priority, the subspecies will generally receive higher priority.

RESOLVING TAXONOMIC UNCERTAINTY
IN CONSERVATION DECISIONS

Choices of what to conserve must often be made with regard to populations that are not completely separate from others, or when information regarding the relationships and degrees of distinction among populations is incomplete. Such decisions, although often difficult because of remaining uncertainties, are similar to decisions made in other contexts in which scientists have imperfect knowledge or where nature does not present clear boundaries (Hey et al. 2003). Thus, taxonomic decisions should not be viewed as fundamentally different from other conservation decisions that must be made regardless of uncertainties.

There are two error types described in conservation decisions (Woods and Morey 2008). The first type is an underprotection error, which in taxonomy means defining too few taxa to effectively conserve biodiversity (Skalski et al. 2008). This is a particular issue with island species for which the true species diversity is underestimated because similar taxa from different islands are lumped (Hazevoet 1996; Pratt and Pratt 2001; Pratt, this volume). The potential consequences of this error include loss of taxa and preclusion of conservation actions before species are critically endangered. The second type of error is overprotection: defining too many taxa, which can create excessive administrative costs and dilute conservation dollars.

There are potentially serious biological risks to flawed taxonomic splitting if it results in deleterious management. For example, erroneous delineation of subspecies or distinct population segments can delay or impede management actions to reestablish gene flow to an inbred population fragment that has become isolated because of habitat loss or other anthropogenic barriers. Although this could be overcome through intercrossing subspecies or distinct population segments (that are erroneously delineated) to rescue the inbred population fragment from local extinction, such intercrosses are rarely used and require the approval of the USFWS director (USFWS and NMFS 2000). In any case, the potential consequences of over- and underestimating the number of subspecies are economic losses, loss of conservation options because funds are misdirected, loss of scientific credibility, and loss of important components of biodiversity. Balancing these errors requires interpreting taxonomic descriptions in a manner similar to that used in other types of conservation decisions, which establish explicit criteria appropriate to the problem (Haig et al. 2006, Gippoliti and Amori 2007). Decisions will ultimately be made in the context of societal goals and the resources available. For example, a country or state with many resources and the strong support of citizens might be able to manage what would be considered dramatic overprotection by an entity that can only afford to try to keep portions of ecosystems from disappearing entirely.

Issues of taxonomic disagreement at the species and subspecies levels are handled on a case-by-case basis under the Endangered Species Act. Although neither the Endangered Species Act nor its implementing regulations dictate how the USFWS or NMFS must resolve taxonomic uncertainty, the preamble to the listing regulations, published in the *Federal Register* on February 27, 1980 (p. 13013), established that the Services "will rely on generally accepted lists of taxa [maintained by professional societies] when they are available." Because the USFWS and NMFS are also directed to make listing determinations solely on the basis of the best scientific and commercial data available, newer data from the scientific literature and information provided by professional taxonomists must be considered when a professional society's taxonomic treatment is not kept current (Center for Biological Diversity v. Lohn 2003, Western Washington District Court). This latter approach can consume considerable amounts of time and resources, place difficult and controversial taxonomic decisions in front of the USFWS and NMFS, and increase litigation risk. Thus, from this perspective, the onus is more appropriately placed on the taxonomic societies to maintain updated species and subspecies lists and criteria. If these lists are not maintained, the USFWS and NMFS may need to expend scarce resources for outcomes that result in either type of conservation decision error, with potential implications for the taxa in question. Recognizing the costs, timeliness, and importance of these taxonomic descriptions, perhaps agencies like the U.S. National Science Foundation could provide grants to major taxonomic societies to regularly update their treatments. This would formally involve independent experts, help resolve disputes in a timely manner, and highlight areas that need further research.

Although critical, taxonomic lists alone are not enough for resolving Endangered Species Act issues. Scientific societies can make significant contributions to policy deliberations by maintaining taxonomic lists and providing the rationale for their listing decisions. The American Ornithologists' Union's (AOU) updates to its check-list, which identify new information and why a certain course is being followed in response to it, are particularly valuable in that they provide the scientific basis for changes in entries (although subspecies revisions are not included at present, except when they are being raised to species status). When there are multiple taxonomic authorities with competing lists, providing such a rationale adds context to the content of the lists. This transparency, in turn, can help reduce the number of unwarranted Endangered Species Act petitions or legal challenges that arise from taxonomic disagreements.

Without input from taxonomic authorities, the USFWS, with its resources already stretched thin on many high-priority conservation needs, is forced to undertake resolutions of taxonomic disputes under regulatory and legal time constraints, sometimes operating in a politically charged atmosphere. Furthermore, without the backing of scientific organizations, the results of USFWS efforts are often called into question in court and are not always afforded the same deference given to taxonomic treatments by groups like the AOU. For example, in 2006, the 9th Circuit Court ruled that the USFWS failed to explain adequately why it reversed course after decades of recognition of the subspecies Western Sage Grouse (*Centrocercus urophasianus phaios*, now *C. minimus*; see Oyler-McCance et al., this volume) in making its 90-day not-substantial finding (Institute for Wildlife Protection v. Norton; 9th Circuit 2006). The decision rested on a conclusion that the Western Sage Grouse is not a valid subspecies. The court ruled that the USFWS did not explain the principles used to determine the validity of a subspecies classification and noted that the only taxonomist whom the USFWS consulted believed that, while the subspecies was weakly characterized, it would be wise to continue to recognize it. However, the court upheld the USFWS's determination that the petitioned population did not constitute a distinct population segment. The court remanded the finding to USFWS to revisit its 90-day finding. In a similar example with the opposite outcome (United States v. Guthrie, 50 F.3d 936; 11th Cir. 1995), the court ruled that the USFWS was not arbitrary and capricious in its finding that the Alabama Red-belly Turtle (*Pseudemys alabamensis*) was a valid species despite uncertainty regarding speciation from the Florida Red-belly Turtle (*P. nelsoni*). The court gave deference to the agency partly because the validity of the species designation was accepted by several taxonomic authorities (and published in their lists or books), notwithstanding a lack of complete certainty. Thus, in the absence of up-to-date taxonomic treatments, the

USFWS not only bears the weight of generating an updated taxonomic synthesis but must also defend that synthesis and, if unsuccessful, may be forced to repeat analyses until the plaintiffs or the courts are satisfied.

Subspecific Taxonomic Rank versus Taxonomic Inflation

Scientific debate over species concepts and the identification of taxonomic groups below the species level continues (e.g., in this monograph). These taxonomic issues can be especially problematic for avian species because dispersal is often orders of magnitude greater than in other vertebrates, leaving differences among groups proportionately smaller (Haig et al. 2006). Although numerous species concepts have been proposed, the biological species concept is the current standard in avian taxonomy, and the phylogenetic species concept is the leading challenger. There are legal, administrative, and conservation costs and benefits to choosing one species definition over another for Endangered Species Act listing decisions. Below, we explore the implications of continuing to use the biological species concept versus adopting the phylogenetic species concept on Endangered Species Act listing activities for birds.

Biological Species Concept

The biological species concept defines a species as a group of interbreeding or potentially interbreeding natural populations separated from other such groups by intrinsic (genetically fixed) barriers to gene flow (Mayr 1942a, 2000a, b). The biological species concept is the most universally accepted species definition, but it is not without its critics. For example, problems arise when there is gradual evolution of reproductive isolation and discrete population entities are difficult to identify (González-Forero 2009, Irwin 2009, and many others). However, there are benefits to using this approach, including the relative ease with which a species can be identified.

The biological species concept includes the concept of subspecies, defined by Avise and Ball (1990) as groups of actually or potentially interbreeding populations (normally allopatric) that are genealogically highly distinctive but reproductively compatible with other such groups. Criteria for determining subspecies have never been universally defined (reviewed in Haig et al.

2006); however, one definition, the 75% rule (Amadon 1949, Patten and Unitt 2002), states that a subspecies is valid if 75% or more of a population is separable from all (or >99%) members of the adjacent population. Although the 75% rule is more quantitative than other definitions, there is disagreement about the 75% threshold and the number of characters that should be used when comparing populations (Patten and Unitt 2002; James, this volume; Haig and Winker, this volume). Another criterion for subspecies is that of reciprocal monophyly (Avise 2000), which has been endorsed by Zink (2004), although the same criterion is used to define species under the phylogenetic species concept. Thus, there is debate about the appropriateness of using reciprocal monophyly for both species and subspecies concepts (Goldstein et al. 2000).

Continued use of the biological species concept in avian taxonomy means that subspecies listings under the Endangered Species Act can be maintained (Fig. 1). No changes to subspecies means that the USFWS can focus on conservation priorities such as reducing the backlog of candidates, rather than administrative corrections to listings. Figure 1 illustrates the implications for implementing (or not) these ideas for avian species. There are two additional outcomes not portrayed in the figure: (1) that both methods define the same species, or (2) that the biological species concept defines more species than the phylogenetic species concept. The latter is particularly likely for recently derived species in which differentiation is driven by disruptive selection. However, the scenario portrayed is what we believe to be most likely for the vast majority of bird species.

PHYLOGENETIC SPECIES CONCEPT

An overarching phylogenetic species concept has yet to be definitively described (Coyne and Orr 2004), but some definitions have stated that species are the least inclusive taxon recognized in a phylogenetic classification (Hennig 1966, Wheeler and Platnick 2000) or that species are the smallest diagnosable clusters of organisms, distinct from other such clusters, within which there are parental patterns of ancestry and descent (Cracraft 1983). Although the phylogenetic species concept can include other criteria, such as morphology and song, molecular markers are

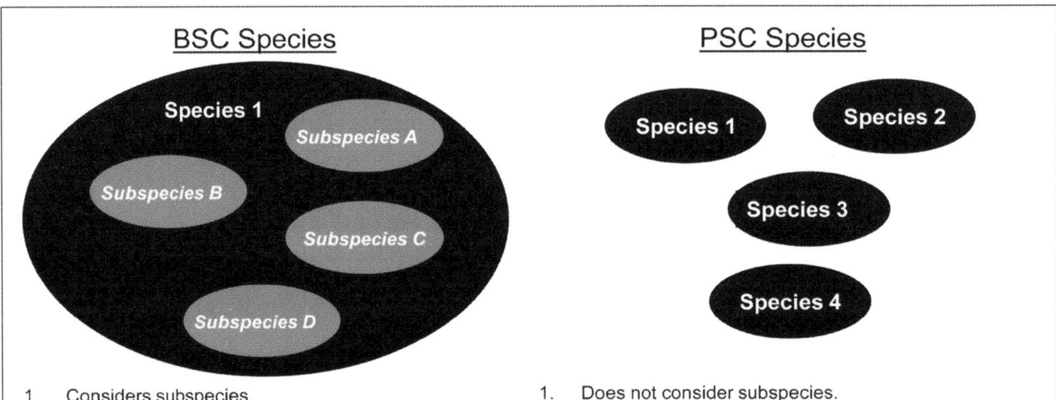

	BSC Species		PSC Species
1.	Considers subspecies.	1.	Does not consider subspecies.
2.	One species listing could contain many PSC species	2.	Might have to list numerous taxa relative to 1 BSC sp.
3.	Avian[1] taxa usually identifiable to law enforcement.	3	Avian taxa not always identifiable to law enforcement.
4.	Considers genetic, morphological, behavioral, ecological and environmental factors.	4.	Primarily considers genetic and morphological traits.
5.	Species may be more difficult to define if more than molecular markers are considered.	5.	Defining species may be easier to quantify based on molecular markers.
6.	Currently accepted by AOU[2], thus no changes to U.S. Endangered Species list.	6.	If accepted by AOU[2], many changes needed to re-list taxa on U.S. Endangered Species list.

[1]Most other taxa as well, however there are some cryptic species/subspecies that could be difficult.

[2]AOU refers to AOU Committee on Classification and Nomenclature.

FIG. 1. Considerations for Endangered Species Act listing of avian taxa under the biological species concept versus the phylogenetic species concept. See text for discussion of further outcomes of either scenario.

currently the prime source used in defining species under the criterion of reciprocal monophyly. The phylogenetic species concept generally addresses issues at the species level; however, the Committee on Phylogenetic Nomenclature currently accepts subspecies designations as long as they are not used as a rank (Dayrat et al. 2008).

Proponents of using the phylogenetic species concept in avian taxonomy argue that the major benefit of adopting it would be long-term benefits in that a relatively simple, arguably objective methodology would be in place to resolve species-level issues. For example, Zink (2004) suggested that this is necessary if we are to move beyond arguing over defining a listable unit and actually do something of conservation value. Although this argument has some merit, the major issue with the phylogenetic species concept is that it mistakes diagnosability for biological importance, something that is meaningful only to humans.

Under the phylogenetic species concept, many species based on the biological species concept would be split into two or more species (i.e., many subspecies would become full species). By some estimates, there would be 48% more species among all taxa (and no subspecies) if the phylogenetic species concept were used over the biological species concept (Agapow et al. 2004). Dillon and Fjeldså (2005) compared the numbers of bird species recognized in sub-Saharan Africa under the biological species concept and phylogenetic species concept and found ~33% more under the phylogenetic species concept, but patterns of endemism and diversity remained unchanged. Zink (2004) stated that for birds there could be as many as 100% more species. This taxonomic inflation results not only in the detection of new species but also in a greater proportion that are endangered because of a reduction in their range (Harris and Froufe 2004, Isaac et al. 2004, Padial and De la Riva 2006). However, taxonomic inflation does not necessarily correlate with inflation in endangered species lists where subspecies and populations (e.g., distinct population segments) are already eligible for protection, as in the case of the Endangered Species Act. Conversely, endangered species lists that are limited to, or biased toward, full species protection (e.g., IUCN Red List) would be affected by taxonomic inflation if the phylogenetic species concept were adopted (Garnett and Christidis 2007). The challenge is to weigh the costs and benefits (scientific, biological,

economic, and political) of these scenarios and understand the consequences of this choice.

Adoption of the phylogenetic species concept by the AOU Committee on Classification and Nomenclature would be relatively benign for many species, such as the Spotted Owl (*Strix occidentalis*), in which subspecies are genetically distinct and would likely be considered full species under the phylogenetic species concept (Haig et al. 2004 Funk et al. 2008). Most island subspecies would likely find support as full species under the phylogenetic species concept, especially those on remote islands such as Hawaii, where separation of the subspecies from conspecifics occurred long ago and where there is little or no interbreeding (Pratt, this volume). This is significant, because island species make up 42.5% of the subspecies currently listed under the Endangered Species Act (30% are from Hawaii or other remote South Pacific islands; Table 2). Some mainland subspecies that are in close proximity (e.g., the three southern California Clapper Rails that are listed; see Fleischer et al. 1995; Table 2) would likely not be recognized under the phylogenetic species concept, because of a recent common ancestor, occasional interbreeding, or both, resulting in low levels of genetic differentiation.

Although the phylogenetic species concept could potentially double the number of recognized bird species, this might not be problematic for Endangered Species Act implementation because subspecies would be eliminated and populations below subspecies are already eligible for protection (provided that they meet distinct-population-segment criteria). Thus, the net change in listable entities under the Endangered Species Act might not change substantially under the phylogenetic species concept. However, there is considerable uncertainty regarding what the taxonomic landscape would look like under the phylogenetic species concept, and uncertainty can cause confusion in agencies tasked with maintaining the lists as well as among conservation partners that rely on these lists for management decisions (Funk et al. 2007). The impact on conservation of threatened and endangered species under the Endangered Species Act could be significant if many species based on the phylogenetic species concept were identified outside the bounds of currently identified subspecies or distinct population segments.

Regardless of which taxa are maintained, split, or no longer recognized under the phylogenetic

TABLE 2. Avian subspecies as listed under the U.S. Endangered Species Act (E = Endangered, T = threatened).

Common name	Scientific name	Listing status	Year listed	Island subspecies?	Listing citation
Akepa, Hawaii (honeycreeper)	*Loxops coccineus coccineus*	E	1970	Yes	35 FR 16047
Akepa, Maui (honeycreeper)	*Loxops coccineus ochraceus*	E	1970	Yes	35 FR 16047
Bobwhite, masked (quail)	*Colinus virginianus ridgwayi*	E	1967	No	32 FR 4001
Caracara, Audubon's crested	*Polyborus plancus audubonii*	T	1987	No	52 FR 25229
Coot, Hawaiian	*Fulica americana alai*	E	1970	Yes	35 FR 16047
Crane, Mississippi sandhill	*Grus canadensis pulla*	E	1973	No	38 FR 14678
Elepaio, Oahu	*Chasiempis sandwichensis ibidis*	E	2000	Yes	65 FR 20760
Falcon, northern aplomado	*Falco femoralis septentrionalis*	E	1986	No	51 FR 6686
Flycatcher, southwestern willow	*Empidonax traillii extimus*	E	1995	No	60 FR 10693
Gnatcatcher, coastal California	*Polioptila californica californica*	T	1993	No	58 FR 16742
Hawk, Puerto Rican broad-winged	*Buteo platypterus brunnescens*	E	1994	Yes	59 FR 46710
Hawk, Puerto Rican sharp-shinned	*Accipiter striatus venator*	E	1994	Yes	59 FR 46710
Kingfisher, Guam Micronesian	*Halcyon cinnamomina cinnamomina*	E	1984	Yes	49 FR 33881
Kite, Everglade snail	*Rostrhamus sociabilis plumbeus*	E	1967	No	32 FR 4001
Millerbird, Nihoa	*Acrocephalus familiaris kingi*	E	1967	Yes	32 FR 4001
Moorhen, Hawaiian common	*Gallinula chloropus sandvicensis*	E	1967	Yes	32 FR 4001
Moorhen, Mariana common	*Gallinula chloropus guami*	E	1984	Yes	49 FR 33881
Owl, Mexican spotted	*Strix occidentalis lucida*	T	1993	No	58 FR 14248
Owl, northern spotted	*Strix occidentalis caurina*	T	1990	No	55 FR 26114
Petrel, Hawaiian dark-rumped	*Pterodroma phaeopygia sandwichensis*	E	1967	No[a]	32 FR 4001
Pigeon, Puerto Rican plain	*Columba inornata wetmorei*	E	1970	Yes	35 FR 16047
Plover, western snowy	*Charadrius alexandrinus nivosus*	T	1993	No	58 FR 12864
Prairie-chicken, Attwater's greater	*Tympanuchus cupido attwateri*	E	1967	No	32 FR 4001
Rail, California clapper	*Rallus longirostris obsoletus*	E	1970	No	35 FR 16047
Rail, light-footed clapper	*Rallus longirostris levipes*	E	1970	No	35 FR 16047
Rail, Yuma clapper	*Rallus longirostris yumanensis*	E	1967	No	32 FR 4001
Shearwater, Newell's Townsend's	*Puffinus auricularis newelli*	T	1975	No[a]	40 FR 44149
Shrike, San Clemente loggerhead	*Lanius ludovicianus mearnsi*	E	1977	Yes	42 FR 40682
Sparrow, Cape Sable seaside	*Ammodramus maritimus mirabilis*	E	1967	No	32 FR 4001
Sparrow, Florida grasshopper	*Ammodramus savannarum floridanus*	E	1986	No	51 FR 27492
Sparrow, San Clemente sage	*Amphispiza belli clementeae*	T	1977	Yes	42 FR 40682
Stilt, Hawaiian	*Himantopus mexicanus knudseni*	E	1970	Yes	35 FR 16047
Swiftlet, Mariana gray	*Aerodramus vanikorensis bartschi*	E	1984	Yes	49 FR 33881
Tern, California least	*Sterna antillarum browni*	E	1970	No	35 FR 8491
Tern, roseate[b]	*Sterna dougallii dougallii*	E/T	1987	No	52 FR 42064
Thrush, Molokai	*Myadestes lanaiensis rutha*	E	1970	Yes	35 FR 16047
Towhee, Inyo California	*Pipilo crissalis eremophilus*	T	1987	No	52 FR 28780
Vireo, least Bell's	*Vireo bellii pusillus*	E	1986	No	51 FR 16474
White-eye, bridled	*Zosterops conspicillatus conspicillatus*	E	1984	Yes	49 FR 33881

[a] Pelagic birds that nest on islands.

[b] The Roseate Tern is listed as two distinct population segments. The north Atlantic Coast population is listed as endangered and the other populations in the Western Hemisphere are listed as threatened.

species concept, removal of subspecific taxonomy would compel the USFWS to reexamine the listing of 33 avian subspecies that are currently listed nationally, 37 that are listed internationally, and 3 that are candidates for protection (Table 2). All subspecies currently listed would have to undergo a technical correction at a minimum, which would be relatively easy for those subspecies that were simply made species. However, there are likely to be cases in which elimination of subspecies would force expanded reviews, cause changes to critical habitat designations, or invite new petitions or litigation. This could be costly, even if it represented a fraction of the subspecies listed, because of the administrative complexities associated with adding a species to the list. Conversely, the AOU's adoption of the phylogenetic species concept would not absolve the USFWS of their responsibility to address avian subspecies, because subspecies are explicitly included among entities eligible for protection under the Endangered Species Act (USFWS and NMFS 1980; Endangered Species Act, section 3). Furthermore, unless the AOU or some other taxonomic authority maintains a standing review process, USFWS will always be liable for evaluating new information for taxa on the AOU's check-list. Absent new information, taxa on AOU lists would not be immune from Endangered Species Act challenges.

Each species listing requires publication of a proposed rule, peer review, solicitation of public comments, and publication of a final rule (USFWS and NMFS 1980). Listing rules published in the Federal Register require review and signatures from approximately 10 to 20 biologists and managers in the USFWS, regional and federal office legal council, and the signature of the regional director, director of USFWS, and high-level officials at the Department of the Interior. Costs of developing and publishing proposed and final listing rules can vary widely depending on the number of offices involved and the complexity of the species (e.g., narrow endemics are generally less costly than wide-ranging species that span several USFWS regions); however, totals of approximately $300,000 are typical, with the cost approaching $400,000 if designation of critical habitat is included and exceeding $1 million for wide-ranging controversial species (USFWS unpublished document entitled *Listing and Critical Habitat Allocation Methodology*). It is important to note that these are the administrative costs associated with assessing a species' status and critical

habitat, publishing documents in the *Federal Register*, holding public meetings, and responding to public input. These costs do not include any on-the-ground conservation.

The added workload associated with adoption of the phylogenetic species concept, without concomitant funding, would be added to the already sizeable backlog of USFWS listing and critical habitat decisions, further delaying listing for ~250 species (or subspecific units) that are on the candidate list (i.e., species, subspecies, or distinct population segments that the USFWS has already determined warrant a position on the List of Endangered and Threatened Wildlife, but for which it did not have funding to propose listing; see http://ecos.fws.gov/tess_public/). Subspecies would have to be reexamined for listing as either a species or some other entity such as a distinct population segment or a significant portion of the species' range. This would have large transaction and opportunity costs, the former because of the enormous administrative task of updating lists, maps, and regulations and the latter because the money used to navigate the administrative updating could have been spent implementing on-the-ground conservation (Garnett and Christidis 2007). Finally, this taxonomic inflation might be perceived by some as motivated solely by conservation purposes and a form of bureaucratic mischief (*sensu* O'Brien and Mayr 1991), scientific dishonesty (Garnett and Christidis 2007), or professional legerdemain (Winker et al. 2007).

A COMPLEX PHILOSOPHICAL DILEMMA

Scientific discourse regarding issues as basic as the definition of a species will continue. Ultimately, the quest to determine and relate biological findings should not be stymied by political implications. No species concept is perfect (Winker et al. 2007); however, it would be useful to decision-makers if scientific uncertainty and the conservation consequences were explicitly stated by scientists when making a specific taxonomic proposal (Haig et al. 2006). Discussing the impact of choosing one species definition over another in view of the conservation implications is fraught with multidimensional philosophical conflicts (Mace 2004). However, as with many complicated situations, not making a decision is a decision. For example, if the phylogenetic species concept were adopted by the IUCN, the Red List would need to be updated to address those newly identified full species. This could take resources away from other high-priority conservation actions, including the protection of currently recognized full species. More subspeciose countries with fewer resources for conservation might be challenged to manage the protection and rehabilitation of an increased number of listed taxa (Table 1).

Conservation efforts will be more effective and less costly if they are initiated before an entire taxon has become critically endangered. Planning ahead is critical from a biological perspective in order to avoid irretrievable losses of heritable adaptations to local and regional conditions. Subspecies groupings, phylogenetic species, distinct population segments, or evolutionarily significant units can help target and prioritize protection for these populations that are geographically disjunct or morphologically unique.

Whatever species concept is accepted, scientists must operationally define the species concept and subspecies definition they used in their work so that USFWS and other conservation agencies understand the criteria by which results were judged (Helbig et al. 2002, Agapow et al. 2004, Haig et al. 2006, Fallon 2007). Clearly stating criteria for taxonomic identification will facilitate comparisons with similar taxa when undertaking listing, recovery, and status assessments. North American ornithologists have been leaders in this effort in the past (AOU 1957, 1998); however, there is a critical need to update this work at the subspecific level on a regular basis.

Acknowledgments

We are grateful to S. Chambers, C. Cicero, M. Fitzpatrick, F. James, C. Phillips, C. Schuler, R. Waples, K. Winker, and an anonymous reviewer for comments on this paper. The findings and conclusions are those of the authors and do not necessarily represent the views of the U.S. Fish and Wildlife Service or the U.S. Geological Survey.

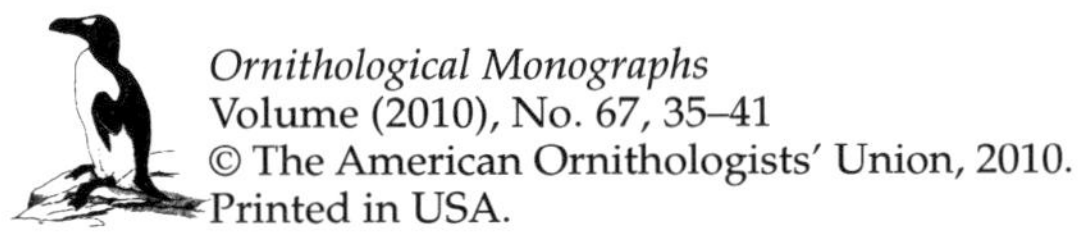
Ornithological Monographs
Volume (2010), No. 67, 35–41
© The American Ornithologists' Union, 2010.
Printed in USA.

CHAPTER 3

NULL EXPECTATIONS IN SUBSPECIES DIAGNOSIS

MICHAEL A. PATTEN[1]

*Oklahoma Biological Survey and Department of Zoology, University of Oklahoma,
Norman, Oklahoma 73019, USA*

ABSTRACT.—The utility of subspecies in studies of evolution and migration and in conservation planning has been debated hotly for a half-century. Inconsistent and sometimes sloppy application of the subspecies concept has led some to deem it a failure, but recent quantitative definitions of subspecies have put the concept on more rigorous footing. Nonetheless, the molecular revolution has added fuel to the fire as researchers attempt to test subspecies by genetic means. Until a sound and defensible null expectation is developed for genetic differentiation of subspecies, genetic approaches will be fraught with problems. A test for monophyly is insufficient, because parapatric subspecies interbreed by definition. Moreover, because much geographic variation may arise via natural selection, tests restricted to selectively neutral genetic data are likewise problematic. Moreover, long-standing charges of subjectivity in the naming and diagnosis of subspecies must be addressed if subspecies are to continue to be accepted as valid taxonomic entities. Statistical advances, including pairwise tests, spline regression, module identification in neural networks, Monmonier's algorithm, and unsupervised, fuzzy k-means cluster analysis offer considerable promise as means of identifying and quantifying geographic variation in an objective yet statistically rigorous manner.

Key words: algorithms, diagnosability, null models, statistics, subspecies.

Expectativas Nulas en la Diagnosis de Subespecies

RESUMEN.—La utilidad de las subespecies en los estudios de evolución y migración, así como en la planeación de la conservación, ha sido debatida fuertemente por medio siglo. La aplicación inconsistente y a veces descuidada del concepto de subepecie ha llevado a que algunos lo consideren un fracaso, pero el desarrollo reciente de definiciones cuantitativas de las subespecies ha puesto al concepto sobre unas bases más rigurosas. Sin embargo, la revolución molecular le ha agregado combustible al fuego en la medida en que los investigadores han intentado poner a prueba la validez de las subespecies usando herramientas genéticas. Mientras no se desarrolle una expectativa nula sobre la diferenciación genética de las subespecies que sea razonable y defensable, los enfoques genéticos estarán rodeados de problemas. Una prueba de monofilia es insuficiente, porque las subespecies con distribuciones parapátricas, por definición, se entrecruzan. Además, debido a que buena parte de la variación geográfica puede surgir como consecuencia de la selección natural, las pruebas basadas en datos genéticos que son selectivamente neutros también son problemáticas. Más aún, las críticas en cuanto a que existe subjetividad sobre la nomenclatura y la diagnosis de las subespecies deben ser abordadas para que las subespecies puedan seguir siendo aceptadas como entidades taxonómicas válidas. Los avances estadísticos, incluyendo las pruebas por pares, las regresiones de "splines", la identificación de módulos en redes neurales, el algoritmo de Monmonier y los análisis de conglomerados difusos y no supervisados de k medias, son altamente promisorios como medios para identificar y cuantificar la variación geográfica de manera objetiva y estadísticamente rigurosa.

[1]E-mail: mpatten@ou.edu

Ornithological Monographs, Number 67, pages 35–41. ISBN: 978-0-943610-86-3. © 2010 by The American Ornithologists' Union. All rights reserved. Please direct all requests for permission to photocopy or reproduce article content through the University of California Press's Rights and Permissions website, http://www.ucpressjournals.com/reprintInfo.asp. DOI: 10.1525/om.2010.67.1.35.

THE SUBSPECIES HAS had a long and contentious history as a taxonomic unit. After its first use—in 1844, by an ornithologist no less (Sibley 1954, Borgmeier 1957)—the subspecies became a staple of museum taxonomy during natural history's great age of discovery (Goetzmann 1986). An overwhelming number of the world's avian subspecies were named during that time (see Sabrosky 1955), and certainly the zeal of that era contributed mightily to the surfeit of trinomials with which any present-day taxonomist must contend. Criticism of the very concept of subspecies was inevitable given the plethora of subspecific names, many based on trivial perceived differences or tiny sample sizes. Wilson and Brown (1953) fired the first serious salvo, spawning a surfeit of their own: some 20 responses and meta-responses were published in *Systematic Zoology* alone over the next 5 years (Starrett 1958), and others weighed in through the pages of influential books (e.g., Inger 1961, Simpson 1961, Mayr 1969). Calls for balance included Smith and White's (1956) observation that constructive criticism of how subspecies are defined or diagnosed could be, and was being, misconstrued to imply that the subspecies concept itself lacked merit. The end result of the fervent debate was not a resolution but an entrenchment of opposing positions. One need look no further than the well-known forum in *The Auk* (Wiens 1982) to see how little the debate had progressed: 11 prominent avian taxonomists and systematists devoted 23 pages to covering the same arguments heard out, but apparently not exhausted, three decades earlier.

Another round of salvos began recently (Zink 2004, Remsen 2005, Haig et al. 2006, Phillimore and Owens 2006), but by this time the tone of the argument had shifted. Zink (2004) concluded that subspecies obscure biological diversity, specifically in birds, and that this intraspecific category of biodiversity misleads conservation efforts—conclusions reached despite an exclusive reliance on genetic surveys, even though it is not clear what a subspecies should look like genetically. Subspecies names have been applied almost exclusively to populations that differ phenotypically. Because of a lack of widely accepted standards, numerous—perhaps as many as half of all (see Patten et al. 2003:71)—avian trinomials will not stand up to scrutiny, but before we can bring modern statistical and genetic techniques to bear on the problem, it is imperative that we have a clear understanding of what a subspecies is and a firm grasp of the null expectations of the underlying genetics.

THE NATURE OF THE SUBSPECIES

Fundamentally, "subspecies are . . . the place at which we stop lumping populations" (Groves 1986). Rensch (*fide* Mayr 1942:106) provided one of the first clear definitions of a subspecies, also known as a "geographical race" or, most commonly, a "race," a loaded term that should be avoided (Patten 2009). Rensch's definition emphasized interfertile individuals that differed in both geographic range and morphology, with morphological differences assumed to have a genetic basis. Mayr himself shortened the definition but again emphasized geography, genetics, and taxonomic distinctness. Four important features about the nature of subspecies can be inferred from these definitions and from others proffered over the years: (1) a subspecies is not reproductively isolated from other subspecies of that species, (2) its defining features have a genetic or developmental basis, (3) it has a unique breeding range separate from that of other subspecies, and (4) it is diagnosably distinct from other subspecies. This fourth feature is what Mayr (1942) meant by subdivisions differing taxonomically.

Patten and Unitt (2002) combined these features into a clear definition of a subspecies: "a collection of populations occupying a distinct breeding range and diagnosably distinct from other such populations." That is, a subspecies represents a level of biological organization below the species that has phenotypic properties that are sufficiently distinct (i.e., separable statistically, as defined below) from other populations. That subspecies are a biological entity whose limits are determined statistically should not deter anyone from making use of this level of biological organization. Subspecies reflect what we can readily observe and have served as the firm basis for decades of literature that has advanced knowledge of distribution, migration, biogeography, ecology, and systematics.

Note that there is nothing in the Rensch-Mayr definition that requires subspecies to be monophyletic. Presence or absence of monophyly is not a relevant criterion for defining or assessing the validity of subspecies, and such a requirement cannot be added. Indeed, this problem exists in defining species, in that monophyly is almost always tested by means of gene trees, particularly using mtDNA (e.g., Zink and Barrowclough 2008), yet as Edwards (2009) observed, monophyly of a gene tree is not a suitable criterion for delimiting a species because natural selection, founder effects, and sampling error affect monophyly but have no bearing on species limits. In the case of subspecies, by the time monophyly could be corroborated, a population may or may not have passed the threshold to become a biological species. Yet reproductive isolation and monophyly evolve independently or are correlated only loosely: reproductive isolation may exist well before monophyly is achieved (e.g., following strong natural or sexual selection) or vice versa (e.g., isolated populations that do not differ phenotypically or behaviorally).

Subspecies are frequently treated as incipient species (e.g., Zink 2004), which is incorrect. Subspecies are, by definition, a stage in the process of allopatric speciation (Mayr 1942b), but even though every species that has evolved in this way must have passed through a subspecies stage, it does not follow that any subspecies that reaches this stage will become a species. Consider a simple abstract model wherein a lineage progresses through different stages of differentiation through time:

$$A \Leftrightarrow B \Rightarrow C$$

in which stage A represents an undifferentiated population, stage B a subspecies, and stage C a species. Given allopatric divergence, to reach C, population A must go through B, but once at B the subspecies might continue to C (i.e., speciate) or revert to A (i.e., differentiation swamped by gene flow in secondary contact). A subspecies thus represents an early diagnosable stage through which a biological species must pass (Mayr 1942), and this stage may or may not be an incipient species.

Phenotypic divergence in allopatry may be the result of natural selection, sexual selection, or drift. The environment certainly plays a role in geographic variation (James 1983), but the suggestion that all geographic variation is solely the result of environmental forces is not supported by scientific evidence. For example, Gloger's rule—the tendency for a species to be paler in warm, arid climates and darker in cool, humid ones—is well documented in birds (Zink and Remsen 1986; cf. Chui and Doucet 2009). Underlying mechanisms of Gloger's Rule are open to debate and may

include background matching, thermoregulation (Zink and Remsen 1986), adaptation to feather-degrading bacteria (Burtt and Ichida 2004), and other factors still to be identified. Regardless of the cause, the result is geographic variation in plumage with a putative basis in the nuclear genome. Let us not forget that Darwin's (1859) successful and extensive artificial selection of domestic pigeons underscores the genetic basis of wide phenotypic variation within a bird species.

PHILOSOPHICAL UNDERPINNINGS

Numerous studies have attempted to apply modern genetic analyses to determine whether one or several subspecies are valid. Many of these studies, across a wide range of organisms, have concluded that a subspecies is not valid because either no genetic difference was detected in some putatively neutral genetic marker or reciprocal monophyly was not found (e.g., Dijkstra and Jellyman 1999, Burbrink et al. 2000, Zink et al. 2000, Daniels et al. 2005, Ramey et al. 2005, Marthinsen et al. 2007). A few authors have drawn more tempered conclusions (e.g., Mock et al. 2002, Bulgin et al. 2003, Swei et al. 2003), and some have even reported new subspecies on the basis of genetic data (e.g., Päckert et al. 2006).

Although a particular species concept may seem to be purely an operational choice that has little bearing on research but says a great deal about one's philosophy, it may be more appropriate to think of species concepts as competing Kuhnian paradigms. Sites and Crandall (1997) intimated that the choice of a species concept should be stated as a hypothesis to be tested because the choice affects interpretation of the results; that is, the concept used affects taxonomic conclusions (Gamauf et al. 2005). On the basis of its definition, the subspecies lies squarely in the purview of the biological species concept. Any diagnosably distinct, geographically circumscribed population may qualify as a subspecies if it is not reproductively isolated from other such populations (and assuming that variation is not merely smoothly clinal). Put simply: if reproductively isolated, the populations are biological species; if not, they are subspecies of a polytypic biological species. It follows, then, that any researcher making use of the subspecies concept in its proper context is making use of the biological species concept to designate phenotypically diagnosable units within the species.

The phylogenetic species concept classifies a population as a distinct species if it is diagnosable by at least one heritable character and if reciprocal monophyly is evident (Cracraft 1983). This definition bears more than a passing similarity to that of a subspecies, but note that it requires 100% diagnosability. Any statistical rule for diagnosing a subspecies at <100%, such as the widely used "75% rule" (see below), renders phylogenetic species and subspecies nonequivalent. Under the phylogenetic species concept, any 100% diagnosable population is elevated to the rank of species, but any population that falls short of 100% diagnosability has no taxonomic status—for example, McKitrick and Zink (1988) advocated not naming subspecies. That subspecies are not recognized in the phylogenetic species paradigm has profound philosophical ramifications for testing a subspecies' validity scientifically. In a hypothetico-deductive framework, the null hypothesis must be "not diagnosable," regardless of paradigm. In the phylogenetic species paradigm, however, the sole alternative hypothesis is that diagnosable taxa are not subspecies but species. These two alternatives exclude the possibility of a subspecies being valid: a population is either undifferentiated or represents a phylogenetic species; there is no middle ground. On purely philosophical grounds, this situation is untenable.

NULL EXPECTATIONS

Most of the papers that I have cited above failed to detect reciprocal monophyly of mtDNA trees and thus concluded that subspecies designations were incorrect. Using monophyly as a criterion meant that these authors were, perhaps unwittingly, equating species and subspecies, in that both taxonomic levels were being held to that same expectation. Recognizing that subspecies are, by definition, a stage in the process of allopatric speciation, the proper null expectation for reciprocal monophyly of a set of subspecies is that it will not exist. Failure to reject this null expectation tells us nothing about a subspecies' validity. Let us further recall that a properly defined subspecies is a diagnosable population not isolated reproductively from other such populations, a definition that allows for the possibility of persistent gene flow between parapatric populations, although not enough to swamp phenotypic diagnosability. Accordingly, a broader question surfaces: What is the null expectation for the genetic differentiation of subspecies, particularly at a neutral locus?

A null expectation of no difference among populations under study is both obvious and unhelpful, yet it is fundamentally difficult to construct a test for H_0 = no genetic differentiation. By definition, a neutral marker is not associated with local adaptation, so this locus may be relatively uniform throughout two distinct populations that hybridize across a shared boundary. Taxa that hybridize even modestly are likely to have at least some introgressed mtDNA, and in some cases mtDNA of one taxon replaces mtDNA in another taxon despite no evidence of introgressed nuclear genes or of mixed morphology (Ballard and Whitlock 2004). Yet ongoing (potential) hybridization characterizes a subspecies—the lack of reproductive isolation is why subspecies, although distinct morphologically, are not classified as biological species.

A further complication arises because subspecies are often posited to be the products of local adaptation, and such divergence may have been geologically recent (see Reznick et al. 2004). Rapid, disruptive selection may yield distinct phenotypes with an underlying genetic basis that is invisible to a neutral marker because differentiation is evident only in non-neutral portions of the nuclear genome. A selective sweep may alter gene frequencies to a point that they no longer reflect phylogeny accurately. Balancing selection (e.g., through frequency-dependent selection) can decouple mtDNA phylogeny and taxon pedigrees (Rand 1996). And strong stabilizing selection on morphology and behavior may produce two allopatric populations that are uniform phenotypically and behaviorally but, if they have been isolated long enough, distinct genetically.

The point about genetic divergence raises another philosophical issue: use of neutral markers assumes that we need only concern ourselves with time. Not to belittle mutation rates, population sizes, sex ratios, or dispersal, but time and natural selection constitute the two pillars of divergence. Emphasizing only time is easier operationally, but doing so ignores our growing understanding of how speciation works. If all speciation occurred via random accumulation of genetic differences when populations are allopatric, the extent to which a neutral genetic marker differs will be in proportion to the degree of divergence between the taxa under study. Any other mechanism that promotes speciation, such as disruptive or directional selection in the face of gene flow (Nosil 2008), would create discordance between the neutral marker and the signal of diverging loci and phenotypic characters under selection (for a review, see Winker 2009).

Any attempt to determine the validity of a subspecies needs to take these caveats into account and use them to construct a meaningful null. This task will not be easy. Whether divergence was recent or the result of strong selection or occurred despite persistent gene flow, we might well expect "no difference" in gene frequencies or haplotypes and expect no reciprocal monophyly. If no difference is the expectation, how can we reject H_0?

STATISTICAL APPROACHES

A working null hypothesis of no genetic differentiation when phenotypic differentiation is evident must, at a minimum, examine a larger portion of the genome. Perhaps the day will come when enough genomes have been characterized to pinpoint specific genes associated with observed phenotypic differences. Another option would be to develop a model-comparison approach (*sensu* Anderson et al. 2000), wherein the weight of evidence for competing alternative phylogenetic schemes—assuming that they could be characterized—could be judged objectively. Until either of these lofty goals is reached, the best position for advocates of the subspecies concept is to opt for statistical rigor in defining what they mean.

It is no secret that numerous named subspecies are ill defined, in that they do not conform to a rigorous, formal definition of subspecies based on diagnosability (here meaning an emphasis on effect size, not on mean differences). One solution would be to discard the notion of subspecies (e.g., McKitrick and Zink 1988); if they do not exist, they cannot be ill defined. That other levels of taxonomic classification suffer from similar problems typically passes unremarked by those who favor discarding subspecies. A rank-free system (e.g., de Queiroz and Gauthier 1992) would eliminate all such problems, but rank-free systems are not without drawbacks (Benton 2000). Accordingly, rather than discard subspecies, we ought to ensure that they are well defined and defensible. If a researcher then chooses to ignore subspecies, at least it will be for philosophical reasons rather than discomfort that a particular taxonomic scheme may be arbitrary or trivial.

Statistical assessment of subspecies dates back to the advent of the 75% rule (Amadon 1949). At its core, this rule was an attempt to emphasize that a large effect size, not whether means differ significantly, is the statistic of interest as well as what should matter biologically. (Effect size is a statistical term for a measure, in terms of standard deviation, of distance between the sampling distributions of H_0 and H_A; i.e., it is a measure of the degree of overlap between the null distribution and that of the alternative.) Mayr's (1969:189) related coefficient of difference (CD),

$$CD = \frac{\bar{x}_b - \bar{x}_a}{SD_b + SD_a}$$

stressed this point further, as we can see in comparing it with a simple formula for effect size (**d**; Cohen 1988:20):s

$$\mathbf{d} = \frac{\bar{x}_b - \bar{x}_a}{\sigma}$$

where σ is the (pooled) standard deviation or the SD for either population (they are assumed to be equal). Yet despite numerous efforts to solidify this foundation (Rand and Traylor 1950, Simpson 1961, Mayr 1969), the emphasis on a large effect size still goes largely unheeded. And from a practical standpoint, the vast majority of subspecies were named before this framework was built (see Remsen, this volume), so the vast majority of subspecies have not been evaluated with modern statistical methods (Patten and Unitt 2002, Remsen 2005).

Two approaches have been proposed in the past decade. The first is a pairwise test of diagnosability (Patten and Unitt 2002), a technique based on the t-distribution that can be applied either to a single morphometric or other quantitative (e.g., scores from a colorimeter) character or to a composite value derived from a multivariate statistic such as a discriminant analysis (see Marantz and Patten, this volume). The pairwise test is limited to testing for differences among predefined groups. The other recent method involves application of spline- and step-regression techniques to identify breaks among characters that vary clinally (Skalski et al. 2008). It is unclear how sensitive this method is for detecting small but potentially biologically meaningful steps, because the simulation results presented by Skalski et al. (2008) have abrupt breaks in character values.

Three additional techniques would be worth developing for use in diagnosing morphologically distinct groups that have separate geographic ranges (i.e., subspecies). The first is a simulated annealing algorithm that detects modules (analogous to clusters) in a network (Guimerà and Amaral 2005). The algorithm is unsupervised, which means that there is no need to define in advance the number of clusters desired. Instead, a measure of degree of modularity signifies how distinct the modules are. Input matrices are a set of links, which could be between spatial location (Carstensen and Olesen 2009) and morphological traits or between individuals and those traits. If the latter were used, then short links would indicate that individuals were close to one another and, thus, potentially in the same module. Software for this algorithm is available (see Olesen et al. 2007).

Monmonier's maximum-difference algorithm (Monmonier 1973, Manni et al. 2004) is a spatially explicit technique that can use any distance matrix to determine where barriers (natural breaks) exist in the data. The algorithm begins with a map of sites using specific coordinates (e.g., latitude and longitude). A distance (dissimilarity) matrix is mapped onto a triangulation created among sites such that each pairwise line between sites has an associated distance. The algorithm builds barriers, beginning with the maximum pairwise distance and continuing until a loop forms or the map's edge or another computed barrier is hit. Distances can be from any semimetric index, such as F_{ST} (Nicholls et al. 2006) or Jaccard's index (Patten and Smith-Patten 2008), and are thus amenable to either genetic or morphometric data. Although Monmonier's algorithm is supervised, there are ways to determine *a priori* how many barriers to select (Patten and Smith-Patten 2008). A key advantage to this technique stems from a recent software implementation that allows input of multiple matrices (Manni and Guérard 2004). As such, either bootstrap matrices can be generated to determine support for specific barriers (Manni et al. 2004, Patten and Smith-Patten 2008) or matrices from multiple data sets can be entered to determine levels of correspondence among them.

On purely heuristic grounds, perhaps the most appealing statistical option is an unsupervised fuzzy cluster analysis, a technique virtually unknown in biology but used widely in pattern recognition and artificial intelligence. In classical set theory, group membership is discrete if sets are exclusive. Object x is an element either of set **A** or of set **B**. In fuzzy set theory, group membership is a probability, which means that x might have $P = 0.73$ of being an element of **A** and $P = 0.27$ of being an element of **B**. Such a probabilistic partitioning seems more naturally in accord with the fuzzy nature of subspecies—they are defined by real or potential gene flow and, therefore, every individual may not be classifiable. Operationally, membership probabilities can be assigned by scaling linearly between minimum and maximum values for the variable in question (Roberts 2008). These assigned P values are then used in a clustering algorithm to build k clusters on the basis of minimizing some predefined objective function (Gath and Geva 1989, Liu and George 2005). Software applications can be found on the Internet (e.g., Kenesei et al. 2006).

Although the foregoing discussion is geared toward morphological data, closely related techniques are already used widely for genetic data. For example, assignment tests (e.g., Piry et al. 2004) classify individuals probabilistically—using a Bayesian approach with priors set by group of origin—among a set of populations, and genetic assignment can correspond well to subspecies (e.g., Pruett et al. 2008a). Algorithms such as STRUCTURE (Pritchard et al. 2000) work like an unsupervised k-means cluster analysis in that genetic data are used, again within a Bayesian framework and this time coupled with Monte Carlo randomization, to determine the number of groups most likely given the data. STRUCTURE further allows genetic data to be constrained by phenotype, and individuals can be assigned a genetic score that can be correlated against a morphological score (e.g., Patten et al. 2004).

Concluding Remarks

Conservation policies and wildlife management decisions often are based on subspecific taxonomy (Stanford 2001, Haig et al. 2006). Practitioners must be able to defend subspecies diagnoses critically and objectively, a move that will doubtless lead to a more cautious and conservative intraspecific taxonomy, which is warranted given the high stakes and limited funding available for biodiversity conservation. There is certainly no point in complaining that others ignore subspecific designations if advocates cannot defend them quantitatively. Conversely, advocates of the phylogenetic species concept ought to be forthcoming about their philosophical inability to test the validity of a subspecies—thus far, no one has explained how such a test can be done within that paradigm.

The utility of subspecies rests on our ability to assign, with a high degree of confidence, individual specimens to a particular geographic population. A properly named subspecies is a surrogate for geographically isolated pools of phenotypic variation whose basis is putatively genetic. Diagnosability of subspecies has been defined statistically for qualitative and quantitative characters (e.g., Patten and Unitt 2002). A roughly ordered expectation of which types of characters are best suited to proper diagnosis of subspecies is pattern > color ≈ shape > size. This ordered expectation may not hold in all cases; it only suggests that size, for example, is more likely than pattern to vary clinally. Smoothly clinal variation has no place in the proper description of subspecies. We instead need to emphasize only those characters that exhibit "breaks" between geographic sites (see Skalski et al. 2008).

To achieve rigor in subspecies diagnosis will require strict adherence to the now 60-year-old recognition that extent of overlap is what matters. Between the statistical and genetic techniques summarized above and recent advances in the quantification and analysis of shape (González-José et al. 2008) and color (Endler and Mielke 2005), there are few excuses to avoid the careful work needed to put subspecies on a firm quantitative and objective footing. Yet there are recent examples of subspecific diagnoses that relied on tests of mean differences (e.g., Engelmoer and Roselaar 1998, Jiguet 2002, Cabot and Urdiales 2005) rather than focusing on effect size. Editors and peer reviewers cannot allow this practice to continue, and the time to integrate approaches is now.

ACKNOWLEDGMENTS

I thank L. R. Bevier, J. V. Remsen, Jr., P. Unitt, and K. Winker for input over the years; each has been instrumental in shaping my thinking on this subject. I further thank K. Winker and S. M. Haig for inviting me to present a version of this paper at the symposium on subspecies. F. C. James, J. V. Remsen, Jr., B. D. Smith-Patten, and K. Winker provided excellent comments on various drafts of this paper.

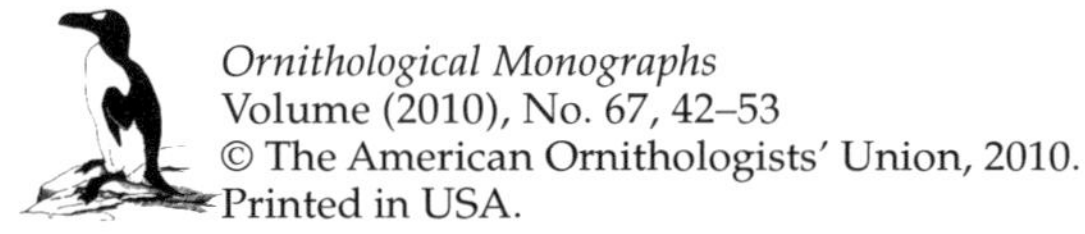

Ornithological Monographs
Volume (2010), No. 67, 42–53
© The American Ornithologists' Union, 2010.
Printed in USA.

CHAPTER 4

SUBSPECIES ORIGINATION AND EXTINCTION IN BIRDS

ALBERT B. PHILLIMORE[1]

*Natural Environment Research Council Centre for Population Biology and Division of Biology,
Imperial College London, Silwood Park Campus, Ascot, Berkshire, SL5 7PY, United Kingdom*

ABSTRACT.—Avian taxonomists have traditionally used subspecies to describe geographic variation in morphology, plumage, and song. A complementary evolutionary perspective is that subspecies are incipient species, representing the first stages of speciation. Here, I review the evidence that subspecies may capture early stages of the speciation process and consider what we have learned about factors that drive subspecies diversification. I apply variants on the birth–death model to species age and subspecies richness data from 1,100 bird species. Clade-wide estimates of species diversification rates correlate positively with subspecies origination (hereafter "subspeciation") rates but not subspecies richness; thus, the evidence for heritable factors promoting speciation and subspeciation is equivocal. Subspeciation rates are higher among insular than among continental species, although this result is highly sensitive to the definition of insularity. *A posteriori* simulations based on the maximum-likelihood constant-rate birth–death parameter estimates reveal model inadequacy. One possible explanation for such model inadequacy is non-homogeneity in diversification rate through time, and I find support for a model that invokes an exponential decline in subspeciation rates through time, with differing rates in continental and insular species. Finally, I discuss some alternative models of subspecies origination and how they might be assessed using population genetic information and geographic range maps.

Key words: birth–death, diversity-dependent, phylogenetic, speciation, subspecies.

Origen y Extinción de las Subespecies de Aves

RESUMEN.—Los taxónomos de aves tradicionalmente han utilizado a las subespecies para describir la variación geográfica en morfología, plumajes y cantos. Una perspectiva evolutiva complementaria es que las subespecies son especies incipientes, por lo cual representan las primeras etapas de la especiación. En este trabajo reviso la evidencia que indica que las subespecies podrían capturar las primeras etapas del proceso de especiación y considero lo que hemos aprendido acerca de los factores que impulsan la diversificación de las subespecies. Aplico variantes del modelo de nacimiento y muerte para analizar datos sobre la edad de las especies y sobre la riqueza de subespecies en 1100 especies de aves. Los estimados de las tasas de diversificación a través de los clados se correlacionan positivamente con la tasa de origen de las subespecies ("subespeciación"), pero no con la riqueza de subespecies. Por lo tanto, la evidencia de que existen factores heredables que promueven la especiación y la subespeciación es equívoca. Las tasas de subespeciación son mayores en especies insulares que en especies continentales, aunque este resultado es altamente sensible a la definición de insularidad. Simulaciones *a posteriori* basadas en los estimados de máxima verosimilitud de parámetros de nacimiento y muerte constante indican que el modelo es inadecuado. Una posible explicación para esto es la existencia de hetereogeneidad temporal en la tasa de diversificación. Encontré evidencia a favor de un modelo que involucra una disminución exponencial en las tasas de subespeciación en el

---

[1]E-mail: albert.phillimore@imperial.ac.uk

Ornithological Monographs, Number 67, pages 42–53. ISBN: 978-0-943610-86-3. © 2010 by The American Ornithologists' Union. All rights reserved. Please direct all requests for permission to photocopy or reproduce article content through the University of California Press's Rights and Permissions website, http://www.ucpressjournals.com/reprintInfo.asp. DOI: 10.1525/om.2010.67.1.42.

tiempo, con tasas distintas en especies continentales y especies insulares. Finalmente, discuto algunos modelos alternos de subespeciación y cómo podrían evaluarse información de genética poblacional y mapas de distribución geográfica.

FOR OVER A century, the designation and utility of subspecies has been a tension point between taxonomists, who aim to describe patterns of geographic variation, and evolutionary biologists, who are interested in the processes of local adaptation and incipient speciation (Mayr 1982a). Mayr (1963:424) defined the subspecies as "an aggregate of local populations of a species, inhabiting a geographic subdivision of the range of the species, and differing taxonomically from other populations of the species." Although this definition is precise in terms of geography, it is vague regarding the taxonomic differences required for subspecies delimitation, which has led to difficulties in the consistent application of the subspecies concept and may partially explain why this rank has drawn such strong criticism from taxonomists (Wilson and Brown 1953, Zink 2004, Rising 2007). The challenge for taxonomists is accentuated because different characters often show discordant divergence among populations, which means that the delimitation of subspecies may differ depending on the characters used (Wilson and Brown 1953). Avian subspecies were traditionally described on the basis of phenotypic characters, in particular morphology and plumage, but in recent years the advent of molecular tools has seen many cases in which neutral genetic information conflicts with subspecies taxonomy (Barrowclough 1980, Zink 2004). From an evolutionary perspective, however, discordance among different characters that are subject to different evolutionary forces is not surprising (Wilson and Brown 1953; de Queiroz 1998, 2005; Fraser and Bernatchez 2001; Winker 2009).

Here, I begin by discussing evidence that avian subspecies may represent the early stages of geographic speciation. I review comparative studies that have sought to explain variation in the numbers of subspecies across different species and highlight the variety of ways that subspeciation has been modeled and the assumptions that have been made. Then, as a case study, I examine the extent to which variants on the birth–death model (in which rates are allowed to differ among clades, between islands and continents, and through time) can explain patterns of subspecies richness using a large sample of bird species. I also test for a relationship between species and subspecies diversification. Finally, I examine the gaps in

our knowledge regarding subspecies diversification and how they might be addressed. I focus on birds in the present study because the delimitation of subspecies in this group has received more attention than subspecies of any other taxon. A corollary of the disproportionate attention given to avian subspecies is that the findings of this study may not be extendable to other groups.

REVIEW

SUBSPECIES AND SPECIATION

Ernst Mayr advocated the idea that subspecies could be incipient species (Mayr 1940, 1942b; O'Brien and Mayr 1991); indeed, he stated that "geographic speciation is thinkable only, if subspecies are incipient species" (Mayr 1942b:155). However, Mayr considered that only geographically isolated subspecies have the potential to become new species (Mayr 1942b, 1982a). Endler (1977) demonstrated that subspecies with abutting ranges are often strongly differentiated, and this finding, combined with evidence that ecological speciation can occur in the face of gene flow (reviewed in Rundle and Nosil 2005), suggests that incipient species status may reasonably be extended to contiguous parapatric subspecies. If subspecies are on the evolutionary trajectory toward being full species (whether or not they ever go on to attain full species status; Patten, this volume), we can reasonably expect them to show divergence in characters that are considered important in speciation (e.g., pre- and postzygotic reproductive isolation, ecological divergence, and genetic divergence).

Subspecies are often delimited on the basis of characters that may play a role in prezygotic reproductive isolation. For example, geographic variation in vocalizations (Slabbekoorn and Smith 2002, Dingle et al. 2008), plumage color, and plumage pattern (Mayr 1942b) may all render members of a population most attractive to members of the same population (Clayton 1990, Uy et al. 2009), and these same traits are routinely used to delimit subspecies.

I am aware of no reports of postzygotic incompatibities (measured as F_1 hybrid fertility and viability) arising between avian subspecies, which is not surprising given that even among full species these barriers are often weak (Price and Bouvier

2002). However, Bensch et al. (1999) suggested that when subspecies follow different migratory routes, hybrids between these forms might have reduced recruitment. In addition, a study of sympatric red and black color morphs of Gouldian Finch (*Erythrura gouldiae*) found that the progeny of between-color-morph crosses suffered greater inviability than those of within-morph crosses (Pryke and Griffith 2009).

Selection drives ecological speciation (Rundle and Nosil 2005). For subspecies to represent the initial stages of ecological speciation, phenotypic differences between them ought to be the result of selection and have a genetic basis. Common garden studies have revealed cases in which differences between subspecies in morphology, life history, phenology, and song-based mate preferences have a genetic component (Price 2008: table 3.1). However, some geographic differences between subspecies, such as carotenoid-based plumage in House Finches (*Carpodacus mexicanus*; Hill 1993) and morphology in Red-winged Blackbirds (*Agelaius phoeniceus*; James 1983), reflect environmental effects. Unfortunately, there are too few common garden and reciprocal transplant experiments on avian subspecies to draw general conclusions regarding the genetic and environmental contributions to phenotypic variation. Field studies have highlighted cases in which divergence between subspecies in bill morphology (Benkman 1993a) and plumage and behavior (Mumme et al. 2006) has an adaptive function and is presumably the result of selection. Further evidence that selection plays an important role in geographic variation comes from a meta-analysis across many taxa (including but not restricted to birds), which found that phenotypic divergence between populations tends to exceed the null expectation under genetic drift (Leinonen et al. 2008).

Because (1) a long period of no (or little) gene flow is required for neutral genes to coalesce (Hudson and Coyne 2002) and (2) under selection, populations may diverge in the face of gene flow, monophyly of subspecies should not be necessary for their delimitation (Patten, this volume). It has been pointed out elsewhere (e.g., Winker 2009) that the divergence between species may take place along two axes, one phenotypic, the other genetic, and the divergence along each axis need not be correlated. The observation that more than a third of subspecies are monophyletic on a mitochondrial DNA (mtDNA) tree (i.e., members of a subspecies are more closely related to one another than to members of another subspecies)

and that monophyly is greater among insular subspecies (Phillimore and Owens 2006) is consistent with geographically isolated populations being on independent evolutionary trajectories, as envisaged by Mayr (1942b).

Darwin (1859) predicted that speciose genera should generally subtend species that possess more varieties. A positive correlation between species richness and subspecies richness would be consistent with heritable or shared factors promoting diversification at both taxonomic levels. Haskell and Adhikari (2009) reported a positive correlation between the number of species in avian genera and the average number of subspecies subtended by each species. However, species and genera both vary in age, which means that variation in species and subspecies richness may reflect variation in genus and species age rather than variation in rates of diversification.

Comparative Studies

Numerous interspecific comparative studies have examined variation in the numbers of subspecies per species (subspecies richness), as a proxy for either phenotypic diversification or incipient speciation (e.g., Belliure et al. 2000, Phillimore et al. 2007, Seddon et al. 2008). Such studies make a multitude of assumptions; perhaps the most important is that different taxonomists consistently delimit subspecies across taxa, regions, and environments (but see Fitzpatrick, this volume). Any systematic departure from consistent subspecies delimitation has the potential to confound comparative analyses. A second important assumption is that the focal statistical models adequately capture the process of subspecies diversification, a subject to which I will return.

There is a long history of comparative research addressing variation in subspecies richness. Rensch (1960:23) noted that migratory passerines tend to have fewer subspecies than sedentary forms. In another classic work, Diamond et al. (1976) defined species as "great speciator[s]" if they are represented by five or more subspecies or allospecies in the Solomon Islands, and they found that the great speciators are predominantly short-distance colonists. In cuckoos, subspecies richness is highest among virulent parasitic species with multiple hosts, which implies a role for coevolution in cuckoo subspeciation (Krüger et al. 2009). In other studies, which vary in taxonomic and geographic scope, subspecies richness has been found to be elevated in species that have larger geographic range sizes (Belliure et al. 2000, Phillimore et al. 2007), in species that inhabit montane areas (Mayr and Diamond 2001) and understory rainforest habitats (Burney and Brumfield 2009), and in species that exhibit a greater degree of insularity (Phillimore et al. 2007), shorter natal dispersal distances (Belliure et al. 2000, Newton 2003), greater plumage ornamentation (Møller and Cuervo 1998), more pronounced plumage dichromatism (Seddon et al. 2008), higher song pitch (Seddon et al. 2008), more complex song (Seddon et al. 2008), smaller body size (Seddon et al. 2008), or larger relative brain size (Sol et al. 2005). Strikingly, even those cross-species studies that considered multiple predictors leave more than half of the variation in subspecies richness unexplained (e.g., Belliure et al. 2000, Phillimore et al. 2007).

On observing that a single authority often described the subspecies of a single species, Martin and Tewksbury (2008) went some way toward addressing the problem of inconsistent taxonomy by making intraspecific comparisons. They found that when they bisected species' latitudinal ranges at the latitudinal midpoint, 65.6% of 710 species with more than five subspecies had a greater number of subspecies at low latitudes than at high latitudes. For comparison, 20.0% of species had more subspecies at high latitudes. High-latitude subspecies also tend to have larger geographic ranges than low-latitude subspecies (Rapoport 1982: table 5.6). An additional reason that Martin and Tewksbury (2008) gave for focusing on intraspecific comparisons was that, if speciation occurs in peripheral isolates, recently formed sister species will tend to differ substantially in both range size and subspecies richness. Consistent with this argument is the observation that subspecies richness has a weak phylogenetic signal; that is, closely related species do not tend to have similar numbers of subspecies (Sol et al. 2005, Phillimore et al. 2007).

Modeling Diversification of Subspecies

Although subspecies richness has been the focus of several comparative studies, the underlying processes have received surprisingly little attention. Some studies make no mention of a model and simply compare predictors of subspecies richness (e.g., Belliure et al. 2000). Studies that tried to model a particular subspeciation process have tended to make conflicting assumptions. For example, Phillimore et al. (2007) modeled subspeciation variously as a pure birth, birth–death, or Poisson process, with the consequence that in some analyses they assumed that subspecies extinction was absent (pure birth and Poisson models), and in others they assumed that all species were of equal age (Poisson and birth–death models).

Analyses at the species level and above routinely use the birth–death model and variants thereof to estimate parameters of interest (Ricklefs 2007). Under the birth–death model, at each point in time, every lineage has a constant probability of splitting into two species and a different constant probability of going extinct (Kendall 1948). Here, I will limit my focus to the application of birth–death models to subspecies diversification.

Most previous comparative studies that focused on subspecies did not account for variation in the age of species. Sol et al. (2005) proposed that species age could be ignored on the basis of weak correlations between species age and ln subspecies richness. However, a high constant rate of extinction erodes the correlation between clade age and ln clade richness even if lineages arise at a constant rate (Paradis 2004, Ricklefs 2006). If the birth-death model provides a reasonable approximation of subspecies origination and extinction then species ages should not be ignored in rate estimation.

Temporal variation in diversification rates can also erode the relationship between clade size and richness (Ricklefs 2006). There is growing evidence that speciation and extinction have not been constant through time and that species diversification may in fact be diversity dependent, slowing toward the present (Ricklefs 2006, Weir 2006, McPeek 2008, Phillimore and Price 2008, Rabosky and Lovette 2008, Rabosky 2009b). It is plausible that a similar process could also govern subspecies diversification; as a species' geographic range is subdivided among a greater number of subspecies, subspeciation rates may decline or subspecies extinction rates may increase, or both. If this were the case, and most species are at their subspecies-richness carrying capacity, then addressing correlates of subspecies richness (i.e., carrying capacity) is more appropriate than addressing correlates of subspecies diversification or origination (Rabosky 2009a).

Case Study

The aims of the following study are to (1) examine whether a constant-rate birth–death model can account for the observed co-distribution of species ages and subspecies richness, (2) estimate subspeciation and subspecies extinction rates under a birth–death model, (3) examine the link between species diversification and subspecies diversification, (4) test whether subspecies diversification differs between insular and continental species, and (5) estimate the likelihood of the data under two temporally non-homogeneous models.

Methods

Species ages.—I estimated species ages from the Bayesian posterior distributions of 44 mtDNA-based, species-level phylogenies (Appendix), all of which included >70% of the species recognized in their clades (Phillimore and Price 2008). These phylogenies had been constructed using a relaxed-clock approach in BEAST (Drummond et al. 2006, Drummond and Rambaut 2007), with the assumption that the nucleotide substitution rate of mitochondrial protein-coding genes has a mean of 1% per million years and is lognormally distributed across branches (for more details, see Phillimore and Price 2008). I adopted the species limits recommended by the original authors of the phylogenetic studies. I estimated the age of each species (i.e., the stem age for the intraspecific clade) as the median time to most recent common ancestor across the posterior distribution of 2,000 trees. Note, however, that estimating time to the most recent common ancestor from gene trees will overestimate the age of species to a degree that depends on the extent of ancestral genetic variation (Edwards and Beerli 2000).

Subspecies richness, insularity, and body-mass data.—I obtained subspecies richness information for the species with phylogenetic data from Phillimore et al. (2007: appendix S1B). This data set was compiled by reconciling information in Clements (2000) with Sibley and Monroe's (1990, 1993) taxonomy. Where species limits in the phylogeny differed from those in the data set, I split or lumped subspecies accordingly, with phylogenetic information taking primacy.

Because species-level phylogenetic studies are biased toward larger clades (Ricklefs 2007; Phillimore and Price 2008, 2009), it is possible that the species ages estimated in the present study are a nonrandom sample of all birds. To test whether the subspecies richness of birds in the study departed from a random sample, I compared the number of sampled species that have 1, 2, 3, 4, 5, 6, 7, 8, 9, 10, or >10 subspecies to the expected numbers from the global data using a chi-square goodness-of-fit test.

The number of insular subspecies per species was obtained from Clements (2000). A subspecies was deemed insular if its range was limited to islands in the United Nations Islands Directory (see Acknowledgments). For each species, I then calculated the proportion of subspecies that were restricted to islands. If more than one subspecies was found on a single island, this was counted as a single case, because I was interested in inter-island rather than intra-island subspecies divergence. For example, five subspecies are described for *Aegotheles bennetti*, all of which are restricted to islands; however, four subspecies inhabit a single island, New Guinea, giving an insularity score of $2/5 = 0.4$. If the proportion of insular subspecies was ≥0.3, I classed the species as having numerous insular subspecies. If the proportion was <0.3, I classed the species as continental. I then examined the sensitivity of results to this definition of insularity by repeating all analyses, replacing 0.3 with (1) 0.5 and (2) 0.8.

Statistical analysis.—If subspeciation follows an exponential pure birth process (Equation 1), a positive linear relationship between species age (t) and ln subspecies richness (N), of slope λ (the subspeciation rate) is expected. I examined the evidence for this using linear regression.

$$N(t) = e^{\lambda t} \tag{1}$$

If some subspecies suffer extinction, a constant-rate birth–death model (Kendall 1948, Harvey et al. 1994, Nee et al. 1994),

$$N(t) = \frac{\lambda e^{(\lambda - \mu)t} - \mu}{\lambda - \mu} \tag{2}$$

where μ represents a constant rate of stochastic lineage extinction, is more appropriate than the pure birth model. A constant rate of speciation and extinction can lead to nonlinear relationships between species age and subspecies richness, because extinction will have had more time to prune the lineages descended from old species (Harvey et al. 1994, Paradis 2004, Ricklefs 2006). If subspecies extinction rates approach subspeciation rates, $\ln(N)/t$ will give a biased estimate of the subspecies diversification rate ($r = \lambda - \mu$; Ricklefs 2006).

Given a subspeciation rate (λ) and subspecies extinction rate (μ), I estimated the likelihood ($l_i = P(n|t)$) of observing n subspecies in species i of age t following Equation 3 and Equation 4 (Bokma 2003; Ricklefs 2007, 2009), where the relative extinction rate $\varepsilon = \mu/\lambda$.

$$E(n) = e^{\lambda(1-\varepsilon)t} = e^{(\lambda-\mu)t} \qquad (3)$$

$$P(n|t) = (1-\varepsilon)\frac{[E(n)-1]^{n-1}}{[E(n)-\varepsilon]^n} \qquad (4)$$

The maximum-likelihood values of λ and ε for a distribution of species ages and their subspecies richness were those that maximized the sum of log likelihoods ($\ln L = \Sigma \ln(l_i)$). I identified the maximum-likelihood values for λ and ε by iterating through values of ε in the range 0–0.99999 in steps of 1^{-5} and using the optimize function (Brent 1973) in R (R Development Core Team 2008) to identify the maximum-likelihood value of λ for each value of ε. The maximum-likelihood values of λ and ε combined were those that corresponded to maximum $\ln L$ across these iterations. Approximate 95% and 99% confidence intervals (CIs) were calculated as those values of λ and ε for which $\ln L(\text{max}) - \ln L(\lambda,\varepsilon) = 2.996$ and 4.605, respectively. Because incomplete species sampling could bias inferences by inflating species ages, I repeated parameter estimation including only phylogenies in which >90% of recognized species were included.

I estimated λ and ε for each phylogeny. Approximate 95% CIs for λ were calculated by fixing ε at its maximum-likelihood value and finding the values of λ that satisfied $\ln L(\text{max}) - \ln L(\lambda,\varepsilon) = 1.92$. A similar procedure was followed to estimate confidence intervals for ε. I tested whether there was any evidence for heterogeneity in rates among clades using a likelihood ratio statistic, T, calculated as follows (Bokma 2003):

$$T = 2(\ln L_1 + \ln L_2 + \cdots + \ln L_{44} - \ln L_0) \qquad (5)$$

The values $\ln L_{1-44}$ represent the maximum log likelihoods for each of the 44 phylogenies, and $\ln L_0$ denotes the global estimate across all trees. T should be approximately chi-square-distributed, with degrees of freedom equal to the difference in number of parameters between the null and alternative hypothesis, in this case $(44 \times 2) - 2 = 86$.

For each of the 44 phylogenies, I estimated the diversification rate among species as follows (Magallón and Sanderson 2001):

$$\hat{r} = [\ln(n) - \ln(2)] / t \qquad (6)$$

where t was the median crown-group age for the phylogeny and n was the number of extant species. Using regression, I then examined the degree to which $\hat{r}$ predicted variation in (1) the mean ln transformed subspecies richness for clades and (2) clade subspeciation (λ) estimates.

Islands are often thought to be promoters of species and subspecies diversification (Mayr and Diamond 2001, Phillimore et al. 2007, Moyle et al. 2009). I tested whether estimates of λ and ε differ between a continental and island setting by estimating the $\ln L(\text{max})$ separately for (1) species classed as continental and (2) species classed as insular, and then calculating T (Equation 5). I used a likelihood ratio test with 2 degrees of freedom to compare the multiple-rate versus fixed-rate models. I also explored the sensitivity of estimates to different definitions of insularity.

To assess the fit of the maximum-likelihood birth–death model to the data, I conducted simulations using the relevant parameter estimates for continental and insular taxa. I used a modified version of the "birthdeath.tree" function in the GEIGER R library (Harmon et al. 2008) to simulate branching and extinction of lineages starting with a single lineage and running the simulation for a fixed duration, after which the numbers of extant tips were counted. If all lineages were extinct before the simulation was completed, the simulation was repeated until ≥1 extant lineages resulted. The distribution of simulation durations followed the distribution of species ages. I conducted 1,000 replicates, each consisting of a separate simulation yielding extant subspecies for each species age observation. For each replicate, I calculated the mean ln('subspecies' richness) in each of the following species age classes (in millions of years): 0–0.9, 1–1.9, 2–2.9, 3–3.9, 4–4.9, 5–5.9, 6–6.9, 7–7.9, 8–8.9, and ≥9. I then calculated the 0.025 and 0.975 quantiles for each of these expected mean subspecies richness values and compared the observed and expected distributions. Good model fit would correspond to a case in which most of the observed data lie within the expected quantiles.

I then assessed the likelihood of the data under a model describing an exponential decline in subspeciation (and subspecies extinction) rate from an initial level λ_0, during the time since species origination (t). This was implemented by replacing r ($\lambda - \mu$) in Equation 3) with r_t, which corresponds to diversification rate at time t; thus (Rabosky 2009b):

$$r_t = \lambda_0 e^{-zt}(1-\varepsilon) \qquad (7)$$

I identified the maximum-likelihood values of λ_0, z, and ε after calculating the likelihood of all combinations of parameters described by $\lambda_0 = 0.5 - 2$ (in units of 0.01), $z = 0 - 1$ (in units of 0.01), and $\varepsilon = 0.5 - 0.999$ (in units of 0.001). I assessed support for the exponential-decline model by comparing the $\ln L(\max)$ to the likelihood of the best model satisfying $z = 0$ (constant rate of diversification) via a likelihood ratio test with a single degree of freedom. I also fitted the exponential decline model separately for continental and insular species.

If subspeciation is diversity dependent, one of the main determinants of subspeciation and subspecies diversification rates is likely the size of species' geographic ranges. Therefore, I replaced λ_0 with species' geographic range sizes (A) and a slope (c) and intercept (d) and replaced r ($\lambda - \mu$ in Equation 3) with $r_{i,t}$ (corresponding to the diversification rate of species i at time t), thus:

$$r_{i,t} = (d + c \log(A_i))e^{-zt}(1 - \varepsilon) \qquad (8)$$

Equation 8 is similar to equation 7 in Rabosky (2009b) but with the addition of an intercept (d). The model allows geographic range to influence the initial rate of subspeciation (and subspecies diversification). An assumption of this model is that species' geographic ranges have been constant through time (Rabosky 2009b).

Breeding-season geographic ranges (km^2) of 1,038 of the species included here were obtained from Phillimore et al. (2007)—these data were extracted from ARCGIS global distribution data (Orme et al. 2005, 2006). The maximum-likelihood values of d, c, z, and ε were identified after calculating the likelihood of the data under all combinations of parameters described by $d = -1 - 2$ (in units of 0.01), $c = -1 - 1$ (in units of 0.01), $z = 0 - 1$ (in units of 0.01), and $\varepsilon = 0.5 - 0.99$ (in units of 0.01). A model in which $c = 0$ corresponds to the initial subspeciation rate being unaffected by breeding-range size, which means that d is then the same as λ_0 in Equation 7. To assess whether geographic range size influences the rate of diversification, I used a likelihood ratio test to compare $\ln L(\max)$ with the log likelihood of the most likely model satisfying $c = 0$. I also tested the support for this geographic-range model separately for continental and insular species.

None of the interspecific comparisons accounted for phylogenetic autocorrelation. However, given that subspecies richness appears to have little phylogenetic signal (Sol et al. 2005, Phillimore et al. 2007), this is unlikely to be a major source of statistical bias. All analyses were conducted in R.

RESULTS

Global birth–death model.—Species age estimates (median = 3.17 MY, 2.5% quantile = 0.17 MY, 97.5% quantile = 10.61 MY) were obtained for 1,100 species from 44 different phylogenies. The median subspecies richness across sampled species was 2 (min = 1, max = 50), and in total there were 3,353 subspecies. Species included in this study were a random sample of all bird species with respect to subspecies richness ($\chi^2 = 7.43$, df = 10, $P = 0.68$).

There was a significant positive linear correlation between species age and subspecies richness ($b = 0.044 \pm 0.008$, $P < 0.001$); however, species age explained only 2.5% of the variance in subspecies richness (Fig. 1A). When a birth–death model was fitted to the same data, subspecies extinction was estimated to have occurred at a rate essentially equal to the subspeciation rate ($\lambda = 0.740$, $\varepsilon = 0.9999$, $\ln L = -2{,}244.16$; Fig. 1A, B), and the likelihood of the data was substantially higher under this model than under a pure birth model ($\lambda = 0.298$, $\ln L = -2{,}510.10$). Restricting the data to 27 phylogenies with good species sampling ($\geq 90\%$ complete at the species level) led to very little change in maximum-likelihood estimates of subspeciation and subspecies extinction ($\lambda = 0.744$, $\varepsilon = 0.99999$, $\ln L = -1{,}083.57$).

Heterogeneity in subspeciation rate.—Maximum-likelihood estimates of λ and ε for different phylogenies revealed considerable heterogeneity in λ rates, from 0.06 in storks (Ciconiidae) to 2.02 in *Cinclodes*, but very little variation in ε, with 42 of 44 phylogenies returning a value of 0.99999 and 35 of 44 returning ε estimates that differed significantly from zero (Appendix). The sum of the maximum-likelihood estimates was $-2{,}122.91$, and the likelihood ratio compared with the global model was -121.25, which strongly supports the alternative hypothesis that λ and ε vary among clades ($T = 242.51$, df = 86, $P < 0.001$).

Species versus subspecies diversification.—Clade-wide species diversification rates did not predict variation in average subspecies richness ($a = 0.80 \pm 0.13$, $b = -0.47 \pm 0.49$, $R^2 = 0.02$; Fig. 2A), but these clade-wide rates were a strong positive predictor of variation in subspeciation rate estimates ($a = 0.19 \pm 0.18$, $b = 2.06 \pm 0.67$, $R^2 = 0.18$; Fig. 2B).

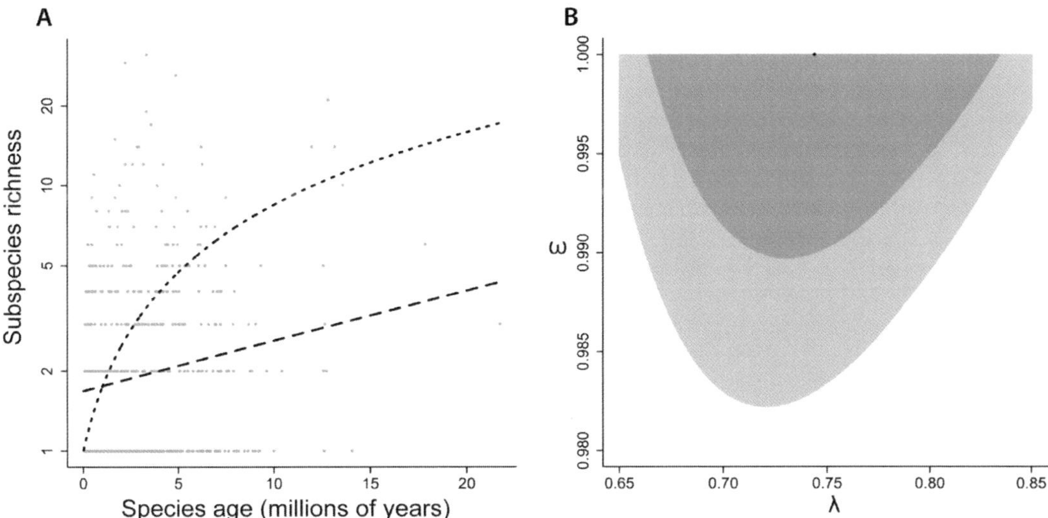

FIG. 1. (A) Species age versus subspecies richness for 1,100 species. The dashed line ($a = 0.564 \pm 0.039$, $b = 0.044 \pm 0.008$) is from a linear regression of $\ln(N)$ on t. The dotted line is fitted using the maximum-likelihood estimate for λ and ε, which are 0.740 and 0.99999, respectively. (B) A contour plot showing the maximum-likelihood values of λ and ε and the corresponding approximate 95% and 99% confidence intervals ($\ln L(\max) = \ln L(\lambda,\varepsilon) = 1.92$ and 3.32, respectively).

Continental versus insular species.—The subspeciation rate among the 212 insular species ($\lambda = 1.029$, 95% CI: 0.869–1.224; $\varepsilon = 0.99999$, 95% CI: 0.969–0.99999; $\ln L = -451.19$) exceeded that of their 888 continental counterparts ($\lambda = 0.672$, 95% CI: 0.615–0.735; $\varepsilon = 0.99999$, 95% CI = 0.993–0.99999; $\ln L = -1,783.34$), despite appearing superficially similar (Fig. 3A, B). This difference in rates between the two models was highly significant ($T = 19.25$, df = 2, $P < 0.001$).

When applied to the distribution of continental and insular species ages and subspecies richness,

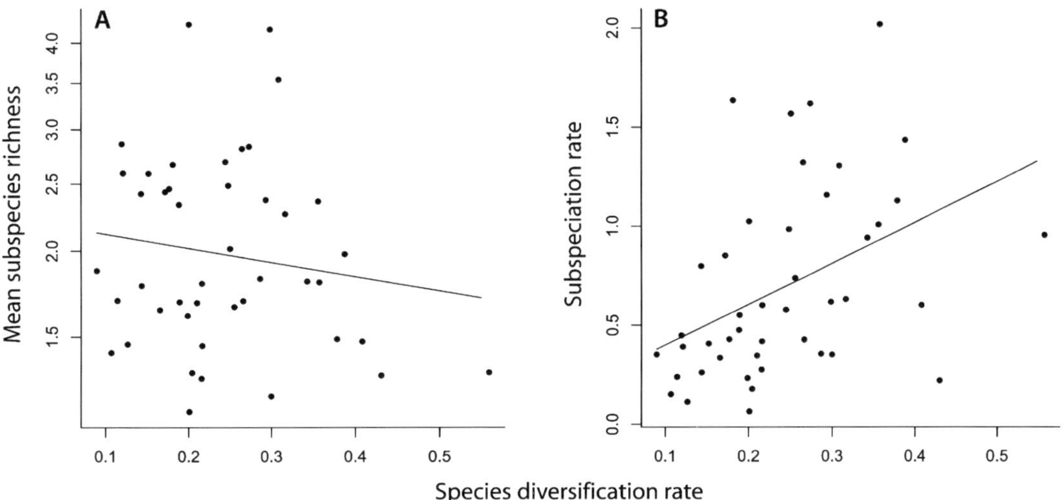

FIG. 2. (A) The linear regression of back-transformed mean ln(subspecies richness) on clade species diversification rate (see Equation 6), $a = 0.80 \pm 0.13$, $b = -0.47 \pm 0.49$, $R^2 = 0.02$. (B) The linear regression of subspeciation rate (λ; see Appendix) on species diversification rates (r), $a = 0.19 \pm 0.18$, $b = 2.06 \pm 0.67$, $R^2 = 0.18$.

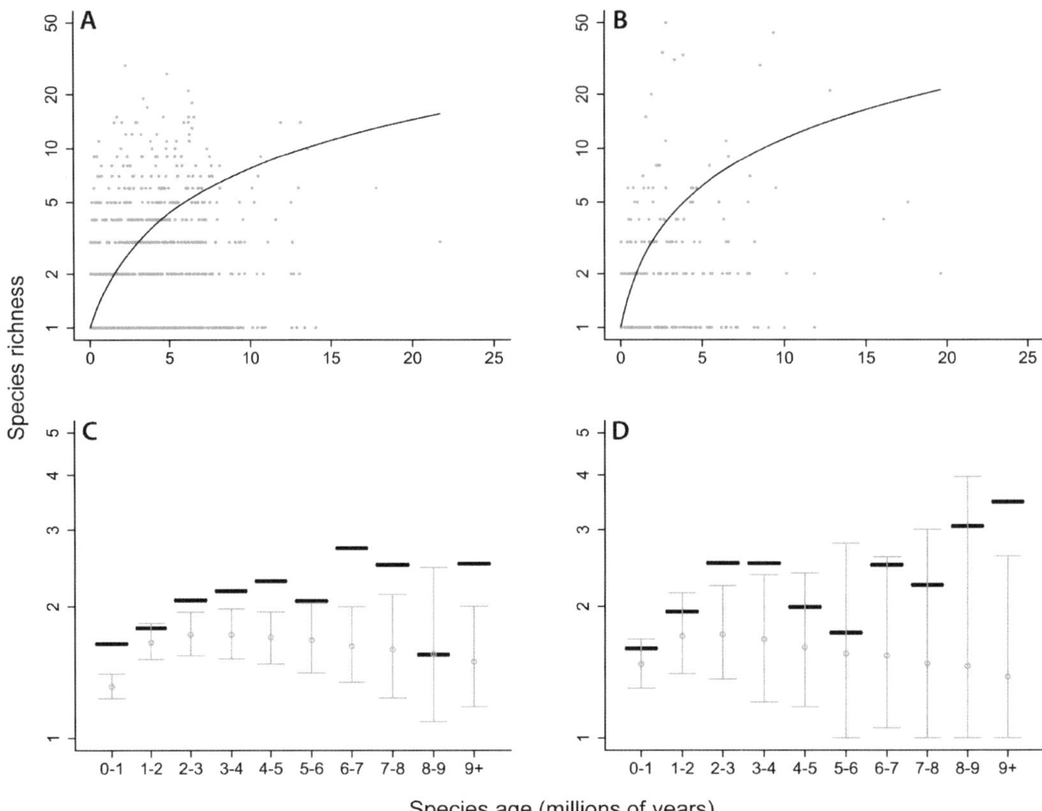

Fig. 3. Plots A and B show the observed relationship between species age and subspecies richness in continental (solid line fitted using $\lambda = 0.67$, $\varepsilon = 0.99999$) and insular species (solid line fitted using $\lambda = 1.03$, $\varepsilon = 0.99999$), respectively. Plots C and D show the average subspecies richness observed (black lines) and expected (gray 95% confidence interval; see text) in different age classes on continents and islands, respectively.

the birth–death model again received greater support than the pure birth model. However, there was a poor fit between the observed distribution and that expected given the maximum-likelihood parameter estimates (Fig. 3C, D). The average subspecies richness observed in each age class consistently exceeded the median that was expected and, in many cases, lay outside the 95% CIs.

Subspeciation rate estimates proved very sensitive to the definition of insularity. Defining insularity at a 0.5 cut-off (see above) resulted in a substantial reduction in the subspeciation rate estimated for insular species (number of species = 185; $\lambda = 0.673$, 95% CI: 0.552–0.823; $\varepsilon = 0.99999$, 95% CI: 0.980–0.99999; $\ln L = -335.25$) to below that of continental species (number of species = 915; $\lambda = 0.753$, 95% CI: 0.691–0.821; $\varepsilon = 0.99999$, 95% CI: 0.994–0.99999; $\ln L = -1,908.40$).

The difference in rates between the two groups was nonsignificant ($T = 1.01$, df = 2, $P = 0.60$). A more stringent definition of insularity ($\geq 80\%$ of subspecies must be island endemics) led to a considerable decrease in the estimated subspeciation rate for insular taxa (number of species = 141; $\lambda = 0.416$, 95% CI: 0.318–0.543; $\varepsilon = 0.99999$, 95% CI: 0.999–0.99999; $\ln L = -200.61$), as compared with continental taxa (number of species = 959; $\lambda = 0.783$, 95% CI: 0.720–0.851; $\varepsilon = 0.99999$, 95% CI: 0.995–0.99999; $\ln L = -2,033.69$). In this instance, a model that allowed rates to differ between the two groups was preferred to the null ($T = 19.72$, df = 2, $P < 0.001$).

Diversity-dependent subspeciation.—A model of exponentially declining subspeciation (Equation 7) performed significantly better than the best constant-rate model ($\lambda_0 = 1.53$, $z = 0.19$, $\varepsilon = 0.999$, $\ln L = -2,142.01$, likelihood ratio = 205.04, df = 1,

$P < 0.001$). When this model was applied separately to insular and continental species (using the 0.3 cut-off; see above), the exponential-decline model was preferred to the constant-rate model in both cases (continental species: $\lambda_0 = 1.54$, $z = 0.21$, $\varepsilon = 0.999$, $\ln L = -1{,}691.91$, likelihood ratio = 184.27, df = 1, $P < 0.001$; insular species: $\lambda_0 = 1.61$, $z = 0.13$, $\varepsilon = 0.999$, $\ln L = -438.95$, likelihood ratio = 24.65, df = 1, $P < 0.001$), and parameters differed significantly between the two ($T = 22.3$, df = 3, $P < 0.001$).

In the maximum-likelihood range-size model (Equation 8), the initial rate of subspeciation was correlated with geographic range size and subspeciation rates declined exponentially thereafter ($d = -0.38$, $c = 0.14$, $z = 0.20$, $\varepsilon = 0.99$, $\ln L = -1{,}995.24$). The data were significantly more likely under this model than one positing no relationship between geographic range size and initial rate of subspeciation ($d = 1.58$, $z = 0.19$, $\varepsilon = 0.99$, $\ln L = -2{,}059.80$, likelihood ratio = 129.12, df = 1, $P < 0.001$). However, when assessed separately across 841 continental species and 197 insular species with range-size information, the slope (c) between geographic range size and initial subspeciation rate did not differ significantly from zero in either case (continental species: $d = 1.36$, $c = 0.01$, $z = 0.22$, $\varepsilon = 0.99$, $\ln L = -1{,}547.95$, likelihood ratio = 0.30, df = 1, $P = 0.32$; insular species: $d = 1.41$, $c = 0.03$, $z = 0.19$, $\varepsilon = 0.99$, $\ln L = -401.68$, likelihood ratio = 0.21, df = 1, $P = 0.65$). The difference in rates between the two settings was significant ($T = 91.22$, df = 4, $P < 0.001$), but the sum of continental and insular log likelihoods did not differ significantly between the maximum-likelihood model and models with c fixed at zero (total $\ln L = -1{,}949.75$, likelihood ratio = 0.22, df = 2, $P = 0.89$).

Discussion

Species diversification rate was not correlated with subspecies richness. By contrast, diversification rates were positively correlated with subspeciation rates, agreeing with a recent study that correlated species richness with subspecies richness in birds (Haskell and Adhikari 2009). If species and subspecies origination and extinction both followed constant-rate birth–death processes, the observed correlation would be consistent with shared factors promoting speciation and subspeciation. However, I found evidence that subspeciation has slowed over time, rather than being constant. This could lead to estimates of subspeciation being higher for younger species (i.e., species originating from clades with rapid diversification rates) than for older species (Rabosky 2009a). Although it remains to be established whether species are at their subspecies carrying capacity, if this were the case then comparative analyses focusing on subspecies richness (the carrying capacity) would be more informative than analyses focusing on subspeciation or diversification rates (Rabosky 2009a).

Subspecies extinction rates approached subspeciation rates in all the models considered in the present study. Extinction of a subspecies may arise because of the extirpation of a population, but it may also arise if the phenotypic differences between two populations collapse (O'Brien and Mayr 1991). A corollary of high subspeciation and subspecies extinction-rate estimates is that many contemporary subspecies should have arisen relatively recently (Nee et al. 1994).

The estimated subspeciation and subspecies extinction rates in the present study were higher among insular than among continental species. Although these results were highly sensitive to definitions of insularity, they persisted under the exponential subspeciation-rate-decline models. An earlier study on biogeographic predictors of subspecies richness also identified insularity as a factor promoting higher levels of subspecies richness (Phillimore et al. 2007). There are several potential explanations for this result. A reduction in gene flow between island populations may facilitate greater phenotypic divergence, either by drift or by selection. In addition, ecological conditions may vary more across islands than across continental regions, making divergence via selection more likely on islands (Price 2008). Alternatively, taxonomic practice may differ between continental and insular species. For instance, Pratt (this volume) suggests that 20th-century taxonomists, such as Mayr and Amadon, tended to relegate what in essence were good insular species to subspecies and lumped these together as polytypic species. If this practice was widespread, the ratio of subspecies to species for insular taxa will have been overestimated in this study.

Subspecies diversification models.—Constant-rate pure birth and birth–death models of subspeciation and extinction are unable to account for the subspecies richness observed across species of different age. Both models describe subspeciation as an exponential process, with each lineage equally likely to subspeciate at each moment in time; this seems unlikely in reality. Both subspeciation-rate-decline models were preferred to the constant-rate models. These models may correspond to a scenario in which the probability of subspecies formation is a function of geographic range size, with larger geographic ranges more prone to subdivision (Rosenzweig 1978) and species' geographic range sizes changing little through time. As subspecies subdivide the geographic range among themselves, the probability that each of these subspecies ranges will themselves be subdivided further is expected to decline.

Although I found greatest support for models invoking exponentially declining subspeciation rates, I considered only two of a wide range of possible temporally non-homogeneous models. Diversification rates may, in fact, decline linearly (Rabosky 2009b) or be dependent on the number of lineages present at a particular time. Perhaps even time-varying birth–death processes do not capture the true subspecies diversification process. For instance, multiple subspecies may already exist at the time of speciation, or one subspecies with a large range may be the parent of many peripheral isolate subspecies (Rapoport 1982).

Comparative analyses that address the topology and temporal dynamics of species-level molecular phylogenies have shed light on modes of speciation (Barraclough and Nee 2001). Unfortunately, below the species level, ongoing gene flow, historical introgression, and incomplete lineage sorting are more pronounced, and these factors may reduce the efficacy of subspecies-level phylogenies for reconstructing patterns of divergence. In this context, population genetic approaches to identifying hierarchical structure may offer greater promise, particularly where some gene flow is likely to have occurred.

In addition to phylogenetic–taxonomic approaches, the support for different subspecies diversification models can be assessed from the geographic distributions of subspecies. Rapoport (1982) found that there were more North American mammalian subspecies at the periphery of species' ranges than predicted under a simple null model and that peripheral subspecies tended to have smaller geographic ranges; both findings suggest a peripheral-isolates model of subspeciation. In recent years, however, most work on geographic ranges has been conducted at the species level, and subspecies distributions have been overlooked.

Model inadequacy.—*A posteriori* simulations of subspecies diversification given the maximum-likelihood birth–death parameters generated consistently fewer subspecies than were actually observed (Fig. 3C, D), which implies that the maximum-likelihood model is inadequate. Two possible explanations are that subspeciation rates vary across subspecies because of one or more unaccounted-for variables and that subspeciation rates vary through time (Rabosky 2009b), both of which appear to be true of these subspecies data. Recent work by Rabosky (2010) demonstrates that if the rate of diversification (r) varies across lineages, then, even in the absence of extinction ($\varepsilon = 0$), application of a likelihood model to clade ages versus richness data will often lead to estimates of ε that approach 1. Rabosky also found that the likelihood surface around the maximum-likelihood estimate of ε can be very steep, yet this feature is not recovered when data are simulated using the estimated parameters. Across all analyses in the present study, I found that estimates of ε were close to 1 and estimated with an apparently high degree of confidence; therefore, I recommend caution in interpreting these high extinction-rate estimates.

Subspecies and evolutionary potential.—Appropriately delimited subspecies may be indicators of future evolutionary potential and speciation hotspots (Fraser and Bernatchez 2001, Winker et al. 2007). If we accept that, like species, subspecies may diverge along two somewhat independent axes, a genetic and a phenotypic axis, this may facilitate greater consistency and agreement among taxonomists (Winker 2009). A sound understanding of the processes that govern the origination and extinction of subspecies will be invaluable in the practical conservation of evolutionary potential. This study is the first to provide evidence that subspecies diversification rates may vary through time.

Acknowledgments

I am very grateful to T. Barraclough, S. Meiri, I. Owens, A. Pigot, T. Price, D. Rabosky, R. Ricklefs, G. Thomas, and J. Tobias for comments or advice that greatly improved this manuscript, and to S. Haig and K. Winker for organizing the coordinated session on subspecies in Portland that led to this volume. I also thank the Royal Society for a conference grant and the Natural Environment Research Council for funding. The United Nations Islands Directory is at http://islands.unep.ch/isldir.htm.

APPENDIX. Parameters estimated for individual phylogenies. For details regarding the phylogenies used, see Phillimore and Price (2009). The maximum-likelihood value of λ was estimated following Equation 4.

Clade	Number of species in phylogeny	Median species age (millions of years)	Median subspecies richness	λ (approximate 95% CI)	ε (approximate 95% CI) [a]	lnL (max)
Acanthiza	13	4.150	2.0	0.47 (0.24–0.97)	1.00 (0.49–1.00)	−24.93
Aegotheles	9	4.434	2.0	0.26 (0.10–0.66)	1.00 (0.00–1.00)	−13.78
Albatross	14	1.032	1.0	0.60 (0.25–1.40)	1.00 (0.00–1.00)	−17.49
Alcinae	22	3.156	2.0	0.42 (0.23–0.77)	1.00 (0.54–1.00)	−34.99
Alectoris	7	2.921	3.0	0.85 (0.37–2.18)	1.00 (0.58–1.00)	−16.01
Amazona	28	1.362	1.0	0.60 (0.32–1.30)	1.00 (0.71–1.00)	−44.03
Anas	45	0.656	1.0	1.13 (0.64–1.99)	1.00 (0.91–1.00)	−79.67
Anthus	37	5.142	3.0	0.58 (0.40–0.86)	1.00 (0.75–1.00)	−79.99
Caciques and oropendolas	17	3.238	2.0	0.35 (0.18–0.70)	1.00 (0.42–1.00)	−26.96
Catharus	12	5.188	4.5	1.02 (0.55–2.07)	1.00 (0.79–1.00)	−29.68
Cinclodes	13	0.868	2.0	2.02 (0.96–4.35)	1.00 (0.77–1.00)	−24.14
Cracidae	14	2.298	1.0	0.22 (0.08–0.56)	1.00 (0.00–1.00)	−13.50
Cranes	15	2.091	1.0	0.18 (0.08–0.32)	0.00 (0.00–1.00)	−14.15
Dendroica, Parula, Seiurus and *Vermivora*	40	2.467	1.0	0.94 (0.64–1.41)	1.00 (0.91–1.00)	−83.78
Empidonax	17	2.249	1.0	0.55 (0.27–1.14)	1.00 (0.67–1.00)	−27.32
Ficedula	27	4.560	1.0	0.33 (0.20–0.57)	1.00 (0.66–1.00)	−45.68
Geositta	11	5.497	1.0	0.15 (0.06–0.36)	1.00 (0.00–1.00)	−14.59
Grackles and allies	36	3.203	2.0	1.01 (0.67–1.55)	1.00 (0.92–1.00)	−84.84
Grouse, turkeys, partridges and tragopans	53	2.665	2.0	1.64 (1.19–2.30)	1.00 (0.97–1.00)	−131.25
Hemispingus	12	5.300	1.0	0.24 (0.10–0.57)	1.00 (0.21–1.00)	−15.30
Icterus	28	2.540	2.5	1.16 (0.72–1.91)	1.00 (0.91–1.00)	−65.66
Laridae	52	0.616	1.0	0.95 (0.57–1.58)	1.00 (0.76–1.00)	−66.98
Meliphaga	12	5.288	2.5	0.39 (0.20–0.81)	1.00 (0.51–1.00)	−23.95
Myiarchus	19	1.507	2.0	1.57 (0.87–2.90)	1.00 (0.88–1.00)	−39.90
Myioborus	12	1.956	1.5	1.43 (0.68–3.18)	1.00 (0.74–1.00)	−24.40
Parus	42	3.967	4.0	1.31 (0.94–1.86)	1.00 (0.95–1.00)	−115.93
Penguins	18	1.390	1.0	0.27 (0.11–0.63)	1.00 (0.00–1.00)	−18.99
Phylloscopus and *Seicercus*	59	4.310	1.0	0.43 (0.30–0.62)	1.00 (0.89–1.00)	−113.43
Pteroglossus	13	1.338	1.0	0.74 (0.32–1.69)	1.00 (0.44–1.00)	−18.59
Puffinus	24	1.464	1.0	0.35 (0.17–0.70)	1.00 (0.00–1.00)	−24.42
Ramphastos	8	2.494	2.0	0.34 (0.11–1.02)	1.00 (0.11–1.00)	−11.53
Sterinae	34	2.583	2.5	0.80 (0.52–1.25)	1.00 (0.93–1.00)	−73.60
Storks	16	4.069	1.0	0.06 (0.02–0.18)	1.00 (0.00–1.00)	−11.63
Swallows	31	3.902	1.0	0.23 (0.14–0.39)	1.00 (0.29–1.00)	−41.05
Sylvia	23	4.725	3.0	0.41 (0.24–0.69)	1.00 (0.75–1.00)	−44.97
Tangara	42	3.768	2.0	0.63 (0.43–0.94)	1.00 (0.85–1.00)	−83.63
Tauraco	13	1.974	2.0	0.35 (0.15–0.83)	1.00 (0.63–1.00)	−19.10
Thamnophilus	30	3.064	3.0	1.32 (0.86–2.08)	1.00 (0.91–1.00)	−72.56
Toxostoma	10	4.361	2.0	0.43 (0.20–1.98)	1.00 (0.30–1.00)	−18.05
Tringa	12	7.073	1.0	0.11 (0.05–0.27)	1.00 (0.00–1.00)	−15.43
Trogons	29	7.335	3.0	0.45 (0.29–0.71)	1.00 (0.88–1.00)	−66.42
Turdus and allies	60	4.224	2.0	0.98 (0.73–1.35)	1.00 (0.94–1.00)	−142.19
Woodpeckers	21	2.773	3.0	1.62 (0.97–2.82)	1.00 (0.92–1.00)	−54.39
Wrens	50	5.382	4.5	0.62 (0.50–0.77)	0.79 (0.66–0.93)	−134.03

[a] The maximum value allowed for ε was 0.99999.

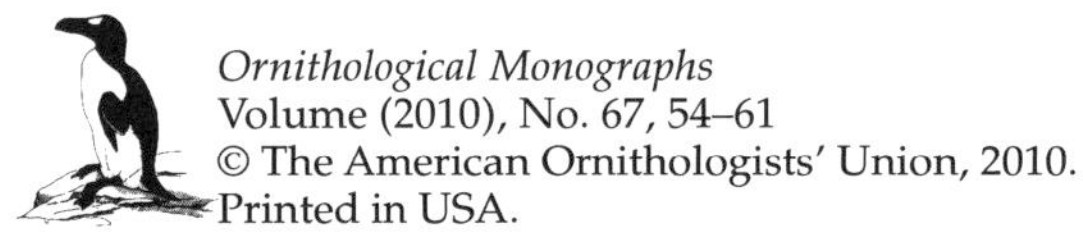

Ornithological Monographs
Volume (2010), No. 67, 54–61
© The American Ornithologists' Union, 2010.
Printed in USA.

CHAPTER 5

SUBSPECIES ARE FOR CONVENIENCE

John W. Fitzpatrick[1]

Cornell Lab of Ornithology, 159 Sapsucker Woods Road, Ithaca, New York 14850, USA

Abstract.—The century-long debate over the meaning and utility of the subspecies concept has produced spirited print but only superficial consensus. I suggest that genuine consensus about subspecies is an impossible goal, because trinomial epithets will inevitably be applied to a heterogeneous mix of evolutionary phenomena, thereby precluding genuine standardization of the concept. Populations that have intermediate levels of phenotypic differentiation and geographic isolation from one another often fall into a region I refer to as the "zone of art," where even skilled experts can disagree about the validity of any one subspecific treatment. The trinomial system cannot accurately represent the kind of information now available about genetic and character variation across space. Instead, ever more accurate tools are being perfected for quantitative, standardized descriptions of variation. These analyses—not subspecies classifications—will keep providing new scientific insights into geographic variation. Even more important, those of us who propose, debate, set, or enforce scientifically based conservation policies need to recognize that trinomial nomenclature survives primarily as a tool of convenience that cannot be viewed as strict science and should not be called on to establish or resolve crucial policy issues such as endangered-species listings. I have described new subspecies myself, and I regard the concept as a useful convenience. However, I submit that art and judgment will always be involved in practice and that no one trinomial treatment can be scientifically proved to be the biologically correct one. In this context, the subspecies concept itself is simply too heterogeneous to be classified as strict science.

Key words: conservation, endangered species, differentiation, geographic variation, subspecies.

Las Subespecies Son por Conveniencia

Resumen.—El siglo de debate sobre el significado y la utilidad del concepto de subespecie ha producido escritos muy animados pero se ha llegado a un consenso sólo de modo superficial. Sugiero que alcanzar un consenso genuino sobre las subespecies es un objetivo imposible, porque los epítetos trinomiales serán aplicados inevitablemente a un conjunto heterogéneo de fenómenos evolutivos, lo que impide una estandarización genuina del concepto. Las poblaciones que tienen niveles intermedios de diferenciación fenotípica y aislamiento geográfico entre sí usualmente se ubican en una región a la que yo llamo como la "zona de arte," donde incluso los expertos más hábiles pueden estar en desacuerdo sobre la validez del tratamiento de alguna subespecie. El sistema trinomial no puede representar de modo preciso el tipo de información disponible actualmente sobre la variación genética y en caracteres a lo largo del espacio. En cambio, las herramientas más precisas están siendo perfeccionadas para brindar descripciones cuantitativas estandarizadas de la variación. Estos análisis—no las clasificaciones de subespecies—van a seguir brindando nuevas visiones científicas sobre la variación geográfica. Incluso más importante, aquellos que proponen, debaten, fijan o hacen cumplir políticas de conservación con base científica necesitan reconocer que la nomenclatura trinomial sobrevive principalmente como una herramienta de conveniencia que no puede ser vista como ciencia estricta y que no debe ser utilizada para establecer o resolver la política de asuntos cruciales como el listado de especies amenazadas. Yo mismo he descrito nuevas subespecies y veo a este concepto como una conveniencia útil. Sin embargo, acepto que arte y juicio siempre estarán

---

[1]E-mail: jwf7@cornell.edu

Ornithological Monographs, Number 67, pages 54–61. ISBN: 978-0-943610-86-3. © 2010 by The American Ornithologists' Union.
All rights reserved. Please direct all requests for permission to photocopy or reproduce article content through the University of California Press's Rights and Permissions website, http://www.ucpressjournals.com/reprintInfo.asp. DOI: 10.1525/om.2010.67.1.54.

involucrados en la práctica y que ningún tratamiento trinomial puede ser probado de modo científico como el correcto biológicamente. En este contexto, el concepto mismo de subespecie es simplemente demasiado heterogéneo como para ser clasificado como ciencia estricta.

LIKE MOST QUESTIONS for which multiple correct answers exist, the century-long debate over the meaning and utility of the subspecies concept has produced spirited print but only superficial consensus. Naming subspecies codifies our recognition (traceable to Darwin) that species are neither static nor unitary across space and time. Few biologists dispute that understanding intraspecific geographic variation remains elemental to understanding evolution, or that naming distinguishable geographic units (whether as species or as subunits of species) facilitates conversations about them. Subspecific names provide convenient handles by which to describe, sort, store, retrieve, and discuss certain kinds of information about phenotypic geographic variation (Mayr 1982a). Beyond these generalizations, however, opinions about meaning and process associated with subspecies remain as diverse as ever (Haig et al. 2006; Haig and D'Elia, this volume; Winker, this volume). Indeed, diversity of opinion on the subspecies question has been amplified, not narrowed, by today's burgeoning access to genetic information about spatial variation within species and near-species.

Here, I suggest that genuine consensus about subspecies treatments—and about the subspecies concept itself—is an impossible goal, because trinomial epithets will inevitably be applied to a heterogeneous mix of evolutionary phenomena. I further suggest that no special reasons exist to expect or demand standardization in the application of trinomial names, because this nomenclatural system cannot possibly represent accurately the kind of information now available about genetic and character variation across space. Ever more accurate tools are being perfected for quantitative, standardized descriptions of variation. Simply put, in the 21st century, we know too much to be bound by a 19th-century nomenclatural convention, however useful aspects of that convention may be in some contexts.

Most important, it is essential that we who propose, debate, set, or enforce conservation policies recognize that trinomial nomenclature persists primarily as a tool of convenience and that it cannot be treated as strict science, because no standardized method for diagnosing and naming discrete units of evolutionary differentiation can be equally meaningful across taxa. In this context, I concur with Crandall et al. (2000) that management policies (including endangered species listings) should be derived directly and exclusively from ecologically and genetically relevant information about population distinctiveness, not from names. It is both tactically and ecologically inappropriate for conservation policies to be determined by subspecific taxonomy, because the latter is so famously subject to heterogeneous, often arbitrary, and inevitably fallible personal conventions of alpha taxonomists, nomenclature committees, reviewers, and editors.

TOOL OF CONVENIENCE: EASTERN TOWHEES AS A CASE EXAMPLE

Across the scrub and pine flatwoods of central Florida lives a distinctive form of Eastern Towhee (*Pipilo erythrophthalmus*). Compared to all other populations east of the Rocky Mountains, this form is smaller, longer-legged, duller, and has much less white on the back, wings, and tail. This Florida form also has pale, cream-colored, or whitish irides, whereas those of all other Eastern Towhees are dark red. The distinctive Florida form has a convenient and unambiguous handle: it is the "pale-eyed" form of Eastern Towhee, *Pipilo erythrophthalmus alleni*.

Greenlaw (1996) followed the American Ornithologists' Union (1957) in recognizing three other subspecies of Eastern Towhee. Interestingly, Greenlaw (1996:4) also noted that the "four-subspecies concept in Eastern Towhee may no longer be defensible," because outside of Florida, character variation is discordant, named taxa are only weakly differentiated from one another, and intermediate populations form broad geographic zones between any two of them. Greenlaw's reservations about the biological validity of these four taxa are justified, because none is fully discrete. Indeed, no two of them even represent the same kind of subspecies. The widespread nominate race has numerous plumage characters that vary gradually—but discordantly—over most of the species' breeding range; *P. e. canaster* "is a Gulf Coast extreme of geographic clines" (Greenlaw 1996:4); *P. e. rileyi* represents "strongly introgressed secondary contact between [the other two races] and pale-eyed *alleni* in Florida" (Greenlaw 1996:4—5); and *P. e. alleni* is a form so different from the others, and so uniform across the Florida peninsula, that Greenlaw suggested genetic investigations to determine whether it is a "separate phylogenetic entity" (Greenlaw 1996:5).

Biologically justified as they are, Greenlaw's reservations do not discount the conveniences achieved by having four names, because these (1) generally distinguish four kinds of Eastern Towhees and (2) generally identify each with a geographic subset of the species' range. The fact that borders of these subsets are variously fuzzy does not negate the utility of lumping variation into a few discrete categories, as long as we recognize that these crude categories are for purposes of overview only. Often, they help us organize conversations about, and even initiate the study of, variation. But these categories are not, in and of themselves, appropriate units for biological analysis. For the latter, we have other tools at our disposal, as discussed below.

The Subspecies as a Term of Art

Few tools in science are applied in as heterogeneous a manner as the subspecies concept, and the reason is simple. Evolutionary changes across space and time develop like snowflakes: no two are identical. The taxonomic consequences of this fact are diabolical. Because histories vary, and no two populations differentiate from one another in precisely the same way, any attempt to apply a single nomenclatural category to the process cannot help but encompass a diverse array of configurations, stages, and degrees of divergence.

Heterogeneity of the subspecies as a taxonomic category has unavoidable consequences, two of which are especially relevant here. First, many different kinds and degrees of variation will always be lumped into this single taxonomic level, the most common of which are illustrated in Figure 1. For example, separate names (Fig. 1A, B) are routinely applied to widely allopatric populations, whether they differ grossly (uppermost) or barely (second from top) from one another, and also to parapatric populations, whether the contact zone is sharp (third from top) or gradual (third from bottom). Intermediate or hybrid populations often differ so much from those on either side that an additional subspecific name or names are proposed for these as well (Fig. 1C, D). The bottom two cases in Figure 1 illustrate two common nomenclatural treatments of a smoothly clinally variable species (not shown is a third option espoused by many: to recognize no subspecific distinctions at all, even to represent extreme points of clines). More

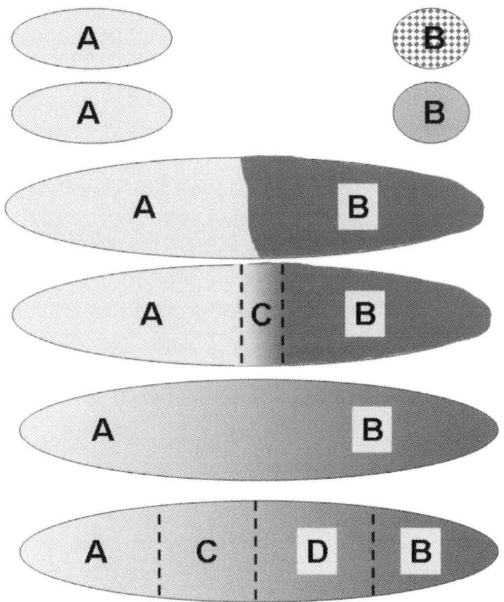

Fig. 1. Schematic examples of heterogeneity in most common applications of subspecies names (denoted here by A, B, C, and D) to geographically variable populations or taxa.

complex variation than is shown here (e.g., multiple discordant characters, variation along two dimensions, males and females showing different patterns, mix of isolated and continuous populations varying in different characters and degrees, etc.) is commonplace among widespread bird species. Taxonomists have dealt with this assortment by naming dozens of different kinds of subspecies. It is unfair to dismiss this historical heterogeneity as old fashioned or shoddy scientific practice (Patten, this volume; Winker, this volume). To the contrary, few such treatments can be proved incorrect, because the subspecies, including any quantitative rules used for "diagnosability" (Patten and Unitt 2002), is a human construct of convenience, not a biological entity that can be identified deductively and unambiguously.

Second, among birds at least, a large proportion of formally and usefully recognized subspecies will always be distinguished along arbitrarily demarcated geographic borders, or on the basis of arbitrarily defined levels of distinctiveness, or both. These features preclude standardization; hence, it will always be difficult (technically, impossible) to compare different species and their subspecies quantitatively and precisely for purposes of biological analysis or regulatory action (e.g., endangered-species listings; see below).

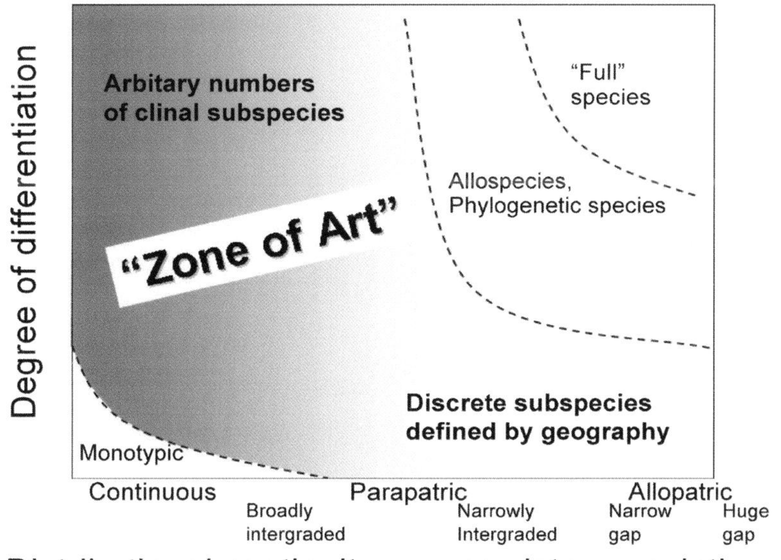

FIG. 2. Nomenclatural space for sister populations or taxa, visualized along two orthogonal axes: (1) degree of differentiation (phenotypic or genetic) and (2) distributional continuity between or among the taxa (adapted from Waples and Gaggiotti 2006).

Even the most careful taxonomists, examining the same data, often disagree with one another about how best to incorporate subspecific nomenclature to describe geographic variation within a widespread species. Disagreement occurs because, in most cases found in nature, no single subspecific treatment can be proven by a standardized algorithm to be the best, let alone the only, depiction of biologically meaningful and informative variation. The most obvious cases in point are subspecies names applied to various transitional stages or endpoints of clinal variation within continuously distributed populations (e.g., the Eastern Towhees discussed above), or even along step-clines that evolve among island or habitat archipelagoes. Both historically and in current practice, class Aves contains tens of thousands of such names (Dickinson 2003, Clements 2007).

Reviewing the question "What is a population," Waples and Gaggiotti (2006) elegantly visualized the well documented inverse relationship between population differentiation and degree of panmixia. I propose that in dividing species into discretely named taxa, taxonomists implicitly or explicitly invoke this relationship, and the exercise can be visualized along two axes (Fig. 2). One axis expresses the degree of divergence or differentiation—phenotypic, genetic, or both—among constituent sister populations. The other

axis describes the degree of geographic or distributional continuity among these populations. Monotypic species have mostly continuously distributed populations that cannot be distinguished from one another (i.e., very low values along both axes). At the other extreme (very high values along both axes), most well-differentiated and widely allopatric sister taxa today are recognized as full species. Just inside the full-species zone are allospecies, the component taxa of superspecies (Mayr 1963, Amadon 1966, Mayr and Short 1970). This zone incorporates most evolutionary units and taxonomic clusters regarded as phylogenetic species by some (e.g., Cracraft 1983, Nixon and Wheeler 1990). In such cases, which are often characterized by conspicuous differentiation among allopatric populations, geography renders moot the question of whether such forms could or would interbreed. Under the biological species concept, these cases are frequently treated as polytypic species (Mayr 1963) using trinomial nomenclature. As noted by many authors (e.g., Winker et al. 2007), neither of these two alternative treatments of well-differentiated allopatric populations is logically superior to the other. Whether they are called subspecies, allospecies, phylogenetic species, or even full species, treatment of well-differentiated sister taxa is biologically uncontroversial when they are allopatric.

Biological Correctness versus the Zone of Art

Often in nature, only slight phenotypic or genetic differences exist among sister populations, or populations differ from one another statistically but with substantial overlap among individuals. Where such populations are parapatric or allopatric, taxonomists historically have been tempted to name them as subspecies, and geography alone is sufficient to identify a given specimen to subspecies unambiguously (Fig. 2, lower right). Where populations are more continuously distributed, geographic differentiation commonly occurs clinally, either smoothly or along one or more steps. As in the Eastern Towhee case discussed above, clinal variation of different characters is often geographically discordant, which leads to a two-dimensional mosaic of phenotypes that grows increasingly complicated as more discordant characters are analyzed. Discordant geographic variation among characters within some widespread taxa may even be configured differently between males and females (e.g., Haffer and Fitzpatrick 1985). No taxonomic formulae or conventions exist for naming subspecies in such complex cases. Instead, the individual taxonomist is on his or her own—and in the case of birds, taxonomists have exercised this liberty for 150 years. This portion of the divergence—continuity spectrum can be referred to as the "zone of art," because imagination, creativity, sample-size constraints, and local geographic idiosyncrasies play as much of a role in delimiting subspecies as could any strict, scientifically based rule. This, fundamentally, was Greenlaw's observation about dividing Eastern Towhees north of the Florida peninsula into three subspecies: populations could be categorized in any number of ways, and no solution is intrinsically better or more biologically correct than any other. As visualized in Figure 2, the zone of art encompasses a broad range of situations within which multiple subspecific treatments are approximately equally defendable. Indeed, some authors argue that no subspecies should be recognized in such cases because broad intergradation or clinal differentiation occurs without obvious breaks or discontinuities. It is impractical to imagine that the zone of art can ever be fully resolved on biological grounds. Such fuzziness, however, does not negate the everyday utility of applying names to extreme variants and easily diagnosable units within this zone.

Fig. 3. Traditional (upper) and recently revised (lower) subspecies treatments of the Ochre-bellied Flycatcher (*Mionectes oleagineus*), a species that is remarkably uniform in phenotype and behavior across most of its range. Upper diagram shows published ranges of 15 poorly defined subspecies (Traylor 1979). Lower diagram identifies seven geographic regions where subtle clinal variation reaches endpoints (Fitzpatrick 2004).

Countless examples of the zone of art exist among widely distributed bird species, from the 200+ purported subspecies of Canada Goose (*Branta canadensis*; Hanson 2006) and 22 named subspecies of Northern Bobwhite (*Colinus virginianus*; Brennan 1999) to the 15 described subspecies of Ochre-bellied Flycatcher (*Mionectes oleagineus*; Traylor 1979). In the latter case—a widespread, dull ochraceous-green tyrant flycatcher of Middle and South America (Fig. 3)—most individuals literally cannot be assigned unambiguously to subspecies without precise data on where the specimen was collected (appropriately, subspecies names in this species translate to such words as "obscure," "similar," "inseparable," "intense," "poor," "pale-bellied," "yellowish," "greenish," and "olive"). I simplified the subspecific taxonomy of this species to seven names (Fitzpatrick 2004), but perhaps a better case could be made that the entire complex represents a broad cline from somewhat darker, greener, and diffusely streaked birds in Middle America south to the paler, ochraceous-bellied, unstreaked examples of the Amazon Basin and southeastern Brazil. Such a treatment might be reduced to just two names, applied to opposite ends of the cline. Even more important in the present context, no phenotypically based nomenclature would capture the complex phylogeographic history suggested by recent molecular analysis of this species in Amazonia (Miller et al. 2008). My points here are (1) that no one subspecific treatment, including mine, can be demonstrated to be the correct one on biological grounds; and (2) that detailed information about genetic variation and population histories, as revealed by modern studies, transcends the scientific utility of any particular naming convention we might apply (see below).

Superior Tools Abound for Analyzing Variation

Trinomial nomenclature entered ornithology in the mid-19th century, before Darwin and Wallace identified natural selection as a basis for geographic variation and a full century before Wright elucidated the role of genetic drift. Identifying bins for conveniently cataloguing intraspecific variation remains in vogue (e.g., Howard and Moore 2003, Clements 2007, del Hoyo et al. 2008, and several of the chapters in this volume), but the range of methods, questions, and outputs in the scientific study of geographic variation have vastly superseded those accommodated by trinomial nomenclature. Today, opportunities abound for detailed analyses of variation using quantitative comparisons of both phenotypic and genetic variation among populations across space. A plethora of tools exists for such analyses, including multivariate and spatial statistics (Maurer 1994), phylogeography (Avise 2000, 2004, 2006), coalescent theory (Wakeley 2006), and even historical demography (Rogers and Harpending 1992, Drummond et al. 2005). These modern tools

promote sophisticated scientific inquiry by providing detailed data at multiple scales, involving geographic variation at all stages of differentiation and speciation. By contrast, subspecies taxonomy provides, at best, a blurry lens through which to attempt to draw inferences about microevolutionary pattern and process. At worst, subspecies are treated as ends in themselves, thereby distracting us from more rigorous and illuminating analyses of this subject. No amount of statistical rigor, including a 75% rule (for discussion of this convention, see Patten, this volume), can change the fact that judgment must always be applied to subspecies treatments. Even the term "diagnosability" (Patten and Unitt 2002) is demonstrably malleable, because we can always add more characters, diversify statistical tools, or debate and change our acceptable criteria for diagnosis. The subspecies applies to such a range of intermediate situations that no single correct algorithm exists to represent these intermediates using the trinomial tool.

Subspecies Status Is a Poor Guide for Conservation Priorities

It is, at long last, axiomatic that long-term conservation of biological diversity demands protection of natural variation at the infraspecific level. The subspecies, however, is at best a very poor proxy for this variation (Wayne and Morin 2004). It is high time that individuals and agencies involved in setting conservation priorities or policies (including local, state, and federal listing decisions) acknowledge the inherent heterogeneity and subjectivity of the subspecies concept and embrace more rigorous analyses of distinctiveness (including ecological distinctiveness) in establishing priorities and setting policies. Although trinomial nomenclature sometimes offers first-order clues about population distinctiveness, this naming system was never designed as a substitute for objective measurement and comparative assessment of the morphological, ecological, behavioral, genetic, and evolutionary data required for pursuing rational conservation policy today (Crandall et al. 2000, Moritz 2002). Despite the fact that the U.S. Endangered Species Act specifically allows for listing of "distinct population segments" that are not described subspecies, listing decisions routinely incorporate, and often hinge upon, debates about the validity of a particular subspecific name (such a case is discussed below) rather than more biologically relevant questions (Haig et al. 2006).

Reliance on subspecies classifications as a proxy for delineating conservation units is misleading on at least two grounds. First, by partitioning complex and often discordant patterns of phenotypic and genetic variation into discrete entities, often using only a limited number of characters, subspecies nomenclature usually simplifies, and often grossly misrepresents, both the amount of variation and its geographic complexity. Modern geospatial statistical tools such as spline, spline regression, and step-regression (e.g., Skalski et al. 2008) allow biologists to detect, demonstrate, and interpret clines, step clines, local peaks of character divergence, and discordant character variation using multiple data sets, objective algorithms, and replicable procedures. Second, when it comes to proposing or recognizing subspecies names within geographically variable species, expert opinions often contradict one another. I suggest that this cannot be avoided, because multiple trinomial solutions—especially within the zone of art—can be approximately equally correct on biological grounds (see above). Because equivocal cases and contradictory treatments are common, conservation agencies run the risk of straying from otherwise warranted findings of population distinctiveness (Haig and D'Elia, this volume) by adhering to any single subspecific treatment. Instead, to ensure a sound biological footing, listing decisions and conservation priorities should be based on explicit and thorough analyses of ecological, behavioral, morphological, and genetic data. Trinomial epithets, whether historical or recent, can help guide these studies, but they should not—as they still so often do—replace them.

Failure to list the virtually isolated "San Diego" population of Cactus Wrens (*Campylorhynchus brunneicapillus*) in southern California provides an example of how undue reliance on the subspecies concept leads to erroneous conservation decisions. In 1990, the U.S. Fish and Wildlife Service (USFWS) received a petition to list as endangered a described subspecies of Cactus Wren (*C. b. sandiegensis*; Rea and Weaver 1990), on the grounds that this rapidly disappearing population represented a "distinct population segment" as defined by the Endangered Species Act (for definitions and discussion of this feature of the Endangered Species Act, see Haig and D'Elia,

this volume). In its published finding not to list (Beattie 1994), the USFWS stated the following:

> The American Ornithologists' Union Committee on Classification and Nomenclature did not recognize the San Diego cactus wren . . . as a subspecies of the cactus wren. . . . Since the conclusion of the committee is that *C. b. sandiegensis* likely only represents an intermediate form between two recognized subspecies of cactus wren, it is not currently under consideration for addition to the Federal List of Endangered and Threatened Wildlife and Plants.

As a member of the referenced AOU committee at the time, I can attest that (1) the committee considered the validity of *C. b. sandiegensis* reluctantly, during a previously schedule meeting at the U.S. National Museum; (2) we spent less than an hour on the question; (3) we examined a total of about 20 Cactus Wren specimens representing several subspecies, including only a few of the focal taxon; (4) we concluded that *C. b. sandiegensis* appeared to differ from both of the subspecies that surrounded it (*C. b. couesi* of the southwestern United States and adjacent mainland Mexico, and *C. b. bryanti* from San Diego County, California, south to northern Baja California); and (5) plumage characters of *C. b. sandiegensis* appeared to be intermediate between the two much more widespread subspecies, each of which appeared to be uniform across their much larger ranges.

Our analysis was not a scientific study of the distinctiveness of a population, and we did not submit our findings to be peer-reviewed or published, as had Rea and Weaver (1990). Our casual conclusion about the intermediacy of *C. b. sandiegensis* was conveyed as such in a letter to the USFWS by the committee chairman. Such determinations represent willing, if informal, participation in the zone of art (Fig. 2), and they abound in the history of ornithological taxonomy. Countless hundreds, perhaps thousands, of the world's named avian subspecies today represent exactly the same level of population distinctiveness as *C. b. sandiegensis*. Identifying such intermediates as subspecies is neither correct nor incorrect but is simply a point of view, given available data—a convenient handle. What was incorrect was citing the committee's conclusion as a fundamental reason for not considering this named population as a candidate for listing. As it turns out, mtDNA studies eventually revealed genetic uniqueness within this named form (Eggert 1996), yet this rapidly disappearing population is still not federally listed. The California Department of Game and Fish recently identified *C. b. sandiegensis* as a "species of special concern" (Unitt 2008).

Most important, subspecies or not, localized Cactus Wren populations are well-documented indicators of a vegetative formation that is unique in North America, the cactus scrubs of southern coastal California (Rea and Weaver 1990, Solek and Szijj 2004). In this context, the status of *C. b. sandiegensis* as a valid subspecies is moot: these populations are well separated geographically and ecologically from other Cactus Wrens, thus fulfilling another of the criteria for designation as a distinct population segment (Haig and D'Elia, this volume). Both the habitat and its wrens have been catastrophically reduced by residential and commercial development and by unnaturally high frequency and severity of wildfires. As a consequence, survival of the coastal Cactus Wren is considered one of the greatest challenges in bird conservation for southern California (Unitt 2004). Section 2(B) of the Endangered Species Act specifies that the purposes of the act include "to protect the ecosystems upon which endangered species and threatened species depend." Protection of coastal cactus scrubs in southern California continues to be impeded by failure of the USFWS to give standing to the habitat's signature bird species. This failure originated from undue focus on the subspecies question. Equivocal validity of the name (about which legitimate debate will continue) became a red herring diverting attention from ecological distinctiveness and conservation importance of this population. Listing should be driven by the latter.

Recommendation: Use Subspecies as a Convenient Tool

By any of the current definitions, species are biological entities. By contrast, subspecies have been human constructions since they were first used in the 19th century. Our innate desire to name things that we see motivates us to apply unique names to each of the variants that we detect within species. This is a useful habit, because recognizing different identifiable types helps draw attention to natural variation. Naming these variants facilitates conversation about them, encourages naturalists to recognize geographically distinct populations in the field, and—as emphasized without embarrassment by Mayr (1982a)—helps museum curators sort specimens to organize variants into categories for inspection and analysis. Persistence of the subspecies as a tool of convenience, however, must neither be confused with, nor stand in the way of, more precise, scientific, and ultimately useful methods of describing, analyzing, interpreting, and conserving geographic variation within species.

When it comes to conservation policies (e.g., prioritizing habitat for preservation, preparing listing petitions, making listing decisions, reviewing listings, or developing recovery plans), I view it as imperative that we refrain from placing too much weight on trinomial nomenclature. Not all biologically important or ecologically informative variation has been described in the form of subspecies, nor could it be. By the same token, described subspecies differ spectacularly in the respective levels of scientific rigor with which they were described, and in the biological meaning that underlies their trinomial names. Subspecific nomenclature is convenient but idiosyncratic. Subspecies should be used as just one, very fallible, clue in triggering more rigorous approaches to the careful evaluation of population distinctiveness, especially for purposes of establishing conservation priorities or actions.

Acknowledgments

I am grateful to S. Haig and K. Winker for inviting my participation in the original symposium and in this volume. I thank F. Gill, J. Greenlaw, S. Haig, F. James, N. Johnson, I. Lovette, M. Morrison, S. Morrison, J. V. Remsen, T. Schulenberg, J. A. Stallcup, D. Stotz, D. Willard, and K. Winker for observations, data, and thoughtful insights that contributed to key points in this paper.

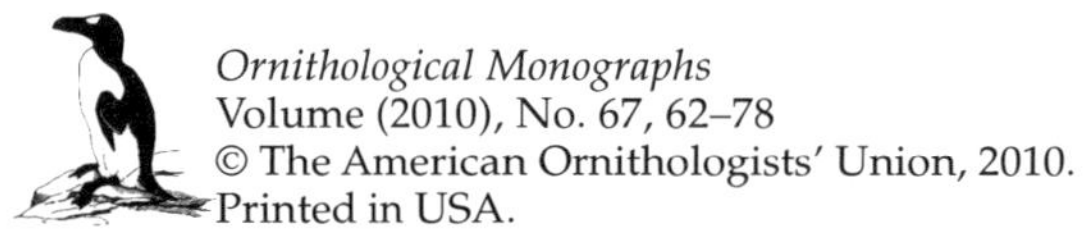

Ornithological Monographs
Volume (2010), No. 67, 62–78
© The American Ornithologists' Union, 2010.
Printed in USA.

CHAPTER 6

SUBSPECIES AS A MEANINGFUL TAXONOMIC RANK IN AVIAN CLASSIFICATION

J. V. Remsen, Jr.[1]

Museum of Natural Science, Louisiana State University, Baton Rouge, Louisiana 70803, USA

Abstract.—Dissatisfaction with the subspecies unit of classification is, in part, a consequence of the failure of many of those who have described subspecies to follow the conceptual definition of the subspecies, namely that it should represent diagnosable units. The antiquity of the descriptions of most subspecies (median year of description of currently recognized subspecies estimated to be 1908–1909) means that the majority predated any statistical tools for assessing diagnosability. The traditional subspecies concept, as originally construed, identifies minimum diagnosable units as terminal taxa, and I suggest that it is thus essentially synonymous with the phylogenetic species concept. Therefore, both must deal with the fundamental difficulties inherent in using diagnosability as a criterion. Application of monophyly as a criterion for taxon rank at the population level has inherent difficulties. An advantage of the biological species concept is that it incorporates, in its classification of taxa, assessments of gene flow and reproductive isolation, which are critical components of the evolutionary process. Critics of the biological species concept persistently overlook the fact that it includes the subspecies rank as a necessary component of that concept for distinct populations within biological species. Analyses that require terminal taxa can, with care, be conducted under the biological species concept using subspecies plus monotypic species. Critics of the biological species concept with respect to its application have missed the biological and political disadvantages of treating minimum diagnosable units as the primary unit of conservation concern. Human perception is in accord with ranking such minimum diagnosable units below the species rank; socially and scientifically, humans consider diagnosable units of other humans as distinct groups but not separate species.

Key words: species concepts, species definitions, subspecies definitions.

Las Subespecies como un Rango Taxonómico Significativo en la Clasificación de las Aves

Resumen.—En parte, la insatisfacción con la unidad de clasificación de subespecie es consecuencia de que muchos de aquellos que han descrito subespecies no han seguido la definición conceptual de la subespecie como una unidad diagnosticable. La antigüedad de las descripciones de la mayoría de las subespecies (la mediana del año de descripción de las subespecies actualmente reconocidas se estima en 1908–1909) significa que la mayoría precedió a las herramientas estadísticas para evaluar la diagnosticabilidad. El concepto tradicional de subespecie, como se concibió originalmente, identifica unidades diagnosticables mínimas como taxones terminales, por lo que sugiero que esencialmente es sinónimo del concepto filogenético de especie. Por lo tanto, ambos deben lidiar con las dificultades fundamentales inherentes vinculadas con el uso del criterio de diagnosis. La aplicación de la monofilia como un criterio para la clasificación de los taxones al nivel poblacional tiene dificultades inherentes. Una ventaja del concepto biológico de especie es que incorpora, en su clasificación de los taxones, evaluaciones del flujo génico y del aislamiento reproductivo, que son componentes fundamentales del proceso evolutivo. Las críticas al concepto

[1]E-mail: najames@LSU.edu

Ornithological Monographs, Number 67, pages 62–78. ISBN: 978-0-943610-86-3. © 2010 by The American Ornithologists' Union. All rights reserved. Please direct all requests for permission to photocopy or reproduce article content through the University of California Press's Rights and Permissions website, http://www.ucpressjournals.com/reprintInfo.asp. DOI: 10.1525/om.2010.67.1.62.

biológico de especie persistentemente pasan por alto el hecho de que éste incluye el rango de subespecie como un componente necesario para poblaciones diferentes dentro de la especie biológica. Los análisis que requieren taxones terminales pueden ser conducidos, con cuidado, bajo el concepto biológico de especie usando subespecies y especies monotípicas. Las críticas del concepto biológico de especie con respecto a su aplicación han pasado por alto las desventajas biológicas y políticas de tratar a las unidades diagnosticables mínimas como las unidades principales de preocupación conservacionista. La percepción humana coincide en clasificar estas unidades diagnosticables mínimas por debajo del rango de especie; social y científicamente, los humanos consideran unidades diagnosticables de otros humanos como grupos distintivos pero no como especies separadas.

WHETHER THE SUBSPECIES rank in classification is considered useful depends on whether one's concept of species includes room for geographically non-overlapping, diagnosable units within a species. Can a species be subdivided into distinct, biologically meaningful units? Should such units be formally named? As noted by all who have written about the classification of organisms, imposing a categorical scheme, such as the Linnaean system of classification, on the pattern of continuous variation produced by evolutionary processes is doomed to be unsatisfactory. As noted by Stresemann (1936:157), "Whoever wants to hold firm rules, should give up taxonomic work. Nature is too disorderly for such a man." Empirical examples can be mustered that defy the tidy either/or demands of any species concept. Yet human perception, dominated by categorical thinking, uses such schemes to produce the vocabulary of labels needed for communication. In short, biological classification attempts to inflict an unrealistic categorical scheme on the patterns produced by a disorderly, fundamentally noncategorical process.

Controversy over the utility and definition of the subspecies rank in such a categorical classification has a long history, with episodic reappraisals (e.g., Wiens 1982), yet the category survives in almost all modern classifications of birds. This survival, since the mid-1800s, is presumably driven by a perception among most humans that the category that we term "species" can include within it named subpopulations to identify nonclinal geographic variation. This, in turn, may follow from our own widespread, long-standing perception of the nature of the species *Homo sapiens*, in which pronounced, nonclinal geographic variation is included within that species rather than each distinct group being considered a separate species.

WHAT IS A SUBSPECIES? CONCEPTUAL DEFINITIONS

To recognize nonclinal intraspecific geographic variation in animals, some taxonomists have applied trinomials as subspecies names since at least

1844 (*fide* Simpson 1961). The concept behind subspecies definitions centers on the existence of separate units or geographic units within the rank of species. Historically, dissatisfaction with ranking every distinctive geographic population as a species was the catalyst for the use of trinomials, which were regarded a century ago as a radical and progressive step in classification (Knox 2007). Definitions of subspecies extracted from standard references and textbooks (Table 1) are founded on the theme that a "species" may consist of subunits that differ from each other in diagnosable ways yet share the characters attributed to the species itself. I combine these ideas into the following definition: "geographic populations diagnosable by one or more phenotypic traits."

The theme that unifies these definitions of subspecies is that subspecific names identify distinct population units: they are phenotypic predictors of past or current genetic continuity, the phenotypic analogue of genetic markers. Hennig (1966:102) stated that the goal of species-level taxonomy was to relegate to subspecies rank "all vicarying reproductive communities."

That the subspecies category has biological meaning is reinforced by the observation that populations known to be reproductively isolated, and thus considered species by any definition, typically differ from close relatives in the same kinds of phenotypic characters and patterns, but to a greater degree (e.g., as seen among subspecies that intergrade where in contact). Whether such characters and patterns represent causation or correlation is an open question.

A definition of subspecies as "geographic populations diagnosable by one or more phenotypic traits" is a simple statement concerning the current geographic distribution of distinct phenotypic traits. This definition makes no assumptions about whether the traits are adaptive or whether they represent populations that are incipient species, and thus makes no predictions concerning the future. Assuming that a phenotypic trait has a genetic rather than environmental basis, subspecies boundaries imply that all individuals of

TABLE 1. Subspecies definitions from textbooks and reference works.

Source	Definition
Mayr et al. 1953, Mayr 1963	Geographically defined aggregates of local populations which differ taxonomically from other such subdivisions of a species
Mayr and Ashlock 1991:43, 430	An aggregate of local populations of a species inhabiting a geographic subdivision of the range of the species and differing taxonomically [differing by sufficient diagnostic characters] from other populations of the species
Futuyma 1979	A set of populations of a species that share one or more distinctive features and occupy a different geographic area from other subspecies
Futuyma 2005:213, 356	A recognizably distinct population, or group of populations, that occupies a different geographic area from other populations of the same species; populations of a species that are distinguishable by one or more characteristics and are given subspecific names
Strickberger 2000	A taxonomic division of a species often distinguished by special phenotypic characters and by its origin or localization in a given geographic region

the subspecies share the genes responsible for the diagnostic trait, which arose in a common ancestor, and, thus, form a monophyletic group with respect to those genes. Whether they also form a monophyletic group with respect to other gene trees is an open question. As I discuss below, however, application of the term "monophyly" at the population level is problematic.

WHAT IS A SUBSPECIES?
OPERATIONAL DEFINITIONS

Mayr et al. (1953) provided objective, quantitative definitions of subspecies based on degree of overlap that can be applied across taxa. They outlined why using simple linear overlap in measurements, for example, overemphasizes extreme individuals in a population and overestimates true population overlap. They also discussed various interpretations of the "75% rule" as the threshold for naming subspecies. Although one interpretation is that only 75% of the individuals of each sample have to be correctly classified, the rule as defined by Amadon (1949), Mayr et al. (1953), and Patten and Unitt (2002) is based on standard deviations from the mean of normally distributed data. Depending on which metric is applied, in essence these definitions mean that 90–97% of the individuals of one population must be distinguishable from the equivalent percentage of the other population to be considered subspecies under the somewhat misleadingly named 75% rule.

As emphasized by Mayr et al. (1953) and Patten and Unitt (2002), defining subspecies solely on the basis of statistically significant differences in population means is an unfortunate misinterpretation of the conceptual definition. Given large enough sample sizes, the means of any two populations likely differ significantly (>95%), even though actual overlap can be nearly complete, and so statistically significant differences in the means alone provide almost no information on how distinctive two populations are in terms of diagnosability, the key theme of the conceptual definitions of subspecies. The problem is that the conceptual definitions emphasize the population as a whole, not the individuals that constitute it, and so statistically significant differences between means can be interpreted as diagnosability if the population is the unit of analysis.

Although the 75% rule has a long history in ornithology, its application has been erratic at best. For example, it is generally not mentioned as a criterion for recognizing subspecies in classifications (e.g., American Ornithologists' Union 1957, Dickinson 2003) or in any of the *Handbook of the World* series (del Hoyo et al. 1992–2008). It is not possible to tell how many of the subspecies currently recognized in such sources would qualify as subspecies under the 75% rule, but it is certain that many subspecies, especially in North America, would not qualify as valid taxa under this rule, particularly those defined by mensural differences. From personal experience in attempting to use subspecies diagnoses, such as the keys in the *Birds of North and Middle America* series (Ridgway and Friedmann 1901–1950), I predict that more than 75% of North American subspecies taxa delimited by mensural data would not survive application of the 75% rule.

Although Patten and Unitt (2002) used 75% as their target level of degree of diagnosability in deference to tradition, they advocated a higher level, 95%, as a standard for diagnosability. I also propose that this level of diagnosability become the operational definition of diagnosable. McKitrick and Zink (1988) gave reasons why aiming for 100% diagnosability for phylogenetic species is conceptually and methodologically unreasonable. Also, geographic sampling to determine diagnosability in the case of parapatric populations must exclude any zones of intergradation in statistical treatments because heavy sampling from that zone would eliminate any potential diagnosability of two populations. To avoid circularity, delimitation of zones of intergradation must be objective (e.g., Harrison 1993), because, hypothetically, one could expand the area considered the zone of intergradation until diagnosability reaches 95% for populations on either side of it. Of course, the amount of intergradation occurring at parapatric zones of contact is of considerable biological and taxonomic interest, especially in determining whether taxa are subspecies or full species (Mayr and Ashlock 1991).

By incorporating a quantitative operational definition into the conceptual historical definition, I produced the following definition, modified from Futuyma (2005), including an explicit statement that diagnosability refers to individuals that comprise the population: A subspecies is a distinct population, or group of populations, that occupies a different breeding range from other populations of the same species; individuals are distinguishable from those other populations by one or more phenotypic traits at the 95% level of diagnosability.

Application of this rigorous definition would result in the synonymization of many subspecies names in North America and elsewhere where broad geographic patterns of smoothly clinal differences in coloration and, especially, morphometrics have been artificially categorized as subspecies. Quantitative analyses of this geographic variation typically found that much or most of this variation, at least in terms of morphometrics, cannot be partitioned into diagnosable units (e.g., Power 1969, Behle 1973, Tacha et al. 1985, Aldrich and James 1991, Wood 1992, Rising et al. 2001, Rising et al. 2009). The implied agenda of much of the work in the first half of the 20th

century was that all geographic variation had to be described in a categorical way, namely by use of subspecies names (see Knox 2007). However, patterns of geographic variation in phenotype provide valuable insights into population structure and the process of evolution, regardless of whether the variation can be apportioned into diagnosable units (James 1970, Zink and Remsen 1986). Note also that this definition is based on phenotypic traits in plumage and morphology with an assumed genetic basis, not on other phenotypic traits such as behavior or physiology, nor on genetic markers not expressed in the phenotype; one could make a case for recognizing any diagnosable, geographically distinct genetic unit with a subspecies label. Note also that this definition is incomplete with respect to distinguishing subspecies from species, which I address below. To refer to populations or individuals that represent extremes of clinal variation or populations that do not meet statistical thresholds of diagnosability, for convenience one could use an informal vocabulary using the formerly recognized subspecies names in quotes, followed by "grade" (e.g., "the *nigrideus* grade" for the darker northeasternmost populations of the American Robin (*Turdus migratorius*). Note that use of a 95% diagnosability criterion applies only to two-way comparisons, as in the 75% rule, not to multiple simultaneous comparisons. It also applies only to populations in the breeding ranges; an empirical consequence of such a rigorous standard (if individuals from intergrade zones were excluded) would also allow assignment of individuals from the nonbreeding range to breeding population with a high level of statistical certainty.

A persistent criticism of the subspecies concept is that analysis of different characters may produce different subdivisions (e.g., Wilson and Brown 1953). In other words, the characters are not distributed in a concordant geographic manner; for example, three characters might show geographic variation, but each character could show three different patterns that would delimit subspecies boundaries in three conflicting ways. The existence of such conflict is inevitable, and application of the subspecies concept in such cases is unwarranted. However, my impression, based on examining many hundreds of primarily Neotropical bird species over the past 30 years with respect to distinct plumage characters, is that such conflicts are greatly outnumbered by examples of concordance. For example, if three traits show nonclinal geographic variation, all three may not show breaks at the same point, but actual conflict in where the breaks are is infrequent. For exceptionally thorough quantifications of patterns of geographic variation in morphometrics and plumage that showed strong concordance, see Johnson (1980) and Cicero (1996). By contrast, smoothly clinal variation regularly shows conflicting patterns, with trends in, for example, tail length showing a different pattern from that of, for example, back color.

Ridley (2004) used lack of concordance in geographic variation as grounds for essentially dismissing the importance of the entire subspecies unit of classification. He used the example of clinal size variation in North American House Sparrow (*Passer domesticus*) to illustrate clinal variation and then noted that there was no reason to expect clines in other characters to match the body-size cline. However, he did not point out that no subspecies have been described or recognized in North American House Sparrows because the variation is smoothly clinal rather than discrete, so his example is inappropriate, and his premise is flawed. If the geographic pattern of variation in distinct characters produces conflicting patterns, then this implies complex underlying population structure that may not be amenable to diagnoses and, therefore, subspecies names. However, Johnston and Selander (1964:549) found that "color differences between samples are both marked and consistent, permitting 100 percent separation of specimens from two localities" but did not name any subspecies because they did not think that the situation was temporally stable.

Subspecies versus Phylogenetic Species

Given that conceptual definitions of subspecies have always emphasized diagnosable units, how do they differ from phylogenetic species other than that, under the biological species concept, many diagnosable units are ranked as species? Cracraft (1983:170) defined phylogenetic species as "the smallest diagnosable cluster of individual organisms within which there is a parental pattern of ancestry and descent." Cracraft did not define how these differ from subspecies but emphasized the heterogeneous nature of the results of applying the subspecies concept. He urged abandoning the subspecies rank in classification, without detailing how it differs from phylogenetic species. The methodological difficulties

that produced the heterogeneity in units called subspecies are assumed to disappear if Cracraft's phylogenetic species concept is adopted—when, in fact, delimitation of "diagnosable clusters" entails all the methodological problems that complicate subspecies delimitation. Any renaming of all minimum diagnosable units as species would require determining what units are actually diagnosable and at what statistical thresholds of diagnosability.

Cracraft (1983) pointed out the biological species concept lacks equivalency among the units called species. However, the same problem pervades species defined under the phylogenetic species concept, in which, for example, species reproductively isolated from all other lineages, including syntopic sister taxa, are treated as the same taxonomic unit as populations that differ only in the possession of a single diagnostic character and cannot coexist syntopically with sister taxa. By contrast, use of the subspecies rank within biological species as the unit of analysis reduces the problems of heterogeneity because population units diagnosed only by minor plumage differences are not treated as the same unit as lineages known or inferred to be sealed from other lineages by reproductive isolation.

Another criticism of the biological species concept is that biological species are not the appropriate unit for biogeographic and speciation analyses (Cracraft 1983). I agree. The appropriate units are indeed minimum diagnosable units— that is, subspecies under the biological species concept. That subspecies can be used productively for such analyses is shown inadvertently by none other than Cracraft (1983), who used subspecies names in outlining his methods for determining areas of endemism. Cracraft (1985) later also used lengthy lists of trinomials to demarcate and name areas of endemism in the Neotropics. Those areas, defined by the terminal taxonomic unit of the biological species, namely subspecies, are still the standard nomenclature for Neotropical biogeographic analyses, thereby demonstrating the utility of the subspecies unit of classification.

Other definitions of phylogenetic species repeat the essence of Cracraft's phylogenetic species concept, with the emphasis on diagnosability and common ancestry, and they do not address how this definition differs from that of the subspecies. Futuyma (2005) and Freeman and Herron (2007) also reported the definition of the phylogenetic species concept without explaining how it differs from the rank of subspecies within the biological species concept. The explicit conceptual definition of phylogenetic species is that they represent monophyletic units, whereas subspecies are not defined explicitly with respect to monophyly. In practice, however, when phenotypic characters define phylogenetic species, the issue of monophyly is often ignored. In fact, the phylogenetic species concept's method of the minimum diagnosable unit, when applied to phenotypes, can be applied to inanimate objects and is not inherently phylogenetic (Johnson et al. 1999). Further, when genetic criteria are used to define monophyly, these criteria are typically just one or two loci, typically non-recombining mitochondrial DNA (mtDNA) genes, and monophyly with respect to other loci is not addressed (see below).

Furthermore, using unique or even multiple characters to identify minimum diagnosable units does not guarantee monophyly of the taxa in question if three or more populations are involved. If rates of character evolution are unequal, then some populations will become diagnosable before others, leading to paraphyletic groupings of populations that have not become diagnosable, similar to the "metaspecies" and "plesiospecies" problems (Donoghue 1985, Olmstead 1995, Willmann and Meier 2000). If three populations have a known history C+(A+B), if B is the first population to acquire a diagnostic character, leaving C and A with nothing but ancestral character states, then even if the true history were known, there would be no way to avoid a paraphyletic taxon A+C if the strict rules of diagnosability are followed. If their geographic ranges are linear, the taxa are sedentary, and the central taxon acquires an apomorphy first, then at least we would be suspicious that the character distribution represents unequal rates of character evolution. In fact, such a linear array provided a clue suggesting that many such cases in Andean birds represented cases of unequal character acquisition that would potentially mislead phylogeny (Remsen 1984). In many cases, however, the populations' ranges are not linear, and in such cases, geography cannot provide hints that the populations without diagnostic characters form a paraphyletic taxon.

Advocates of the phylogenetic species concept often promote its adoption because it makes the fundamental unit of classification "historical taxa" (Zink and McKitrick 1995), whereas in the biological species concept non-sisters can be treated as a single species. As noted above, morphology-based applications of the phylogenetic species concept do not necessarily produce historical taxa. History is a continuum, and the exercise of recognizing which historical units within this continuum are named taxa is inherently arbitrary. Worse, at the population level, defining historical units depends on which characters or which loci are thought to represent the true history. As acknowledged by Zink and McKitrick (1995), it is well known that use of any one set of markers can lead to misrepresentations of history (Tateno et al. 1982, Neigel and Avise 1986, Pamilo and Nei 1988). Only by knowing the gene trees of a large number of polymorphic loci can the true population history be reconstructed, and even then, incomplete lineage-sorting may complicate resolving a single history even if entire genomes are sequenced (Pollard et al. 2006). Further, for all populations with topographically and climatologically heterogeneous ranges, this history likely dates no farther back than the most recent pulse in the cycle of fragmentation and secondary contact.

Using diagnosability as a criterion for naming taxa has inherent methodological problems that affect phylogenetic species and subspecies (under the biological species concept) equally, for four reasons. (1) Any diagnosability level is arbitrary. Because diagnosability is a continuum, from 0 to 100%, any cutoff is inherently arbitrary and cannot be defended conceptually (Johnson et al. 1999). Setting the threshold at 95% is a reasonable level because of the widespread use of that arbitrary level for statistical "significance." Nonetheless, the consequence is that two populations that are, for example, 95% diagnosable are given taxon status, whereas those at 94% are not and are included in the same unnamed category as those population samples diagnosable at 0%. (2) A corollary of arbitrary diagnosability is that the outcome is driven in part by sampling. The closer the diagnosability approaches the threshold, the higher the chance that an increase of one additional individual in the sample will determine the outcome; thus, such an addition to the sample could change the ranking from unnamed taxon to phylogenetic species or subspecies without any true change in the biology and history of the populations. (3) The geography of sampling is critical to the outcome if the character assessed shows any geographic variation (Zink and Remsen 1986). Past gene flow or residual geographic variation in the once-continuous populations makes it essential that sampling be focused on geographically proximate populations. And (4) diagnosability is driven by the resolution of the technique used (see Collar 1997, Avise 2004).

Zink (2006) objected to my definition of subspecies and phylogenetic species as synonyms (Remsen 2005) because, in essence, some minimum diagnosable units under the biological species concept would be ranked as species if reproductively isolated from other such units. Zink's (2006) argument is largely semantic, because a biological species that is monotypic (contains no units ranked as subspecies) would still be treated as equivalent to the subspecies unit in those analyses for which minimum diagnosable units are the appropriate unit of analysis. In other words, an analysis using minimum diagnosable units under the biological species concept would include all taxa ranked as subspecies plus all monotypic species. The difference between the biological species concept and the phylogenetic species concept is not in defining minimum diagnosable units but in the ranking of some of those units as species. Under the biological species concept, 4,677 (48%) of the 9,722 species in Dickinson are monotypic (D. Lepage pers. comm.)

Monophyly at the Population Level?

The original conceptual theme of the phylogenetic species concept is that minimum diagnosable units are not only diagnosable, but monophyletic (Cracraft 1983). As is now well known, the problem is that at the population level, monophyly is difficult to define and determine (e.g., de Queiroz and Donoghue 1990, Wheeler and Nixon 1990, Davis and Nixon 1992). Only if all gene trees within a series of populations that share a common ancestor have topologies that do not conflict can a single population be labeled unambiguously monophyletic. Genetic data (e.g., Avise 1989) confirm what common sense predicts: the turbulent history and complex population genetics of real-world situations are often unlikely to produce true monophyly because of incomplete lineage-sorting and gene flow among populations that are not reproductively isolated. Gene tree topologies, superimposed, probably look more like a tangled net than a tree (Degnan and Rosenberg 2009). Further, it is now well understood that under some circumstances the gene trees of independently segregating loci are not expected to recover the true species tree (Rosenberg

and Tao 2008) and that postdivergence gene flow may make reconstructing species trees from gene trees particularly problematic (Takahata and Slatkin 1990, Eckert and Carstens 2008). Add to this the historical likelihood of repeated phases of expansion, range fragmentation, and secondary contact, and the use of the term "monophyly" becomes problematic. For a particularly well documented example of how a single gene tree can misrepresent species trees of buntings in the genus *Passerina*, see Carling and Brumfield (2008). In part because of this, Hennig (1966: fig. 4) recognized and illustrated this problem graphically, did not apply the term "monophyly" below the species level, and used the reasoning of the biological species concept in his definition of species. Although Hennig used characters to label species in his diagrams illustrating cladistic methodology and is thus widely cited as an advocate of the phylogenetic species concept, Hennig clearly considered reproductive isolation the essential component of speciation (e.g., Hennig 1966:54). Reproductive isolation is the necessary first step toward true monophyly.

Even with respect to a single gene, monophyly at the population level differs fundamentally from monophyly at higher levels because it can be ephemeral, perhaps typically persisting only during the refugial phase of range expansion and contraction cycles, and even then being vulnerable to dispersal-generated gene flow. Therefore, the objection to subspecies or biological species because they are not monophyletic (e.g., McKitrick and Zink 1988) is not condemning. Paraphyly and polyphyly at the population level are predicted, and empirically demonstrated, to be widespread (Funk and Omland 2003). For example, Hull et al. (2008) showed that Swainson's Hawk (*Buteo swainsoni*) is paraphyletic with respect to Galapagos Hawk (*B. galapagoensis*) in terms of mtDNA; however, there is no other biological support for merging *B. galapagoensis* into *B. swainsoni* or for recognizing two or more species within traditionally defined *B. swainsoni*. Further, possession of a diagnostic character, the criterion needed for phylogenetic species rank, is no guarantee of monophyly with respect to other genes. For example, Swainson's Hawk has a suite of diagnostic phenotypic characters despite its being a paraphyletic unit with respect to Galapagos Hawk. Labeling clusters of populations as species on the basis of monophyly with respect to single gene trees indicates monophyly only with respect to that gene tree, and not necessarily with respect to the the population or species tree (Edwards et al. 2005, 2007). Additionally, in practice, far too few individuals are typically sampled to determine whether two populations are monophyletic with respect to the loci surveyed (for an example of how differences in sample size can affect conclusions concerning population monophyly, see Brumfield 2005). In fact, an earlier study of Swainson's and Galapagos hawks (Riesing et al. 2003) had reported that the two were reciprocally monophyletic only because too few individuals had been sampled (Hull et al. 2008). In summary, the putative advantage of the phylogenetic species concept in establishing monophyletic units as the fundamental unit of taxonomy is appealing rhetoric but elusive reality. Hennig's (1966) restriction of the term "monophyly" to levels of classification above that of species under the biological species concept reflects remarkable wisdom given the state of knowledge of population genetics at that time.

What Is a Species?

One cannot discuss subspecies without also defining species. The controversy over species concepts is obviously too large and complex to treat here; see Coyne and Orr (2004) for a comprehensive review. De Queiroz (2005a, b) pointed out that all species concepts share the property, explicit or implicit, that the unit called "species" represents the uniquely biological property of a separately evolving metapopulation lineage. The problem is how to apply that concept and which criteria are used to delimit species. Although de Queiroz (2005a, b) tried to present his broadly defined species concept as a solution, he offered no real operational definition with respect to explicitly defining such a unit; in fact, at one extreme, a pair of individuals colonizing an island and successfully reproducing could fit the definition of "separately evolving metapopulation lineage" after a single generation. Nonetheless, de Queiroz's (2005b) simple diagram of the splitting and subsequent divergence of populations crisply illustrates the underlying problem of setting criteria to demarcate species boundaries. His use of continuous shading aptly emphasizes the continuum of degrees of divergence and the inherently arbitrary decisions necessary. The island example above would represent the first point past divergence on his time axis. Therefore, some

level of subjectivity inevitably influences one's choice of criteria.

As expressed more fully elsewhere (Johnson et al. 1999, Remsen 2005), I favor definitions of species based on a fundamental process of evolution at the population level, namely gene flow or lack of it; that is the essence of the biological species concept. My support for process-based definitions—rather than being "blind allegiance" to the biological species concept, the accusation leveled by Peterson et al. (2006)—is based on recognition that severe diminishment or cessation of gene flow is clearly critical to diversification. Personally, I regard the biological species concept as an imperfect attempt at inflicting a typology on a continuum; however, I dislike even more any other categorical scheme proposed so far (e.g., various versions of the phylogenetic species concept). Rather than become disillusioned at the failures, I recommend rejoicing in the underlying complexity that the failures reveal.

The primary operational problem of the biological species concept, as emphasized by Ernst Mayr from the outset (e.g., Mayr 1942b), is in dealing with ranking allopatric differentiated populations. Here, I note that human cognition deals directly with this problem in recognizing differentiated but reproductively fully compatible units within *Homo sapiens* as conspecifics. This predates science, much less the Modern Synthesis, in that even the earliest historians treated allopatric differentiated populations of humans as "people," rather than as some other type of species. Territoriality and combat, typical manifestations of intraspecific competition but relatively rare in interspecific competition, were expressions of that cognitive framework. Therefore, in treating distinct interpopulational differences as part of the same species, the biological species concept has a subjective appeal that the phylogenetic species concept lacks. The phylogenetic species concept could also produce some unknown number of species within *Homo sapiens*, a result refuted by human behavior long before modern societal influences.

The problem of assigning rank to differentiated allopatric populations is not as intractable as is often portrayed. By placing the degree of differentiation in a comparative phylogenetic framework, namely comparing degree of differentiation in the allopatric form to that seen in closely related sympatric or parapatric populations, a reasonable and testable hypothesis can be made concerning whether the allopatric form has or has not differentiated to the degree shown by related forms that do or do not freely interbreed (Miller 1955; Mayr 1969, 1996; Mayr and Ashlock 1991; Helbig et al. 2002; Futuyma 2005). Any arbitrariness involved in assignment of taxon rank through this process is no greater than that inherent in assessing minimum diagnosable units under the phylogenetic species concept.

Reproductive Isolation

The importance of reproductive isolation in guaranteeing independent evolutionary lineages has been emphasized by many authors, including Hennig (1966) and Cracraft (1983). Proponents of the phylogenetic species concept can seem schizophrenic toward reproductive isolation, first acknowledging its importance, then dismissing its importance. For example, McKitrick and Zink (1988:6) stated that "the 'closure' or sealing of a gene pool is therefore an important evolutionary event." Yet they explicitly denied a role to reproductive isolation in ranking taxa because interbreeding is a "primitive trait" or "ancestral character" (e.g., Zink 2006). The ability to interbreed could perhaps be construed as an ancestral character, but empirical evidence in birds suggests a severe limit to interbreeding in terms of time since divergence: Price and Bouvier's (2002) survey indicated that postzygotic incompatibilities begin to originate by ~2 million years after divergence. Moreover, free interbreeding (i.e., nonassortative mating with hybrids having equal fitness to pure parental) provides a highly reliable indicator of a close relationship. Empirically, it is limited in birds to populations that have diverged to a limited degree; if not sisters, such populations are members of a lineage that abruptly replace each other geographically. In other words, the ancestral component of free interbreeding is highly restricted to parapatric representatives of a single lineage. Zink and McKitrick (1995) reiterated the importance of reproductive isolation and considered studies of it valuable, but they also argued that it should not have a role in delimiting species. Missed altogether is that reproductive isolation or its absence governs the distribution of characters that delimit the phylogenetic species concept's minimum diagnosable units in sympatric and parapatric taxa; therefore, the pattern of diagnosability is a product of the process dismissed as an "ancestral character." As noted

previously (Avise and Wollenberg 1997, Remsen 2005), denying a role in classification to the most important threshold in the history of a lineage seems incongruous if that classification is supposed to be based on the history of a lineage.

Zink and McKitrick (1995) implied that some proponents of the biological species concept place theoretical emphasis on reproductive isolation because the lack of it, namely hybridization, means that the two populations may eventually homogenize. Similarly, Zink (2006) portrayed the biological species concept as placing importance on the "potential future outcome of current interbreeding." Rather than making such predictions, the classification of two differentiated, freely interbreeding populations as one biological species represents only a statement concerning the current interaction of the two populations, namely that in terms of mate selection and recognition, individuals of both populations treat each other as equivalents, regardless of any previous history of differentiation. It does not necessarily predict the future (although considered by some a hallmark of a mature research field, not speculation), nor does it necessarily group historical taxa. However, it represents important information concerning the current situation in terms of individual behavior and its consequences for population genetics. In summary, such population interactions provide taxonomist-free data on whether (or to what degree) two populations consider themselves "the same" or "different."

At least some of the controversy over the importance of reproductive isolation is caused by disagreement over, or misrepresentation of, the definition of reproductive isolation. Mayr's definition of the biological species concept emphasizes free interbreeding, widely interpreted as nonassortative mating in contact with no reduction in hybrid fitness. In contrast to any criteria based on diagnosability, the advantages of these criteria are (1) that ranking depends on the biological behavior of the individuals involved and (2) that any change in that behavior has consequences for gene flow. That patterns of mate choice may change temporally or geographically is inevitable, and these differences will generate problems for anyone who expects a typological categorization scheme to nimbly handle all real-world variation. Populations that interbreed but still mate assortatively (e.g., no hybrid swarm in contact zone) are treated as separate species under most interpretations of the biological species concept. If new data reveal that existing taxa classified by the biological species concept are actually freely interbreeding, that classification should be changed.

SUBSPECIES AS STRAW MEN
AND PHYLOGENETIC SPECIES

Finding that existing classifications of subspecies are defective at some level is not an indictment of the subspecies concept itself, no more than a reanalysis of a phylogenetically based classification that found problems with a previous classification would be an indictment of the phylogenetic species concept. Those who attack subspecies as a taxon rank consistently miss the distinction between a concept and the correct application of that concept. An everyday analogy would be to blame the car, not the mechanic, for a botched repair job. For example, McKitrick and Zink (1988:11) advocated abandonment of the subspecies rank largely because of the historical inconsistency in its application and admitted that properly characterized subspecies—namely, in their words, those "distinct from other populations in one or more characters"—"would be called [phylogenetic] species by our criteria" (as echoed by Zink 2006).

Using existing subspecies classifications as an indictment of anything is disinguous. The vast majority of such classifications has not been subjected to a modern, quantitative analysis since their original presentation, often the *Check-list of the Birds of the World* series (Peters 1934–1987), many dating as far back as the 1930s. More recent synopses, such as Dickinson (2003) and the *Handbook of the Birds of the World* series (del Hoyo et al. 1992–2008), largely repeat the initial classifications in Peters's *Check-list* unless subsequent studies have altered them. Although the 1960s and early 1970s saw a wave of quantitative studies, particularly in North America, few such studies have been published since then. Thus, the vast majority of subspecies-level classifications remain mostly unchanged from those of Peters's *Check-list* and are maintained largely by historical inertia, a diminishment in this type of biodiversity science, and a lack of adequate material to readdress historical hypotheses.

However, many critiques of the subspecies concept seem to assume that these classifications undergo some sort of constant, modern, quantitative scrutiny. As pointed out previously (Remsen 2005), the majority of subspecies were described in a prestatistical era. In fact, the term "statistics" and even the simplest statistical analyses, such as the t-test, postdate the majority of subspecies descriptions. The percentage of subspecific classifications in the Peters's *Check-list* that have ever been subjected to statistical evaluation is minute, perhaps <1%. Therefore, the chances that any of these classifications would not require modification after a modern reanalysis are also minute. I am unaware of any quantitative reanalyses of existing subspecies designations that have not produced modifications of existing subspecies classifications. For example, see Cicero (1996), who found that 4 of the 10 subspecies in the *Baeolophus inornatus* complex were not diagnosable, and Patten and Pruett (2009), who found that only 25 of 51 subspecies of *Melospiza melodia* represented diagnosable units.

To illustrate these points, I plotted (Fig. 1) the date of the type descriptions of all subspecies currently recognized by Dickinson (2003) for two bird families, Parulidae and Pycnonotidae, of similar size but contrasting features. The family Parulidae is restricted to the New World, much of its diversity is at temperate latitudes, and many species are highly migratory. The family Pycnonotidae is restricted to the Old World tropics and includes no highly migratory species. Despite the differences, the chronology and pattern of subspecies descriptions are remarkably similar. Fifty percent of all descriptions predate the first publication of Student's t-test (1908), much less its widespread use in ornithology, 70% predate Fisher's (1930) seminal work on population genetics, and 79% predate Huxley's (1942) book on the Modern Synthesis. Therefore, to use such classifications as ammunition to attack subspecies as a concept is a classic straw-man approach that is counterproductive to elucidating the patterns of diversification and the processes that produce them. Any critique of the subspecies unit as a concept using empirical results should start by determining which named subspecies fit the conceptual definition. Failure to apply such a conceptual definition to subspecies designations over the past century has, in my opinion, directly catalyzed the origin of the phylogenetic species concept.

In contrast to subspecies designations, the phylogenetic species concept benefits from having few empirical applications to examine on any large scale. A reasonable prediction is that if all

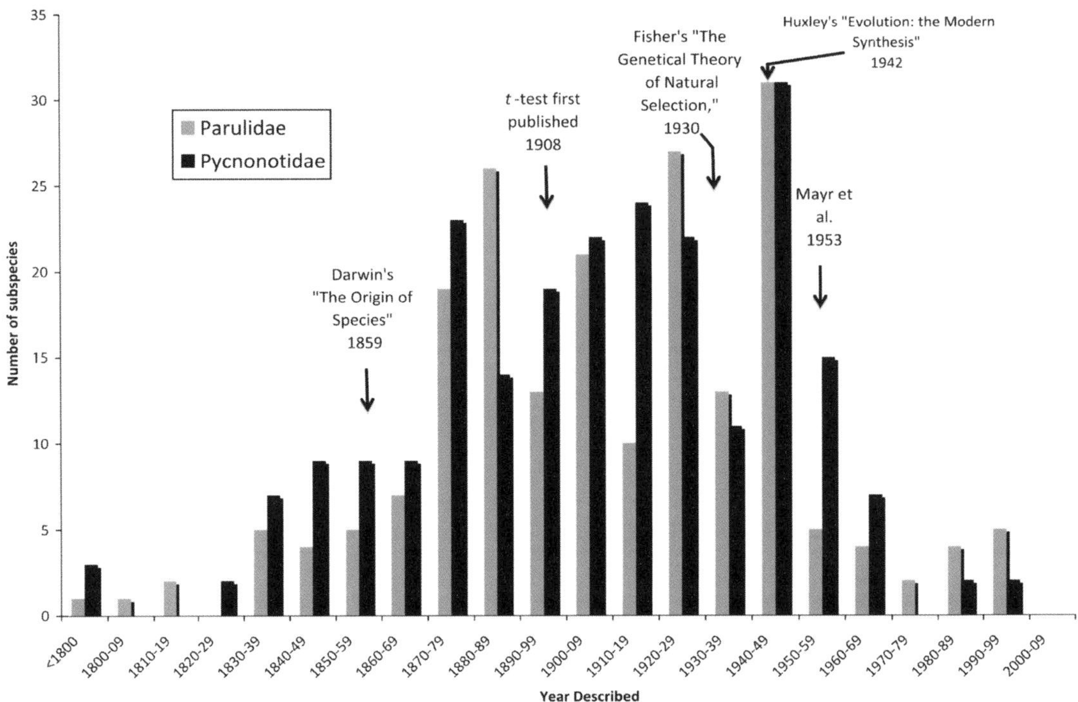

Fig. 1. Historical pattern of dates of descriptions of subspecies in the Parulidae and Pycnonotidae. The data plotted are the publication years of the type descriptions for all subspecies currently recognized by Dickinson (2003), not including, of course, the type description of the species.

classifications started using a phylogenetic species classification as of today, then 100 years from now those results would be viewed with the same disdain directed at current subspecies classifications. Published analyses using phylogenetic species as units already provide ample fodder for criticism, with a near absence of quantitative rigor in determining whether their units actually represent minimum diagnosable units. Given the importance in many analyses of using minimum diagnosable units, whether called subspecies or phylogenetic species, the first step requires a rigorous determination of what those units are.

However, Navarro-Sigüenza and Peterson's (2004) listing of bird species for Mexico based on diagnosable units is quantitatively inferior to that of Robert Ridgway's volumes from the early 1900s that cover the same area (Remsen 2005), although Peterson and Navarro-Sigüenza (2006) assured us that unpublished analyses supported their designations. Peterson and Navarro-Sigüenza (2006:886) also assured us that their 2004 classification was a "consistent taxonomy" that was

"based on the same criteria as in all other clades," yet their methodology and criteria remain unspecified. Likewise, Peterson's (2006) synopsis of diagnosable units in Philippine birds rests on unspecified sample sizes (noted as "woefully small" for many populations) and qualitative assessments. See Collar (2007a) for a full critique of Peterson's (2006) approach. Simply dismissing all trinomial nomenclature and then labeling as species all populations that by qualitative inspection appear diagnosable is not an acceptable program for assessing biodiversity; see Collar (1997) for similar comments on fundamental problems with Cracraft's (1992) revision of a single family, the Paradisaeidae, based on the phylogenetic species concept. In fairness to all these attempts, reevaluation of the diagnosability of currently described subspecies, especially for a rich avifauna such as that of Mexico, is a daunting, monumental task that will require detailed research, and Navarro-Sigüenza and Peterson have made a noble start. Unfortunately for biodiversity assessment, these kinds of baseline analyses of geographic variation are not considered groundbreaking research. Nonetheless, casual, qualitative inspection of study skins is no longer an acceptable practice for taxonomic revisions, whether the taxa are labeled "subspecies" or "phylogenetic species."

Some biogeographic analyses using phylogenetic species as units start with the assumption that certain described subspecies accurately represent minimum diagnosable units, declare them to be species, and then proceed with the analysis. These analyses typically do not report sample sizes or the geography of their sampling distribution, much less character analyses, diagnosability indices, or anything else that would permit replication. Notable recent exceptions are the analyses of McKay (2008) and D'Horta et al. (2008). Prior to any phylogeographic analysis, they began with a quantitative analysis of geographic variation in plumage characters to define minimum diagnosable units. As noted previously, the antiquity of most subspecies names makes it inevitable that many will fail diagnosability tests. However, analyses that do not include the characters used to diagnose the taxa are unlikely to address diagnosability adequately. For example, Drovetski et al. (2009) quantitatively analyzed geographic variation in breast plumage in the currently recognized species of North American rosy-finches in the genus *Leucosticte* but omitted those plumage characters (face pattern) formally (e.g., Ridgway 1901, MacDougall-Shackleton et al. 2000, Johnson 2002, Johnson et al. 2002) used to diagnose the taxa. Snow (1997) pointed out that incomplete geographic sampling and small sample sizes for many taxa make it necessary to study geographic variation and taxonomy in detail before determining what constitutes minimum diagnosable units.

Environmental Induction

In his widely used textbook, Gill (2007:575) stated that "geographical differences in size or color may be due directly to environmental differences rather than evolved genetic differences among populations" but provided no further details or citations. One likely source of such statements is a tiny number of studies that have documented minor environmental effects on body size and shape in relation to the genetic component (James 1983, Larsson and Forslund 1992, Leafloor et al. 1998), although most such studies have not found an environmental component

(see Merilä and Fry 1998). Many subspecies have been described on the basis of measurements that reflect overall body size. Regardless of whether such differences have an environmental component, I suspect that many or most of these subspecies will be shown to fail diagnosability tests. The vast majority of such subspecies have not been analyzed using any test of degree of overlap, and their validity often rests on differences between means and various qualitative assessments of the ranges.

The other potential source of Gill's (2007) statements concerning environmental effects is the relationship between diet, or other measures of condition, and feather pigmentation or structure. Environmental effects on the ability to express appropriate coloration are widely known, in that poor condition or disease may effect the coloration of individuals within a bird population. Coloration based on carotenoids can be affected strongly by diet because carotenoid pigments must be acquired from food (reviewed by McGraw 2006a), and the expression of carotenoid-based coloration can be altered by environmental conditions, such as parasite load (reviewed by Hill 2006). As for melanin-based coloration, the most widespread source of coloration in birds, documentation of environmental effects is not as clear-cut, and some experiments have failed to find an effect of diet on melanin production or expression (e.g., Gonzalez et al. 1999, Buchanan et al. 2001, McGraw et al. 2002). Nonetheless, because melanin is synthesized from amino acid precursors using metabolic energy and their deposition is influenced by at least four classes of hormones (McGraw 2006b), the potential remains for an effect on its production owing to general health and nutrition. As for structural colors, limited experimental data suggest that nutrition during molt may affect their expression (Hill 2006), but such experiments are limited mainly to glossy black species, are largely correlational, and have not addressed potential confounding influences of age (Prum 2006). Nonetheless, given the complex pathways involved and the extraordinary structural precision required to produce normal coloration (reviewed by Prum 2006), the potential for environmental effects would seem large. Whether coloration is based on nanostructure or pigments, environmental effects on individuals within a population are highly likely. (And this is discounting the ways in which the environment can alter plumage through time, such as through fading and wear.) That environmental differences might also produce region-wide effects on certain populations of a species remains a potential explanation for geographic variation in coloration and body size in birds. However, data that document a link between environmental effects and among-population differences in coloration are lacking.

Evidence for natural selection on color shades and patterns is reasonably strong (e.g., Burtt 1981, Rohwer and Ewald 1981, Prum 1997, Negro et al. 1998, Dumbacher and Fleischer 2001, Mumme 2002, Tickell 2003). Certainly, the strong associations between patterns of coloration and various ecological and social factors (for reviews, see Bortolotti 2006, Dale 2006) imply that natural selection on the underlying genetic basis of these patterns is widespread. Dramatic seasonal changes in plumage coloration in some species, typically associated with changes in social system, also imply strong selection (although environmental effects caused by a seasonal shift in food supply are not necessarily ruled out by such correlations).

SUBSPECIES AND CONFLICTS
WITH GENE-BASED PHYLOGENIES

Several recent papers have attacked the utility of subspecies by comparing current subspecies-level classification with patterns of diversification shown by mtDNA (e.g., Zink et al. 2001). A mismatch between units defined by mtDNA versus subspecies is then proclaimed as evidence that subspecies mask patterns of diversity or obscure analyses of the process of historical diversification (Zink 2004).

Conflicts between mtDNA trees and subspecies units may result from faulty delineation of subspecies boundaries, because most have not been critically or quantitatively examined (see above). However, if the subspecies boundaries represent diagnosable units, then I am unaware of any model of evolution that predicts perfect concordance between diagnosable phenotypic units and any single gene tree, particularly those of presumably neutral loci. Lost in the discussion of such conflicts is that (except for populations without any history of fragmentation and secondary contact) gene trees and population trees not only differ, but also are expected to do so because of the influences of incomplete lineage-sorting and gene flow (for review, see Coyne and

Orr 2004). This is especially true for the most frequently analyzed genes, those of mtDNA, which are matrilinearly inherited as a single linkage unit. Empirically, mtDNA markers may do as well as any in tracking population history (Zink and Barrowclough 2008), but to uncritically treat an mtDNA gene tree as equivalent to the true population history should be termed "mtDNA myopia." For example, Zink (2004:563) stated that "subspecies should be judged to fail as meaningful units if they do not predict the evolutionary history of the populations they represent," but in Zink's view mtDNA phylo-groupings represent the only history worth recognizing taxonomically, without recognizing that an mtDNA phylogeny is merely a gene tree. A population marked by a phenotypically diagnosable character, provided that character has a genetic basis, also shares a common history but on a different time-scale. For subspecies units to show perfect concordance with an mtDNA gene tree, each subspecies would also have to have a unique haplotype (or haplotype lineage), an unrealistic expectation. Even so, Phillimore and Owens (2006) showed that Zink's estimates were an order of magnitude too low because of sampling bias and that broader sampling indicated that more than a third of the taxa ranked as subspecies were monophyletic even by the highly restrictive and unrealistic criterion of mtDNA haplotypes. The title of Zink's (2004) paper proclaimed that subspecies obscured biological diversity; however, one could also make a case that using mtDNA phylogroups as taxonomic units obscures biodiversity because it ignores biologically important, phenotypic markers of recent population history. Under Zink's extreme view, some diversity even at the species level would be erased, with most Galápagos finches merged into a few monotypic species because their mtDNA gene trees are not reciprocally monophyletic (Zink 2002). Described as "an unfortunate reliance on a single, potentially misleading molecule" by Grant and Grant (2006), such a treatment as single species would ignore the reproductive isolation and divergence of multiple lineages within this radiation.

Researchers who do not find concordance between genetic data and subspecies boundaries often proclaim that such subspecies are not genetically distinct. Two fundamental problems beset such statements. First, such studies typically analyze one or two genes, often mitochondrial—that is, a tiny fraction of the genome. The appropriate qualifier for such statements would be that a subspecies is not genetically distinct with respect to whatever number of genes was analyzed. Second, if a subspecies is diagnosable by phenotypic characters (external manifestations of genetic characters), then indeed it is also likely genetically distinct, but the gene(s) that control those characters have not been located or analyzed. If two or more populations share the same phenotypic characters that have arisen by common selection pressure, then their grouping into a single taxon would mislead phylogenetic classification; this is where mtDNA or other genetic markers can elucidate the true population history that phylogenetic classification requires.

SUBSPECIES AS IMPEDIMENTS TO CONSERVATION

Some (e.g., Hazevoet 1996, Sangster 2000, Peterson 2006) have claimed that ranking diagnosable units as species under the phylogenetic species concept or a similar concept rather than as subspecies under the biological species concept benefits conservation. See Collar (1996, 1997, 2007a), Garnett and Christidis (2007), and Winker et al. (2007) for opposing views. A benefit of the biological species concept to conservation is that it provides a degree of triage in terms of prioritizing resources at the global level. Restricting the species rank to populations known to be reproductively isolated or to have diverged to a level comparable to that shown by reproductively isolated populations (i.e., species by anyone's definition) allows limited conservation resources to be concentrated on those populations. For example, if one had a limited amount of funding to be divided evenly among Caribbean parrot species in the genus *Amazona*, using the classification based on the biological species concept would divide those funds among species that all differ strongly from one another and are species by any reasonable criterion. By contrast, elevating all diagnosable subspecies to species rank under the phylogenetic species concept would give equivalent taxonomic rank and funding, for example, to *Amazona leucocephala hesterna* (endemic to Cayman Brac and differing from nearby *A. l. caymanensis* of Gran Cayman only in having a larger patch of red in the belly plumage) as to the bizarrely plumaged, highly distinctive *A. guildingii* of St. Vincent. Advocates for conservation on Cayman Brac naturally would be pleased

with such an outcome, and so it is no surprise that among the most vocal advocates for the phylogenetic species concept are those devoted to the conservation of small areas or islands (e.g., Hazevoet 1996), whose cause benefits from raising every endemic subspecies to species rank. A more global view, however, would be that a prioritization scheme based in part on taxon rank is beneficial in that populations diagnosable only by characters that do not impede on gene flow, i.e., taxa ranked as subspecies under the biological species concept, do not receive the resources allocated to taxa ranked as species under the that concept.

The other criticism of the use of subspecies in defining conservation units is that many do not correspond to "historically significant groups" (Zink 2004). However, these groups are typically delimited only by patterns of shared mtDNA haplotypes (e.g., Zink et al. 2001). Whether such groups are the only historically significant groups, however, is open to discussion. Because these genetic markers are assumed to be neutral, by definition they have no biologically meaningful manifestation. Further, because of their matrilineal pattern of descent and because of the widely recognized problem of incomplete lineage sorting, these haplogroups represent only the history of perhaps one or two non-recombining genes (Edwards and Bensch 2009). Although such markers are useful tools for tracking aspects of population history, phenotypic markers also have the potential to do the same. Moreover, in contrast to haplotype differences, phenotypic markers have the potential to be biologically meaningful and should thus be of greater conservation concern (Crandall et al. 2000). Differences in pattern and coloration, for example, frequently correspond to abrupt discontinuities in gene flow in birds, a taxonomic class in which sexual selection has played a key role in diversification; their more subtle manifestation as diagnosable characters that mark subspecies boundaries gave rise to the phrase "incipient species" for some subspecies. To ignore this aspect of geographic variation and population biology only because of lack of correspondence to neutral mtDNA markers in vogue today should be regarded as myopic by those interested in patterns of biodiversity or the identification of units of conservation concern—or, indeed, the process of evolution.

A particularly disingenuous criticism of the biological species concept as an impediment to conservation is the claim that it masks biodiversity. For example, Peterson (2006) denounced the biological species concept for overlooking numerous distinct populations but did not mention that under this concept all of those populations are named, as subspecies, and overlooked only if one restricts an analysis to the species rank. Thus, Peterson (2006) found much higher levels of species richness and unrecognized or underappreciated patterns of endemism by application of a diagnosability-based species concept; however, he did not point out that an analysis that included subspecies would have revealed the same patterns that he "discovered."

Application of the phylogenetic species concept produces two potentially severe problems for conservation. First, opponents of conservation would quickly discover that the definition of species had been changed to elevate more taxa to higher threat levels, with accusations of manipulation of the rules. Changing the definition would only fuel the suspicions of conservation opponents that scientists have abandoned objectivity in favor of a pro-conservation agenda. Second, elevating to species rank many taxa diagnosable only by characters that conservation opponents, the general public, and most biologists would justifiably label as trivial could diminish confidence in conservation science, undermine the credibility of taxonomists, and erode support for programs to protect threatened species.

Subspecies Are Overlooked as a Component of Biological Species

De Queiroz and Donoghue (1988:334) concluded that "no one species concept can meet the needs of all comparative biologists." I suggest that use of a biological species concept that identifies minimum diagnosable units as subspecies spans more of those needs than is appreciated. Debates over the merits of species concepts based on whether they emphasize reproductive isolation or minimum diagnosable units overlook that subspecies, an integral part of the biological species concept, are its minimum diagnosable units. Criticizing the biological species concept for not allowing analyses of basal evolutionary units overlooks that the subspecies rank is an integral part of the concept. The biological species concept encompasses units that fit the conceptual definition of phylogenetic species but calls these minimum diagnosable units subspecies rather

than species (if they are not ranked as biological species). Proponents of the phylogenetic species concept would point out that regardless of conceptual definitions, in practice many subspecies are not diagnosable units. As discussed above, this (1) is largely the consequence of incorrect application of the definition and (2) has to be dealt with regardless of whether these units are called subspecies or species. In fact, Phillimore et al. (2007) showed that analyses of subspecies as an index of intraspecific geographic differentiation within a species yield sensible results with respect to biogeographic influences on intraspecific variation. Under the biological species concept, classification with diagnosable units provides two levels of information: one that emphasizes genetic discontinuities (species) and another that emphasizes geographic units within the species identified by diagnostic characters (subspecies). Analyses that require terminal taxa can use populations ranked as subspecies (e.g., Cracraft 1985), whereas analyses based on active or potential barriers to gene flow can use the species rank. Geographic variation not partitioned into diagnosable units may occur within taxa ranked either as species or subspecies under the biological species concept.

A recurring misconception in some recently published papers is that the phylogenetic species concept reveals diversity and the biological species concept obscures it. For example, Reddy's (2008) application of the phylogenetic species concept to *Pteruthius*, currently considered to consist of 5 species under the biological species concept, first required determining which of the 23 recognized subspecies were diagnosable units; that is the same procedure that would be necessary under a modern reevaluation of the genus. Although the geography of sampling and sample sizes were not reported, Reddy found that 19 taxa were diagnosably distinct. She then claimed that this was "almost a four-fold increase in recognized diversity"; in fact, all of that diversity was recognized under the biological species concept, 5 as species and the other 14 as subspecies of those species. In terms of overall taxonomic diversity, this application of the phylogenetic species concept actually reduced the number of recognized taxa by some 15%. Once nondiagnosable taxa are identified and eliminated (a problem shared by all species concepts), the differences are not in diversity per se but in the ranks assigned to those units of diversity.

In summary, the biological species concept provides two levels of information, whereas the phylogenetic species concept provides one. The biological species concept incorporates the acquisition of diagnostic characters into its classification by ranking diagnosable populations minimally as subspecies. The biological species concept also incorporates reproductive isolation, acknowledged even by many proponents of the phylogenetic species concept as an important evolutionary step in the history of any lineage, by ranking such populations as species.

Subspecies and Human Perception

What would happen if the phylogenetic species concept's minimum diagnosable units were applied to *Homo sapiens*? Certainly, until recent decades, humans classified one another into racial groups thought to have diagnostic characters (e.g., Hall and Kelson 1959), and even today, one's race is a data field in many nonscientific categorization schemes. Research has shown that such schemes fail to classify individuals reliably and that, at the genetic level, ≤95% of all genetic variation is among-individual, not among-group. Nonetheless, despite rampant ongoing gene flow and the relatively recent origin of *Homo sapiens*, the residual variation may accurately predict region of origin and show strong geographic structuring. For example, even different groups of Native Americans differ strongly in haplotype frequencies (Malhi et al. 2003). Research on the genetic basis of human diseases has spawned an interest in ancestry-informative markers that predict the geographic origin of individual humans. Although complex computations are required to identify unique combinations of alleles, the geographic structure of this variation can identify individuals with respect to continent of origin (Rosenberg et al. 2002, Collins-Schramm et al. 2004, Mao et al. 2007, Li et al. 2008) and subregion (Tian et al. 2008a). Recently, Tian et al. (2008b), using a sample of European Americans categorized according to "self-reported" region of European descent, showed that principal component analysis of single nucleotide polymorphisms allowed accurate discrimination of individuals as either northern vs. southern European ancestry and found further evidence of structure within the northern European sample.

If geographic variation in *Homo sapiens* were sampled in the same limited way that it is in most birds, then application of the phylogenetic species concept to *Homo sapiens* would certainly produce "minimum diagnosable units" that are neither biologically nor socially acceptable as "species." However, the detailed structure of this variation, both phenotypic and genotypic, is sufficiently well studied that we can be sure that few if any character states analogous to those used in bird taxonomy would unambiguously diagnose any subpopulations of humans. Even today, after much global movement and genetic mixing, our own genetic and morphological (e.g., Shriver et al. 2003) diversity could be partitioned into an unknown number of diagnosable units by use of unique combinations of characters and allele frequency differences. By contrast, although cultural barriers prevent full application of the biological species concept to humans, this concept would consider all humans conspecific (*Homo sapiens*). In terms of perception and the absence of biologically based reproductive isolation, humans clearly think of themselves as belonging to one species, as defined by the biological species concept, despite marked geographic variation within *Homo sapiens*. Given that species definitions are scientifically untestable matters of taste (Brookfield 2002), human perception has spoken with resounding clarity that "species" are not minimum diagnosable units.

Common Ground

The debate over species and subspecies concepts is healthy, particularly in forcing a reevaluation of currently recognized subspecies names. I strongly agree with critics of the biological species concept that terminal taxa should be used in analyses of, for example, biogeography and biodiversity. The uncertainty of the diagnosability of many subspecies, especially in temperate North America, requires that anyone undertaking an analysis using terminal taxa must carefully scrutinize their diagnosability. Empirically, however, using named subspecies from Peters's *Check-list* series, even without critical evaluation (e.g., Cracraft 1985), successfully demarcates areas of endemism. So, if the sample is large enough and the error rate (nondiagnosable taxa) small enough, real patterns should emerge even if current subspecies names are taken as is.

I also strongly concur with McKitrick and Zink (1988) and others that subjective notions of whether a character is too trivial to use to diagnose a taxon are unscientific. What matters is whether that character is a marker for a cohesive evolutionary unit, regardless of any known functional significance. If that character is "one extra hooklet on a barb of the seventh primary" (McKitrick and Zink 1988:9), and it passes the 95% diagnosability test, then it defines an entity worthy of a name, in my opinion.

Some defenders of the biological species concept worry that adoption of the phylogenetic species concept would lead to too many species (e.g., Mayr [1993] as cited by Zink and McKitrick [1995]). Preconceived notions of how many species there ought to be are scientifically indefensible. I echo McKitrick and Zink (1988) and Zink and McKitrick (1995) on the importance of letting the data determine the number of populations ranked as species. Even under the biological species concept, the number of species is increasing dramatically, particularly in the tropics, where many taxa formerly ranked as subspecies are elevated to species rank through careful study of vocalizations and population interactions at contact zones. For example, field studies of polytypic species of antbirds (Thamnophilidae), many using the comparative framework of Isler et al. (1998), have already elevated 31 taxa previously treated as subspecies to species rank. These 31 species, ranked as subspecies either by Peters (1951) or by Meyer de Schauensee (1970), were subsumed under 16 species names, including one, *Myrmeciza castanea*, considered a synonym of an existing subspecies. They include *Frederickena fulva, Cymbilaimus sanctaemariae, Thamnophilus zarumae, T. tenuepunctatus, T. cryptoleucus, T. atrinucha, T. stictocephalus, T. sticturus, T. pelzelni, T. ambiguus, Thamnomanes schistogynus, Dysithamnus leucostictus, Epinecrophylla spodionota, Myrmotherula ignota, M. multostriata, M. pacifica, Herpsilochmus atricapillus, H. motacilloides, H. dugandi, Drymophila rubricollis, Cercomacra laeta, Hypocnemis flavescens, H. peruviana, H. subflava, Hypocnemis ochrogyna, H. striata, Schistocichla humaythae, S. brunneiceps, S. rufifacies, S. saturata,* and *Myrmeciza castanea* (for references, see Zimmer and Isler 2003, Remsen et al. 2009). Species richness has thus increased by 88% in the 18 cases studied so far. Ongoing studies of other groups of thamnophilids will undoubtedly increase this tally, perhaps by as many as 50 species (M. L. Isler pers. comm.). If the thamnophilid results can be extrapolated to tropical avifaunas as a whole, many hundreds of subspecies will be elevated to species rank under the guidelines of the biological species concept when critical data become available. This does not represent a shift toward the phylogenetic species concept but, rather, an increase in data on reproductive isolation.

McKitrick and Zink (1998) provided a protocol, based in part on Zink and Remsen (1986), for applying the phylogenetic species concept to real-world situations. I suspect that they would also share my concern that few studies undertake the necessary steps to determine diagnosability. I disagree only semantically. I call the diagnosable units revealed by such analyses subspecies, not species.

Acknowledgments

I thank M. Batzer, R. Brumfield, N. Collar, E. Dickinson, K. McGraw, M. Patten, F. Sheldon, and K. Winker for comments on sections of the manuscript.

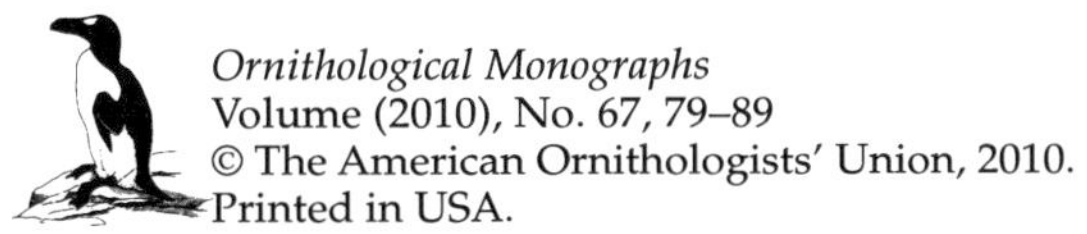
Ornithological Monographs
Volume (2010), No. 67, 79–89
© The American Ornithologists' Union, 2010.
Printed in USA.

CHAPTER 7

REVISITING SPECIES AND SUBSPECIES OF ISLAND BIRDS FOR A BETTER ASSESSMENT OF BIODIVERSITY

H. DOUGLAS PRATT[1]

North Carolina State Museum of Natural Sciences, 11 West Jones Street, Raleigh,
North Carolina 27601, USA

ABSTRACT.—Outdated and overly lumped alpha taxonomy among the world's island birds has serious consequences for scientific research and conservation. The underestimation of biodiversity on islands obscures their role as speciation laboratories, distorts sampling in genetic studies, biases research planning, leads to neglect of endangered island species mistakenly classified as subspecies, and reduces potentially valuable information that might be gathered by recreational birders. Suggestions such as abandoning the biological species concept and the subspecies category in favor of the phylogenetic species concept create new problems and disrupt widely understood terminology. I review avian taxonomic history in the Hawaiian Islands, speciation patterns in Pacific island pigeons and doves, and patterns of variation in the widespread Polynesian Starling (*Aplonis tabuensis*) to demonstrate that the biological species concept, if applied with consideration of potential isolating mechanisms, vagility, and degree of geographic isolation, along with the judicious use of subspecies, produces hypotheses of island biodiversity that meet research and conservation needs. I suggest a thought process for evaluating biological species limits in island birds that is less subjective and more repeatable than previous methods, and use the Fiji Shrikebill (*Clytorhynchus vitiensis*) as a working example. A review of taxonomic history in the Bridled White-eye (*Zosterops conspicillatus*) complex in Micronesia shows that while genetic data are useful for testing hypotheses of species limits based on other data, alone they are insufficient for the purpose and should not be considered essential in species revisions.

Key words: biological species concept, island birds, isolating mechanisms, speciation, subspecies, taxonomy.

Revisitar las Especies y Subespecies de Aves de Islas para una Mejor Evaluación de la Biodiversidad

RESUMEN.—La taxonomía alfa desactualizada y exageradamente agrupada de las aves isleñas del mundo tiene serias consecuencias para la investigación científica y la conservación. La subestimación de la biodiversidad de las islas oscurece su papel como laboratorios de especiación, distorsiona los muestreos en los estudios genéticos, sesga el planeamiento de las investigaciones, lleva a desatender especies isleñas amenazadas clasificadas erróneamente como subespecies y reduce la cantidad de información potencialmente valiosa que puede ser recolectada por los observadores de aves. Las sugerencias como el abandono del concepto biológico de especie y de la categoría de subespecie a favor del concepto filogenético de especie crean nuevos problemas y alteran la terminología ampliamente utilizada. Revisé la historia taxonómica de las aves de las Islas de Hawái, los patrones de especiación en las palomas de las islas del Pacífico y los patrones de variación en la especie ampliamente distribuida *Aplonis tabuensis* para demostrar que el concepto biológico de especie, si se aplica considerando los mecanismos potenciales de aislamiento, la capacidad de dispersión y el grado de

[1]E-mail address: doug.pratt@ncdenr.gov

Ornithological Monographs, Number 67, pages 79–89. ISBN: 978-0-943610-86-3. © 2010 by The American Ornithologists' Union.
All rights reserved. Please direct all requests for permission to photocopy or reproduce article content through the University of California Press's Rights and Permissions website, http://www.ucpressjournals.com/reprintInfo.asp. DOI: 10.1525/om.2010.67.1.79.

aislamiento geográfico, junto con el uso juicioso del concepto de subespecie, genera hipótesis sobre la biodiversidad de islas que contemplan las necesidades de investigación y conservación. Sugiero un proceso razonado para evaluar los limites biológicos de las especies en las aves isleñas que es menos subjetivo y más repetible que los métodos anteriores, y empleo a la especie *Clytorhynchus vitiensis* como un ejemplo de trabajo. Una revisión de la historia taxonómica en el complejo de *Zosterops conspicillatus* en Micronesia muestra que mientras los datos genéticos son útiles para evaluar hipótesis de los límites entre especies basados en otros datos, por separado son insuficientes para este propósito y no deben ser considerados esenciales en las revisiones de las especies.

WE WHO STUDY the world's island birds are burdened with an outdated and overlumped taxonomy that has serious consequences for the assessment and conservation of biodiversity (Collar 2005). Many island endemics that would qualify as biological species by modern standards remain subsumed in what might be called "megaspecies" that reflect the biases of mid-20th-century taxonomy (Collar 1997, 2005; Pratt and Pratt 2001; Chikara 2002; Rheindt and Hutchinson 2007). Despite our increasing knowledge of these birds, most species limits among most island taxa have not been reassessed in the light of new information. A taxonomy that would not pass muster by modern standards remains entrenched, and efforts to alter it often meet resistance. This problem has serious consequences for both science and conservation. Using a series of examples, I discuss the history of species-level taxonomy among island birds and the special nature of allopatry on islands, and suggest a revised methodology for evaluating biological species and subspecies limits among oceanic island birds. I also make a plea for authors and editors not to denigrate species revisions that are based solely on phenotypic characters.

THE PROBLEM AS ILLUSTRATED BY MICRONESIAN FLYCATCHERS (*MYIAGRA*)

In his influential field guide *Birds of the Southwest Pacific*, Mayr (1945) lumped four previously recognized species of flycatchers in Micronesia (*Myiagra erythrops*, from Palau; *M. freycineti*, from Guam; *M. oceanica*, from Chuuk; and *M. pluto*, from Pohnpei) as subspecies of *M. oceanica*, but he observed that they were "so distinct that they might also be considered 4 different species" (Mayr 1945:296). Mayr had no knowledge of these birds in life, and he presented no evidence that they form a monophyletic group, let alone one species. Baker (1951) noted differences in overall size, color, and bill size but accepted Mayr's one-species taxonomy uncritically and thus set the pattern for decades.

In *A Field Guide to the Birds of Hawaii and the Tropical Pacific*, Pratt et al. (1987), drawing on considerable field experience in the region, reversed Mayr's (1945) equivocal lumping and recognized the original four species, which exhibit color variation as broad as that of the entire genus (Burn 2006) and also differ strikingly in size, voice, and, to a lesser extent, habitat. These four inhabit the four high islands found along an east–west (Pohnpei–Palau) axis. With interisland distances ranging from 765 km (Pohnpei-Chuuk) to 1,912 km (Chuuk-Palau), the chance that any of these now highly sedentary birds will ever encounter each other in the wild is almost nil, despite the fact that their ancestors must have crossed large water gaps. Except for Guam, which lies north of the main axis, the Micronesian high islands are closer to potential colonization sources in northern Melanesia than to each other, so origins from different ancestral species for the four forms seem at least as likely as interisland dispersal by a common ancestor. The four-species classification has been used for decades within the region (Pyle and Engbring 1985, Pratt et al. 1987, Wiles 2005), probably because it makes sense to those who know the birds in the field, and it is also followed in some major world check-lists (Sibley and Monroe 1990, Clements 2000, Gill and Wright 2006).

However, in the authoritative *Handbook of Birds of the World*, Gregory (2006) reverted to Mayr's (1945) taxonomy and Burn (2006) depicted three of the nominal subspecies as far more similar than they really are (*M. freycineti* was by then extinct and not included, further distorting the view presented). The third edition of the Howard and Moore checklist (Dickinson 2003) also considered these disparate birds conspecific. Dickinson (2003) stated that those involved in preparing the checklist were unaware of a detailed review of the taxonomy, despite citing Pratt et al. (1987). The original decision to lump these species (Mayr 1945) was based on their occurrence in the same region and the author's preference for polytypic species, not on any review of evidence for conspecificity. The obvious potential isolating mechanisms among these birds should justify restoration of the original four species, but over-lumped taxonomy in island avifaunas remains exasperatingly entrenched (Rheindt and Hutchinson 2007). Examples of polytypic bird species in need of revision can be found throughout the tropical Pacific and include, but are not limited to, *Ptilinopus porphyraceus*, *Todiramphus chloris*, *T. cinnamomina*, *Coracina tenuirostris*, *Pachycephala pectoralis*, *Clytorhynchus vitiensis*, *Chasiempis sandwichensis*, *Myiagra azureocapilla*, *Rhipidura rufifrons*, *R. spilodera*, *Cettia ruficapilla*, *Turdus poliocephalus*, *Aplonis tabuensis*, *Myzomela cardinalis*, *Foulehaio carunculata*, *Gymnomyza viridis*, *Zosterops cinereus*, and *Erythrura cyaneovirens*.

Resistance to changes in species limits reflects (1) a widespread belief among non-systematists that species-level taxonomy is irrelevant, (2) an understandable desire for list stability (Sangster 2000), and increasingly (3) the reluctance of editors to publish revisions that do not include genetic data (even though such data may be irrelevant at the species level, as I discuss below). The failure to recognize that many island subspecies are actually species has several unfortunate consequences that are far more serious than simply misleading illustrators or inconveniencing list-makers.

REPERCUSSIONS OF RANKING
SPECIES AS SUBSPECIES

From a scientific perspective, the most damage occurs when authors of theoretical studies wrongly assume that published species lists for taxonomically long-neglected island regions are essentially equivalent to those of well-studied ones. Underestimation of species-level diversity on islands obscures their important role as speciation laboratories and their importance in the preservation of biodiversity. Modern DNA studies, especially those that require complete taxon sampling, are especially vulnerable to overlumped taxonomy because they may include only one representative of an overlumped polytypic species and assume, falsely, that including others would not change the result. Especially egregious is the conflation of data from subspecies that later turn out to be separate species. For example, Amerson et al. (1982) combined data from the two forms of *Aplonis tabuensis* in American Samoa that, as discussed below, are probably different species, in which case the statistics become meaningless. Research planning may also suffer from

underestimation of species-level biodiversity. For example, Fiji's four largest islands are relatively close together but have distinctive avifaunas. The three islands that have endemic species (Viti Levu, Taveuni, and Kadavu) receive most of the attention of both professional and amateur observers. Vanua Levu, the second largest, whose endemics are all currently ranked as subspecies, is relatively poorly known and visited much less often. Incredibly, I have heard several biologists say that the loss of Guam's entire avifauna (as documented by Savidge 1987) is less regrettable because most of the island's endemics were subspecies.

Unfortunately, given our uncertainties about species limits among island forms, the species category also holds an iconic status among conservationists (Sangster 2000), recreational birders (Pratt 1990), and the general public. Many authors (e.g., Collar et al. 1994, Hazevoet 1996, Myers et al. 2000) have decried the fact that nongovernmental organizations such as World Wildlife Fund and BirdLife International often focus only on endangered species and ignore even highly distinctive endangered subspecies that might turn out to be species. Recreational birders also rely on existing species lists, whether the limits are well constructed or not. These dedicated amateurs are often our only source of new information on remote island endemics, yet they routinely ignore distinctive island subspecies, despite advice to seek out those that are potential splits (Pratt 1990). Popular writers such as Cokinos (2000) usually concern themselves only with species, even when subspecific examples may be equally important and instructive for conservation. The Newfoundland Red Crossbill (*Loxia curvirostra percna*) was driven to extinction by the ill-advised introduction of Red Squirrels (*Tamiasciurus hudsonicus*) to the island (Benkman 1989, 1993b; Parchman and Benkman 2002), but its demise was largely unnoticed at the time because it was only a subspecies. Now that the Newfoundland bird appears to have been an endemic island species (Benkman 1993b), its lessons will perhaps be better appreciated.

CONCEPTS, SPECIES, AND SUBSPECIES

Hazevoet (1996) suggested that species chauvinism actually promotes extinction of island endemics, and he advocated abandoning the biological species concept in favor of the phylogenetic species concept because the latter would reclassify nearly all island subspecies as species.

Peterson (2006) suggested that adoption of the phylogenetic species concept would not only increase the number of species but would reveal such information as previously overlooked centers of endemism. Although I agree that many endemic island species are being neglected because of faulty taxonomy and that current classifications obscure important information, I also agree with Collar (1997:133) that we should "not allow frustration with the misapplication of one concept to result in complete dependence on another." Sangster (2000) suggested that the problem was not a faulty species concept, but rather the fact that avian taxonomy is not as well documented as non-systematists seem to believe. Also, the subspecies concept is quite useful for showing varying levels of differentiation among an array of allopatric populations. Adoption of the phylogenetic species concept would produce a degree of taxonomic inflation that would be just as problematic for conservationists (Collar 1996, Sangster 2000, Pratt and Pratt 2001, Isaac et al. 2004) as current practice under the biological species concept. Noteworthy is that Hazevoet's (1995) phylogenetic reclassification of birds of the Cape Verde Islands was rejected by BirdLife International (Collar 1996).

The biological species concept is fundamentally operational rather than typological or evolutionary in its application. It is not based on degrees of difference, whether morphological, behavioral, or genetic, but rather on how such differences affect (or, in the case of allopatric forms, might affect) the ability of two forms to interbreed. In other words, the criterion is whether the differences are, or are likely to be, isolating mechanisms (I use this widely understood term despite Mallet's [1995] objections). The biological species concept has always had difficulty with allopatric, but obviously related, populations because operational tests are usually unavailable. Mayr (1969) suggested comparisons with related sympatric species pairs as a way of evaluating degrees of difference in allopatric forms, a method Rheindt and Hutchinson (2007) called the "yardstick approach" and used effectively to evaluate some Moluccan birds. Unfortunately, that method is often unavailable. Johnson et al. (1999) suggested a modification of the biological species concept that considers some aspects of the phylogenetic species concept such as diagnosability and genealogy, and stressed the importance of what they termed "independent evolutionary trajectories." Helbig et al. (2002)

provided practical guidelines that supported, more or less, Johnson et al.'s (1999) proposals, but they focused on continental species, or continental species with island populations, rather than on archipelagic taxa with multiple allopatric populations that differ in varying degrees. For these, Helbig et al.'s (2002) guidelines need some modification because, as Steadman (2006:415) stated, "oceanic islands . . . and continental islands or continents . . . have some fundamental differences in geologic development, evolutionary histories, and barriers to colonization." The following examples will show, as did Pratt and Pratt (2001), how an updated application of the biological species concept to island taxa, including effective use of subspecies, can accomplish Hazevoet's (1996) desired goals and more accurately represent the biodiversity of island birds without undermining a long-established and widely understood species definition and without overwhelming endangered species lists with trivially differentiated nominal species. I do not advocate adjusting taxonomy to accommodate attitudes that value species over subspecies, nor do I advocate treating species and subspecies equally when it comes to preserving biodiversity in a world with priorities to set, but we should strive to recognize all biological species as such because their survival may depend on it.

The Hawaiian Islands: A Well-studied Example

The Hawaiian avifauna is particularly instructive in this context (Pratt and Pratt 2001) because it is arguably the most thoroughly studied archipelagic fauna, and our knowledge of it is enlightened by both a rich subfossil record (Olson and James 1982, 1991; James and Olson 1991; Burney et al. 2001) and an ever-growing body of genetic data (Fleischer and McIntosh 2001; Fleischer et al. 1998, 2008) with some studies that combine both lines of evidence (Fleischer et al. 2001, Paxinos et al. 2002). Amadon's (1950) classification, which exhibits the overuse of polytypic species typical for its era, was the standard for many decades. Beginning in the 1980s, most of Amadon's polytypic species were dismantled (Pratt and Pratt 2001) on the basis of new behavioral, ecological, and morphological (Pratt 1982, 1989, 1992), as well as paleontological (Olson and James 1995) and genetic (Tarr and Fleischer 1994, Fleischer et al. 2007), information. As a result, his 25 passerine species comprising 56 named forms have become 51 biological species with only 5 forms remaining as subspecies (as reviewed by Pratt and Pratt 2001). Not counted in this tally are three intra-island subspecies of *Chasiempis sandwichensis* (Pratt 1980) of which Amadon was unaware. If, as ongoing genetic studies (R. C. Fleischer pers. comm.) suggest, all three subspecies of *Loxops coccineus* are elevated to species rank as Pratt (2005) suggested might happen, only three subspecies will remain among Hawaiian passerines. Encouraging is the fact that DNA studies have, to date, corroborated species limits based on phenotypic characters in every case, although they have revealed some strikingly misleading examples of convergence at generic (Reding et al. 2008) or higher (Fleischer et al. 2008) levels. As numerous authors have noted, populations on islands are more strongly isolated than allopatric mainland populations (Phillimore and Owens 2006, Steadman 2006) and the severely restricted gene flow can drive rapid speciation (Moyle et al. 2009). We should expect island birds to exhibit a greater ratio of species to subspecies than continental avifaunas. Even though the Hawaiian example is the extreme, it suggests that Mayr's (1942b, 1969) clearly articulated when-in-doubt-lump precept is the wrong approach when applied to islands. Indeed, as Pratt and Pratt (2001:69) stated, the opposite bias "is more likely to result in a species list that will stand up to independent corroboration." Interestingly, by proper use of the biological species concept, Hawaiian species limits are now nearly the same whether we use the biological or the phylogenetic species concept (Pratt and Pratt 2001), but that will not likely be the case in less isolated archipelagoes.

The Importance of Geography

The Hawaiian Islands are so remote that successful colonists are immediately isolated from their source populations. In other parts of the tropical Pacific, distance from a mainland or island source plays an important role in the degree of differentiation possible, with remote populations likely to become species while those closer to colonization sources may only differentiate to the level of subspecies because of episodic or continuing gene flow. The avifaunas of Micronesia and Polynesia have many large polytypic species whose component taxa occupy islands in more than one archipelago scattered over vast expanses of ocean (e.g., the aforementioned Micronesian

Flycatcher). These species can exhibit many levels of differentiation among several allopatric populations. Amadon and Short (1976) introduced the term "megasubspecies" in an effort to improve the description of such variation, but only a few recent studies (e.g., Mayr and Diamond 2001) have used it extensively, and subspecies on oceanic islands are still too often regarded as essentially equivalent within a species (Phillimore and Owens 2006, Phillimore et al. 2008), especially by non-systematists.

For Helbig et al. (2002), all allopatry was essentially the same regardless of distances involved, but the dynamics of island biogeography clearly modify evolutionary trajectories. Uniformity across a large oceanic region can indicate a recent expansion and colonization, or ongoing gene flow, or a combination of the two. Deciding the role of each of these processes can be difficult, but environmental, behavioral, geographic, historical, and paleontological information can provide inferences. Evaluating the degree of isolation of a population involves the interplay of vagility and distance, and such judgments are subjective because vagility cannot be measured precisely and birds differ widely even within taxa. Paradoxically, selection against dispersal begins immediately upon successful colonization (Carlquist 1974, Moyle et al. 2009), producing the seeming contradiction that although rails (Rallidae) are highly vagile colonizers of even the most remote islands, most endemic island rails are flightless (Steadman 2006). Pigeons and doves (Columbidae) are excellent island colonizers, distributed throughout Polynesia and Micronesia to some of the most remote islands (Pratt et al. 1987, Steadman 2006). Because columbids live on both atolls and high islands, they can take advantage of intervening stepping stones that many land birds cannot. Both the Pacific Imperial Pigeon (*Ducula pacifica*) and Micronesian Imperial Pigeon (*D. oceanica*) apparently move across large water gaps frequently enough to prevent genetic differentiation across vast regions. One observer in Fiji (V. Masibalavu pers. comm.) reports seeing pigeons flying seaward from Viti Levu in large numbers after passage of a particularly devastating typhoon that destroyed the fruit crop. Perhaps dispersal after such storms drives regional genetic homogenization and slows population differentiation in large pigeons. Archeological evidence indicates that *D. pacifica* is a post-human arrival in Tonga and the Cook Islands (Steadman 2006), perhaps because it was able to colonize only after anthropogenic extinction of other *Ducula* spp., so its lack of geographic variation results from both high vagility and recency of dispersal. Pacific fruit doves (*Ptilinopus* spp.) appear to be somewhat less vagile because their species limits tend to coincide roughly with archipelagoes rather than regions, but the most remote forms, such as the Henderson Island Fruit Dove (*P. insularis*) and Rapa Fruit Dove (*P. huttoni*), are distinctive single-island endemics (Pratt et al. 1987). Clearly, geographic remoteness plays a role in speciation, even in highly vagile birds.

Archipelagoes sometimes sample variation in a way that resembles a series of snapshots taken along a cline. Perplexingly, a trend across an island chain may result from an environmental gradient, as in a true cline, but without any continuing interisland gene flow. Cline-like archipelagic variation is infrequent (none of the former polytypic species in Hawaii resembled fragmented clines). Geographic variation in the Polynesian Starling resembles a fragmented cline in some characters but not in others (Mayr 1942a). This small starling is distributed on high islands from the Santa Cruz group (eastern Solomons) eastward through Fiji to Samoa and Tonga. Western populations have brown eyes, eastern ones yellow, with the shift occurring within Fiji, where some populations have both eye colors. Overall coloration varies from mostly brown in the west to mostly gray in the east, but several populations break the flow of this trend. The prominence of pale shaft streaks on the breast feathers also varies, as does overall size, but with no discernible directional trends. One of the largest and most prominently streaked forms is *A. t. tutuilae* on Tutuila, American Samoa. Immediately to the east, on the isolated Manu'a Islands, *A. t. manuae* represents the end of the line for the species. It is much smaller and darker than *tutuilae*, lacks breast streaks altogether, and has pale feather edges that impart a scaly look unique in the complex. Such sudden shifts in characters between neighboring forms, especially if one is a geographic outlier, may signal the existence of previously unappreciated species.

REEVALUATING SPECIES AND SUBSPECIES
AMONG ISLAND BIRDS

Collar (2006a, b; 2007b) used a numerical scoring system for phenotypic characters to determine species limits, similar to Rheindt and Hutchinson's (2007) "yardstick approach," apparently trying to accomplish the same goals I am advocating here (splitting of distinctive allopatric subspecies) and bring some objectivity to the process. I agree with Peterson and Moyle (2008) that Collar's method is essentially a phylogenetic species approach used in a biological species context. Furthermore, Collar has failed to factor in such things as the role of characters as potential isolating mechanisms and the degree of geographic isolation. Peterson and Moyle (2008) also decried the amount of subjectivity in what is supposed to be an objective process. But Collar (2008) rightly pointed out that all species-limit judgments that involve allopatric forms are, at some level, subjective. The model I offer is an attempt to add geographic and biological dimensions to the process and reduce the inevitable subjectivity so that the decisions reached will be repeatable by other disinterested scientists, but setting biological species limits among allopatric taxa can never be a mindless or mechanical exercise under the biological species concept.

I recommend a thought process wherein any oceanic island bird population is considered a species (or allospecies) if (1) at least one age or sex class is distinct from sister taxa in at least one qualitatively discrete phenotypic character (populations that differ only quantitatively are more likely to be subspecies unless measurements show no overlap or proportions are very different, as in one population having a proportionally larger bill, in which case other criteria come into play); and (2) the population is so isolated geographically that present or future gene flow between it and another related population is nearly impossible (i.e., the likelihood of phylogenetic reticulation is extremely low); and (3) it possesses one or more obvious potential isolating mechanism; or, if not strongly isolated geographically, it possesses two or more functionally independent potential isolating mechanisms (i.e., a plumage difference plus a vocal or morphological difference).

This thought process is not operationally different from Mayr's (1942b) earliest suggestions, but it differs philosophically by placing the burden of proof on the lumper rather than the splitter. Mayr's (1969) comparison method is a valuable tool, although underused in the past, for determining whether a difference is likely a potential isolating mechanism, but when no closely related sympatric species pairs exist, that technique cannot be applied. However, one can use such an approach with more distantly related species to infer the kinds of isolating mechanisms likely to operate in a given taxon. For example, the kinds of isolating mechanisms that separate species of nocturnal burrow-nesting petrels are likely to be very different from those among diurnal forest passerines. In practice, Mayr and his followers rarely considered potential isolating mechanisms among island taxa, perhaps because, at the time, these birds were not well known biologically. The most frequently observed potential isolating mechanisms among terrestrial island birds are differences in appearance, vocal differences, morphological differences, differences in breeding biology, other behavioral differences, and ecological differences.

Differences in appearance.—Though often denigrated by earlier taxonomists (e.g., Amadon 1950), color differences in plumage and soft tissues remain the most obvious and predictive indicator of species limits in island birds (Pratt and Pratt 2001). So far, genetic studies have shown that remote island birds that look different to humans in the field usually are different species. Appearance also includes the presence or degree of sexual dimorphism (Pratt 1989, 1992), variation in maturational stages (i.e., distinctive juvenal or immature plumages), or variation in molt timing or sequence (Banks and Laybourne 1977), all of which can indicate species boundaries.

Vocal differences.—As with coloration, birds that sound different to humans, in song or call notes, often are different species. Darwin's finches are a good example of birds that are not highly variable in color but distinguish themselves with different songs (Grant and Grant 2008). Slabbekoorn and Smith (2002) have shown that song can play a prominent role in speciation even in birds whose songs are not innate, but vocal differences are less significant among birds that learn their songs (e.g., oscine passerines) than among those that inherit them. Island birds have been in the forefront of historical playback studies among birds that look similar but sound different (e.g., Lanyon 1967, Pratt 1982), but note that such experiments do not address the important issue of female choice. Recent studies of crossbills (Snowberg and Benkman 2007, Edelaar 2008, Benkman et al. 2009) suggest that call notes as well as songs can serve as isolating mechanisms.

Morphological differences.—Variation in bill shape and relative size may indicate differences in diet and foraging behavior (Benkman 1989; Pratt 1992, 2005; Smith and Benkman 2007) that

are potential isolating mechanisms. The Hispaniolan Crossbill (*Loxia megaplaga*) was recently split almost entirely on the basis of differences in bill size and shape that indicated distinctive food sources (Benkman 1994), and such differences were the first clue that the Kauai Amakihi was a separate species (Pratt et al. 1987, Tarr and Fleischer 1995). Such different physical attributes may also produce differences in appearance (above).

Differences in breeding biology.—Even if two birds can form an initial pair-bond, they will not breed successfully if their nesting habits are incompatible. Important considerations include nest composition and location, different laying and hatching schedules, and differences in roles of the sexes. Nest placement (terminal leaf clump vs. cavity), along with vocal and visual potential isolating mechanisms, were important in splitting the Akekee (*Loxops caeruleirostris*) from the Akepa (*L. coccineus*; Pratt 1989).

Other behavioral differences.—These can be anything from the numerous well-documented examples of differing mating displays to differential response to predators (mobbing vs. hiding; Pratt 1992) and variation in flocking behavior (Smith et al. 1999).

Ecological differences.—These can be such obvious things as differing habitats or differential response to disturbance, as in the case of white-eyes (*Zosterops* spp.) on Saipan and Rota in the Mariana Islands (Fancy and Snetsinger 2001) or the Elepaio (*Chasiempis sandwichensis*) on Kauai and Oahu (VanderWerf et al. 1997, VanderWerf 1998).

A WORKING EXAMPLE:
THE FIJI SHRIKEBILL COMPLEX

The Fiji Shrikebill (Monarchidae: *Clytorhynchus vitiensis*; Fig. 1), with a dozen allopatric subspecies, provides a good model for the reassessment of species limits in a large, widely distributed complex (but I do not regard this exercise as an actual revision because the data have not yet been completely analyzed). Shrikebills are skulking denizens of the forest understory that forage for insects in dead vegetation such as leaf clumps, vine tangles, or tree bark (Watling 2001). Their bills are more or less wedge-shaped and laterally compressed, with a slightly upturned look produced by the shape of the lower mandible, and resemble those of Neotropical antshrikes (*Thamnophilus* spp.). Shrikebills are generally solitary, but they join mixed-species foraging flocks on some

FIG. 1. Representative geographic variation in the Fiji Shrikebill (*Clytorhynchus vitiensis*) complex: (A) *C. v. fortunae*, Futuna and Alofi; (B) *C. v. powelli*, Manu'a Islands, American Samoa; (C) *C.v. keppeli*, Niuatoputapu, Tonga; (D) *C. v. vitiensis*, Viti Levu, Fiji; (E) *C. v. compressirostris*, Kadavu, Fiji; and (F) *C. v. layardi*, Taveuni, Fiji.

islands (Watling 2001). The Fiji Shrikebill is plain and rather featureless in gray and russet, but intensity and hue vary geographically and, less so, individually. The presence or extent of broad pale tips to the tail feathers and a white stripe along the side of the bill also show geographic variation. The characteristic song is a long, quavering, descending whistle usually described as plaintive or melancholy (Pratt et al. 1987, Watling 2001). Seven subspecies are found among the main islands of Fiji, and three more are found on the neighboring islands of Rotuma (*C. v. wiglesworthi*; ~360 km northwest), Futuna and Alofi (*C.v. fortunae*; ~220 km northeast), and Tonga (*C. v. heinei*; ~250 km southeast). The other two subspecies are isolated outliers: *C. v. keppeli* on the remote northern Tongan islands of Niuatoputapu and Tafahi, >300 km

from the next nearest population; and *C. v. powelli*, ~400 km east of Niuatoputapu on the Manu'a Islands at the far eastern end of the Samoan Archipelago (shrikebills are unknown on the geographically intervening, larger Samoan islands). Within Fiji, many characters vary within and between shrikebill taxa in a bewildering mosaic that makes it "rather difficult to work out subspecies that are well defined and geographically restricted" (Mayr 1933:6). Watling (2001) considered most of the subspecies unidentifiable in the field, and some are connected by intermediate populations (Mayr 1933). The Rotuma and Tonga forms are not strikingly different from most of those in the core range.

On the other hand, at least four forms are megasubspecies with consistently distinctive characters.

Clytorhynchus v. fortunae is the smallest and palest form, with the most prominent and sharply defined white tail tips, unique faint gray streaking in the throat, a nearly white belly, contrasting bright tawny flanks, and a thinner, less wedge-shaped and only slightly compressed bill with a very bold white stripe mostly on the lower mandible. Its song, imitated in the local name *tikilili*, comprises metallic notes (Guyot and Thibault 1987) that are apparently very different from shrikebill songs in Fiji, which could not, in the broadest sense, be called metallic (H. D. Pratt pers. obs.). On Kadavu, the southernmost of Fiji's larger high islands, lives *C.v. compressirostris*, a form with plumage, including the pale tail tips, strongly tinged tawny throughout and a long, thin, and very strongly compressed bill as reflected in its epithet. This distinctive bill shape suggests that this population has rather different feeding habits, but no direct observations of such have been reported. Vocally, *compressirostris* generally resembles other Fijian taxa (H. D. Pratt pers. obs.), but I have not made direct comparisons.

Two remote outliers are even more distinctive than *fortunae* and *compressirostris*. Both *keppeli* and *powelli* are much darker than the core group of subspecies, *powelli* being nearly black on the crown, and both have very restricted white tail tips, but otherwise they do not closely resemble each other. On Niuatoputapu, *keppeli* is nearly uniform dusky gray, slightly paler below, with a prominent white base to the bill that is the most noticeable field character (M. LeCroy pers. comm.). The bill is as large as those of Fiji/Tonga birds but not strongly wedge-shaped and only slightly compressed laterally. The only behavioral information available comes from field notes made by M. LeCroy (pers. comm.) in 1997. She described a flock of 8–10 birds "calling, whistling, giving a trill and squawking." Such a large conspecific flock has never been reported for any other shrikebill, and the vocalizations seem quite different, although in an unexpected social context, from those in the species' core range. The Samoan *powelli* is more colorful than *keppeli*, with a strong tinge of russet in the flanks and a pale gray throat that contrasts sharply with the very dark crown and cheeks. The bill is black, with only a thin white line along the tomia (H. D. Pratt pers. obs.), and a strikingly different shape compared to the bills of other populations: relatively shorter without the upturned look, resembling the bills of more typical monarch flycatchers.

The different shape suggests distinctive feeding behavior, but comparative studies have not been done. Importantly, the Samoan bird's songs are only vaguely similar to those given by shrikebills in Fiji (H. D. Pratt pers. obs.; details to be published elsewhere).

Under the guidelines proposed here, the Fiji Shrikebill would be broken up into several allospecies. The Samoan Shrikebill (*C. powelli*), Dusky Shrikebill (*C. keppeli*), and Futuna Shrikebill (*C. fortunae*) qualify as species unequivocally, but the case of *C. compressirostris* is not so clear-cut. Its plumage differences approach those seen in other nearby populations, although most individuals would be identifiable on that basis alone, so whether coloration is a potential isolating mechanism in this complex is questionable. Likewise, its vocalizations may not be sufficiently different to be a potential isolating mechanism (they have not been thoroughly analyzed). Its different bill shape is quite striking, however, and suggests ecological differences that might affect the survival of hybrids should it become sympatric with a neighboring subspecies. So it is a borderline case, best regarded as a megasubspecies until we have more data on additional potential isolating mechanisms. The other subspecies in Fiji (including Rotuma, although its isolation suggests the need for further investigation) and Tonga seem clearly to be conspecific, and some probably do not warrant recognition even as subspecies. The small islands in the Lau Archipelago of eastern Fiji are numerous and close together, which suggests that gene flow may be producing a true fragmented cline in that region. As this case demonstrates, while many island species are wrongly classified as subspecies, the category is still valuable in describing diversity on islands.

THE PARAPHYLY DILEMMA

The influence of phylogenetic thinking has recently set back the cause of island species revision. Many taxonomists are reluctant to recognize well-differentiated peripheral isolates of large complexes, even when they are obviously good species, because doing so might render the remaining complex paraphyletic. Such thinking allows the perfect to become the enemy of the good. Funk and Omland (2003) showed that more than one in five currently recognized species are paraphyletic, so avoidance of paraphyly is hardly a reason to obstruct progress. In my opinion, the fact that we do not yet understand the evolutionary patterns within a large complex should not deter us from recognizing that some peripheral isolates have clearly diverged to the level of species. If that leaves a paraphyletic group, which Rheindt and Hutchinson (2007) called a "Swiss cheese lump" because some forms have been removed from the complex, leaving holes as in Swiss cheese, it may reflect genuine biological processes and the fact that we have more work to do, but at least the island endemics will receive proper conservation attention in the meantime.

An example of the paraphyly problem is the Rufous Fantail *Rhipidura* [*rufifrons*] complex (see front cover), a huge conglomerate with 30 named forms (Mayr and Moynihan 1946, Schodde and Mason 1999), mostly on islands but with a few on continental Australia. Variation in this group is complex, with many forms that look rather similar found throughout the range but with very distinctive ones imbedded within it or on the periphery. Because two of the rather similar-looking forms are sympatric in northern Australia, the complex was split into two species distinguished mainly on tail shape rather than color pattern, *R. rufifrons* with 19 subspecies and *R. arafura* with 11 subspecies (Schodde and Mason 1999). The very distinctive peripheral form *kubaryi* on Pohnpei has been long recognized by many (Pratt et al. 1987, Sibley and Monroe 1990, Clements 2000, Wiles 2005) as a separate species. It is the most isolated of the forms in the *rufifrons* complex (1,625 km from nearest other member of the group), and the most distinctive in color. Nevertheless, according to Boles (2006:231), it is

> sometimes considered a separate species, based on geographical isolation, vocalizations, and lack of rufous in plumage [i.e., exactly the criteria outlined herein]; however, almost certainly derived from other populations within the *rufifrons* cluster, and separation at species level presents complications.

Yet in the same publication, he split the much less distinctive Manus Island form *semirubra* without comment, apparently solely on the basis of a report that its vocalizations were distinctive! That form is the nearest neighbor to *R. kubaryi* and lies between it and other subspecies of *R. rufifrons* and thus presents all the same complications and more. Application of the steps outlined above would alleviate such inconsistencies. In my opinion, recognition of all the strongly differentiated peripheral isolates (the aforementioned plus *ugiensis* [Ugi, Solomon Islands] and *utupuae* [Santa Cruz Islands]) as allospecies, along with the split of *rufifrons* and *arafura*, would be the most informative interim taxonomy for the Rufous Fantail group.

The Role of DNA in Determining Species Limits

Recent genetic studies suggest that the methodology I recommend would hypothesize species limits too conservatively, the large number of resulting splits notwithstanding. The technique cannot reveal species that have differentiated genetically to a level usually found in species but have not differentiated sufficiently in obvious phenotypic traits such as plumage and voice (Cibois et al. 2007, Rheindt and Hutchinson 2007, Phillimore et al. 2008). To date, the few genetic studies of archipelagic birds have consistently broken up large polytypic species, often yielding more species splits than were apparent on phenotypic grounds (Freeland and Boag 1999; Cibois et al. 2004, 2007; Filardi and Moyle 2005; Filardi and Smith 2005). On the other hand, effective isolating mechanisms can result from only slight genetic changes, and thus populations can remain close genetically but still be reproductively isolated as good biological species (Freeland and Boag 1999, Rheindt and Hutchinson 2007, Grant and Grant 2008, Moyle et al. 2009). No measurement of genetic distance can determine whether two populations are species or subspecies under the biological species concept, although large distances suggest that speciation has occurred. How the genes express themselves phenotypically can drive speciation even in cases of limited genetic divergence. Genetic evidence is therefore a "single-edged sword," as characterized by R. Fleischer (pers. comm.). When DNA reveals huge genetic differences or branching patterns that are inconsistent with current taxonomy, we can use it to modify species limits. But when it reveals only slight genetic differentiation, we cannot then say automatically that the taxa in question are conspecific. Therefore, genetic data should not be regarded as a deal-breaking essential feature of species-level revisions.

The "typical" white-eyes (*Zosterops*) of Micronesia provide an example of both the use and misuse of DNA data for setting species limits. While dividing them into three groups, Baker

(1951) followed Stresemann (1931) in combining all seven taxa, from the Marianas in the north to Palau in the southwest, to Pohnpei in the east, as the Bridled White-eye (*Z. conspicillatus*). Every high island has its own form, and they vary in plumage almost as much as the genus varies worldwide (Pratt 2008). Not only do they look different, they sound different in both calls and songs, and some forms apparently lack territorial songs (Pratt et al. 1987, H. D. Pratt pers. obs.). As with the "Micronesian Flycatcher," Pratt et al. (1987) began the process of dismantling this conglomeration by splitting it into three species along geographic lines that corresponded to the three groups mentioned by Baker (1951), except that they considered the Rota form *rotensis* conspecific with the other two Mariana Islands taxa (*Z. c. conspicillatus* on Guam and *Z. c. saypani* on Saipan and Tinian). The Rota bird resembles the birds of Palau (*Z. s. semperi*), Chuuk (*Z. s. owstoni*), and Pohnpei (*Z. s. takatsukasai*) in having all-yellow underparts, but it differs from them strikingly in vocalizations and in colors of soft parts. This classification was tested in a pioneering DNA study by Slikas et al. (2000), who largely upheld Pratt et al.'s (1987) species limits. However, on the basis of genetic distance that indicated a divergence time of 2 million years, they suggested that the Rota White-eye be given full species status. They detected a much shorter period of separation (~10,000 years) between *conspicillatus* and *saypani*, which bracket Rota geographically, and considered them conspecific. This arrangement has now been widely accepted (Stattersfield and Capper 2000, Dickinson 2003, van Balen 2008). Though not as different from each other as from the Rota White-eye, the two other Mariana Island forms differ in size, color pattern, and especially in voice to the same degree (using Mayr's comparison approach) as many sympatric white-eye species (Pratt et al. 1987). These differences are, in my opinion and in the context of white-eyes worldwide (van Balen 2008, Moyle et al. 2009), sufficient potential isolating mechanisms to warrant species status for each, inasmuch as white-eyes have been shown to speciate more rapidly than most birds (Moyle et al. 2009). Although not yet widely accepted, a newly described crossbill species may have diverged from its closest relative as recently as 5,000 years ago (Benkman 2007, Benkman et al. 2009). Slikas et al.'s (2000) argument that the Saipan and Guam birds are not different enough genetically to be separate species

misses the point. Biological species have no minimum number for either time of divergence or genetic distance. On the basis of classic Mayrian criteria, these two birds' variety of potential isolating mechanisms can be expected to keep them on their separate evolutionary trajectories, and greater genetic divergence would develop in due course. Sadly, we can never test this hypothesis because the Guam bird is extinct (Savidge 1987).

The Rota White-eye was rare and restricted to habitat remnants on the island's central plateau by the 1970s (Pratt et al. 1979, 1987). Later, it experienced a precipitous population decline (Craig and Taisacan 1994, Fancy and Snetsinger 2001, Amar et al. 2008). In the meantime, BirdLife International, which maintains the world's Red List of endangered birds (Collar et al. 1994), made no mention of it because, as a subspecies, it was not within their purview. Only after publication of Slikas et al.'s (2000) study was the bird included in BirdLife International's listings (Stattersfield and Capper 2000), and it is now regarded as one of the world's most critically endangered birds (Hirschfeld 2008). The Rota White-eye was overlooked primarily because my own team (Pratt et al. 1987) was overly timid in making splits. I will not make that mistake again. Island birds worldwide are poised for a splitting spree, and we should get to it. Time is not on our side.

ACKNOWLEDGMENTS

Although the ideas presented herein are solely my own, their development benefited greatly from discussions and correspondence with others involved in the species debate, including W. Boles, N. Collar, S. Conant, G. Dutson, R. Fleischer, F. Gill, R. Moyle, T. Pratt, R. Schodde, E. VanderWerf, G. Wiles, K. Winker, and the late Burt Monroe and Robert L. Pyle. M. LeCroy generously shared her field notes from Niuafo'ou. My field work in American Samoa was sponsored by the Division of Marine and Wildlife Resources, under the supervision of J. O. Seamon. Research in independent Samoa was facilitated by T. Foliga of the Division of Environment and Conservation. Research in Fiji was made possible by M. Abbott of Naturalist Journeys (Portal, Arizona), informed by D. Watling, and aided in the field by V. Masibalavu. Field assistance was provided by P. Arthur, D. Buden, M. Etpison, M. Falanruw, and N. Johnson in Micronesia, and, in Hawaii, by D. Kuhn, R. Pacheco, and personnel of the U.S. Fish and Wildlife Service Pacific Islands Ecosystems Research Center. None of my years of work in the tropical Pacific would have happened without the early assistance and companionship of my friend and coauthor P. L. Bruner.

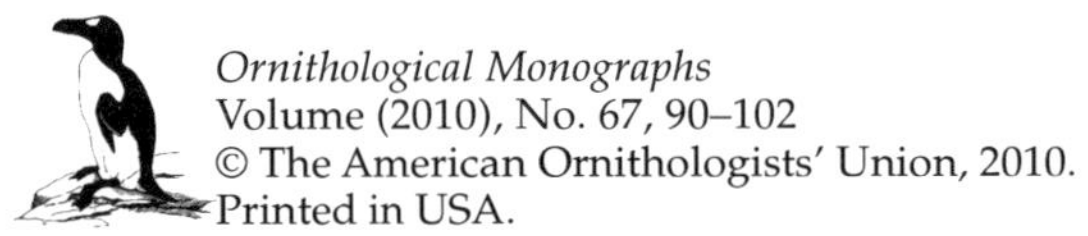

Ornithological Monographs
Volume (2010), No. 67, 90–102
© The American Ornithologists' Union, 2010.
Printed in USA.

CHAPTER 8

PHYLOGEOGRAPHY AND ADAPTIVE PLUMAGE EVOLUTION IN CENTRAL AMERICAN SUBSPECIES OF THE SLATE-THROATED REDSTART (*MYIOBORUS MINIATUS*)

JORGE L. PÉREZ-EMÁN,[1,2,6] RONALD L. MUMME,[3] AND PIOTR G. JABŁOŃSKI[4,5]

[1]*Department of Biology, University of Missouri-St. Louis, 8001 Natural Bridge Road,
St. Louis, Missouri 63121, USA;*
[2]*Instituto de Zoología y Ecología Tropical, Universidad Central de Venezuela, Avenida Los Ilustres,
Los Chaguaramos, Apartado Postal 47058, Caracas 1041-A, Venezuela;*
[3]*Department of Biology, Allegheny College, 520 North Main Street, Meadville, Pennsylvania 16335, USA;*
[4]*Laboratory of Behavioral Ecology and Evolution, School of Biological Sciences, Seoul National
University, Seoul 151-742, Korea, 56-1 Sillim-dong, Gwanak-gu, Korea; and*
[5]*Center for Ecological Studies, PAS, Lomianki 05-092, Poland*

ABSTRACT.—The Slate-throated Redstart (*Myioborus miniatus*) is a common warbler of montane forests from northern Mexico to Argentina. We examined phylogenetic structure and plumage pattern in relation to subspecific taxonomy across the broad geographic range of this species. Phylogenetic analysis of two complete mitochondrial protein-coding genes (subunits 2 and 3 of NADH dehydrogenase) from 36 individuals, representing 10 of the 12 subspecies, revealed four clades, three of which showed general concordance with subspecific classification. However, in a Central American clade, four subspecies (*hellmayri*, *connectens*, *comptus*, and *aurantiacus*) could not be resolved by the molecular phylogenetic analysis, even though populations of *hellmayri* and *connectens* are currently geographically isolated from those of *comptus* and *aurantiacus* by the Nicaraguan lowlands. The genetic homogeneity within this clade suggests a late Pleistocene range expansion at a time when today's montane forest types existed at lower elevations. The Pleistocene hypothesis is supported by both paleoecological reconstructions of Central America and the mismatch distribution of pairwise nucleotide differences among haplotypes in our data set. Despite the genetic homogeneity within the Central American clade, the four subspecies are differentiated in plumage pattern, particularly in the extent of white in the tail. Subspecific variation in the tail pattern is of particular interest because the tails are used in animated displays to startle potential insect prey that are pursued and captured in flight. Field experiments conducted with birds of the Costa Rican subspecies *comptus* and their key prey indicated (1) that a contrasting black-and-white tail is critical to flush-pursuit foraging success and (2) that subspecific variation in the extent of white in the tail reflects evolutionary adaptation to regional prey or habitat characteristics that maximizes flush-pursuit foraging performance. Thus, even though the subspecies of the Central American clade are genetically homogeneous with respect to the mitochondrial genes, analysis of tail pattern and its effect on foraging performance suggests a recent adaptive evolutionary divergence. Our findings serve as a reminder that mitochondrial DNA (mtDNA) gene trees will not always succeed in capturing all evolutionarily significant genetic change, and that manipulative field experiments can provide crucial information on the selective factors that lead to evolution of subspecies-specific morphological traits even in the absence of mtDNA diversification.

[6]E-mail: jorge.perez@ciens.ucv.ve

Ornithological Monographs, Number 67, pages 90–102. ISBN: 978-0-943610-86-3. © 2010 by The American Ornithologists' Union. All rights reserved. Please direct all requests for permission to photocopy or reproduce article content through the University of California Press's Rights and Permissions website, http://www.ucpressjournals.com/reprintInfo.asp. DOI: 10.1525/om.2010.67.1.90.

Key words: foraging ecology, mitochondrial DNA, mtDNA, *Myioborus miniatus*, phylogeography, plumage evolution, Slate-throated Redstart, subspecies.

Filogeografía y Evolución Adaptativa del Plumaje en Subespecies Centroamericanas de *Myioborus miniatus*

RESUMEN.—*Myioborus miniatus* se distribuye en bosques montanos desde el norte de México hasta Argentina. Examinamos la estructura filogenética y el patrón de plumaje y su relación con la taxonomía subespecífica a lo largo de la distribución geográfica de la especie. Análisis filogenéticos de dos genes mitocondriales (subunidades 2 y 3 de la Nicotinamida Adenina Dinucleótido Dehidrogenasa, NADH) en 36 individuos representativos de 10 de las 12 subespecies actualmente reconocidas mostraron cuatro clados, tres de ellos congruentes con la clasificación subespecífica actualmente aceptada. Sin embargo, el clado centroamericano compuesto por cuatro subespecies (*hellmayri*, *connectens*, *comptus* y *aurantiacus*) mostró un sorprendente grado de homogeneidad genética, aún cuando las poblaciones de *hellmayri* y *connectens* hoy están geográficamente aisladas de las de *comptus* y *aurantiacus* por las extensas tierras bajas en Nicaragua. La homogeneidad genética de este clado sugiere una expansión demográfica durante el Pleistoceno tardío a lo largo de Centroamérica durante un período cuando los bosques montanos característicos de hoy día existieron a menores altitudes que en la actualidad, una hipótesis congruente con reconstrucciones paleoecológicas y análisis genético-demográficos. Sin embargo, a pesar de la homogeneidad genética de este clado, las cuatro subespecies se diferencian en los patrones de plumaje, particularmente en la extensión del blanco en la cola. La variación subespecífica en estos patrones es de particular interés porque todas las especies del género *Myioborus* efectúan movimientos rápidos con la cola para ahuyentar presas potenciales (insectos) que luego son perseguidas y capturadas en vuelo. En Costa Rica, experimentos de campo realizados con aves de las poblaciones reconocidas como *comptus* y sus presas clave indicaron que: (1) los patrones contrastantes de negro y blanco en la cola son críticos para el éxito de la estrategia de búsqueda de alimento, y (2) la variación subespecífica en la extensión del blanco en la cola refleja adaptación evolutiva a características regionales de las presas o del hábitat, lo cual maximiza la eficiencia de la estrategia de búsqueda de alimento. Por lo tanto, aún cuando los datos de ADN mitocondrial sugieren que las subespecies del clado centroamericano son homogéneas genéticamente, los patrones del plumaje y su efecto sobre la eficiencia de búsqueda de alimento sugieren divergencia evolutiva adaptativa reciente. Nuestros resultados muestran que los análisis de ADN mitocondrial no siempre reflejan exitosamente cambios genéticos evolutivamente significativos y que los experimentos de campo generan información crucial sobre los factores selectivos que pueden determinar la variación geográfica en caracteres morfológicos, aún en ausencia de diversificación en el ADN mitocondrial.

THE UTILITY AND evolutionary significance of avian subspecies has been a subject of enduring ornithological controversy (Rising 2007). Although the focus of the debate has shifted periodically through the years, it has included considerations of the distinctiveness and diagnosability of subspecies (e.g., Amadon 1949, Wilson and Brown 1953, Patten and Unitt 2002, Remsen 2005, Cicero and Johnson 2006) and whether the concept of subspecies has evolutionary validity in the context of the debate over biological and phylogenetic species concepts (e.g., Cracraft 1983, McKitrick and Zink 1988, Remsen 2005, Zink 2006, Rising 2007, Winker et al. 2007).

In recent years, with the advent of phylogeographic studies (Avise 2000), discussion has often focused on the genetics of subspecies. Genetic data, usually based on mitochondrial DNA (mtDNA) sequence variation, frequently suggest that subspecies originally recognized on the basis

of plumage or morphological characters may not represent discrete evolutionary genetic units (Ball and Avise 1992). For example, Zink (2004) examined the degree to which molecular phylogenies and genetic structure within recognized species correspond to traditional subspecific taxonomy, and he found that only 3% of continentally distributed avian subspecies represent discrete evolutionary genetic units based on the population genetic signature in mtDNA. Although a subsequent analysis (Phillimore and Owens 2006) has shown that the degree of concordance between mtDNA phylogenetic structure and traditional subspecific taxonomy increases to 36% when data from insular and tropical taxa are included, it is clear that many subspecies originally designated on the basis of plumage or morphological characteristics are not supported by the pattern of mtDNA genetic structure. However, even in cases where subspecific taxonomy is not concordant with underlying mtDNA phylogenetic structure, subspecific distinctions can still be valuable if subspecific geographic variation reflects important evolutionary adaptations to regional environmental conditions that have occurred in the apparent absence of mitochondrial genetic divergence (Zink 2004, Remsen 2005, Rising 2007, Winker et al. 2007). Here, we describe an example of such a situation in Central American subspecies of the Slate-throated Redstart (*Myioborus miniatus*).

Redstarts in the genus *Myioborus* comprise 12 species of sexually monomorphic Neotropical warblers that range from the mountains of the southwestern United States to the southern Andes (Curson et al. 1994). The Slate-throated Redstart is the most widely distributed species in the genus, occurring in lower montane forests from northern Mexico to northern Argentina (Curson et al. 1994, Di Giacomo 1995; Fig. 1). Recent phylogenetic analysis indicates that *miniatus* is monophyletic and the sister taxon to a clade of 10 primarily South American *Myioborus* that inhabit higher-elevation upper montane forests (Pérez-Emán 2005).

Within its broad geographic range, 12 subspecies of *miniatus* have been recognized historically (Paynter 1968, Curson et al. 1994; Fig. 1) on the basis of geographic variation in several plumage characteristics. Although subtle subspecific differences exist in coloration of the crown, head, and undertail coverts, the most striking geographic variation is in belly color and the extent of white in the tail. Belly color generally varies along a north–south gradient. Birds of the nominate subspecies *miniatus* of Central Mexico and *molochinus* of the Sierra de Los Tuxtlas of Veracruz have intense vermillion bellies. However, bellies grade from light red, salmon, and red-orange in subspecies of northern Central America (*intermedius, hellmayri, connectens*) to yellow-orange in southern Central America (*comptus, aurantiacus*) and yellow in all five South American subspecies (*sanctamartae, pallidiventris, ballux, subsimilis*, and *verticalis*; Ridgway 1902, Curson et al. 1994).

The pattern of variation in the extent of white in the tail is somewhat more complex (Fig. 1). Two subspecies of northern Central America, *hellmayri* of the coastal mountains of Guatemala and El Salvador and *connectens* of interior El Salvador and Honduras (van Rossem 1936), have the least white in the tail, and the extent of white generally increases both to the north and to the south, reaching its maximum extent

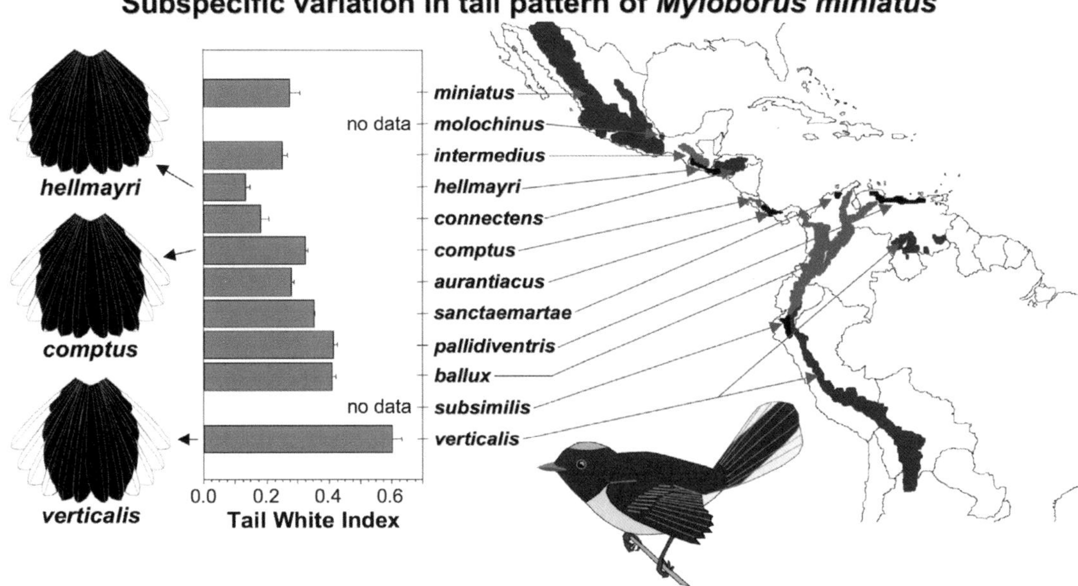

FIG. 1. Range and geographic variation in the extent of white in the tail of subspecies of *Myioborus miniatus*. The tail white index is the average of the maximum linear extent of white in the outer four rectrices divided by rectrix length (based on specimens from the Carnegie Museum of Natural History). Error bars reflect the standard error of the mean.

in *verticalis* of the southern Andes and Guiana Highlands (Fig. 1).

The goals of our study are to (1) examine phylogeographic patterns within *M. miniatus* using sequence information from two mitochondrial genes and focusing mainly on the seven subspecies distributed in Mexico and Central America; (2) contrast the molecular phylogenetic structure to phenotypic variation in plumage characteristics and subspecific taxonomy; (3) illustrate the role of field experiments in understanding the evolution of tail plumage, which is inextricably linked to foraging ecology in this species; and (4) discuss the implications of our results for the long-standing controversies that surround the recognition of avian subspecies

METHODS

Specimens.—We obtained tissues of 23 *M. miniatus*, covering most of the distributional range of the species in North America and Central America (from Mexico to Panama), representing all Central American morphologically differentiated populations (Paynter 1968, Curson et al. 1994). We completed the data set with a previously published group of samples (Pérez-Emán 2005) for a total of 36 specimens of this species, including 10 of the 12 currently recognized subspecies (Appendix). We based subspecific designations on locality information, and most tissues were represented by voucher specimens. Only one individual (01541, *M. m. verticalis*) from the Zoological Museum of Copenhagen was not vouchered.

DNA methods.—We isolated whole genomic DNA from small tissue plugs using the Pure-Gene (Gentra Systems, Minneapolis, Minnesota) extraction kit, or the DNeasy Tissue Kit (Qiagen, Valencia, California), following the animal-tissue protocols provided with the kits. We amplified two complete mitochondrial protein-coding genes, subunits 2 and 3 of the NADH (ND2, 1,041 base pairs [bp]; ND3, 351 bp), using the polymerase chain reaction (PCR) and a set of primers previously used in the genus (Pérez-Emán 2005). The PCR conditions were an initial denaturation cycle at 94°C for 2 min, followed by 35 cycles of denaturation at 94°C for 45 s, annealing at 52°C for 30 s, and an extension phase of 72°C for 60 s. The PCR reactions contained 1–2 µL of DNA template, 0.625U of *Taq* polymerase (PROMEGA), 10 mM Tris–HCl (pH 9.0), 50 mM KCl, 1.5 mM MgCl2, 0.48 µM of each primer, and 80 µM

dNTPs, in a total volume of 25 µL. We purified PCR products in 1% low-melting-point agarose gels, excised from the gel under ultraviolet light, with GeneClean III Kit (BIO 101 Systems, Solon, Ohio), and used them as templates for PCR sequencing reactions. We performed sequencing reactions in both directions using Dye Terminator Kits of ABI and the same set of primers used for amplifications. We precipitated PCR products with an ethanol-sodium acetate solution and ran them on an ABI 377 automated sequencer.

Phylogenetic analyses.—We aligned sequences by eye using the program SEQMAN II (DNASTAR, Madison, Wisconsin). To test for incongruent phylogenetic signal between genes, we used the incongruence-length difference (ILD) test (Farris et al. 1995) as implemented in partition-homogeneity tests in PAUP, with 100 replicates, using only informative characters (Cunningham 1997). We examined phylogenetic relationships among haplotypes using both maximum-parsimony and maximum-likelihood methods in PAUP*, version 4.0b10 (Swofford 2003). We performed an equal-weighted parsimony analysis using heuristic searches with starting trees obtained via random stepwise addition of taxa with the tree bisection reconnection (TBR) branch-swapping algorithm for 1,000 replicates. We estimated support for tree nodes with 100 bootstrap pseudo-replicates (Felsenstein 1985) with the full heuristic search and similar previous settings. We estimated a model of molecular evolution that best fits the data using the program MODELTEST, version 3.7 (Posada and Crandall 1998). This software implements a hierarchical comparison of different nested models, based on likelihood ratio tests, using an initial neighbor-joining tree generated with a Jukes-Cantor model of evolution. The model selected is the simplest model that cannot be statistically rejected in favor of a more complex model. Data used in this study fit a general time-reversible model with a substitution rate for variable sites following a gamma distribution (GTR + G) with a shape parameter alpha = 0.1202. Consequently, we implemented this model in a maximum-likelihood analysis, conducting 10 replicate heuristic searches with random stepwise addition of taxa using the TBR branch-swapping algorithm. We estimated bootstrap support with 100 iterations using previous settings. We tested for departures from a molecular clock using a likelihood ratio test comparing likelihoods of phylogenetic hypotheses with and without a clock assumption. Phylogenetic reconstructions included sequences of *M. pictus* and four other warbler species (*Wilsonia pusilla, W. canadensis, Ergaticus ruber,* and *Cardellina rubrifrons*) as outgroups, based on a hypothesis of phylogenetic relationships obtained in a previous study (Pérez-Emán 2005; Appendix).

We also performed a Bayesian analysis with all sequences using the program MRBAYES, version 3.1 (Ronquist et al. 2005). We set Bayesian search to four chains running simultaneously for 2 million generations, with trees sampled every 100 generations for a total of 20,000 trees. We conducted four independent runs initiated from different random trees to avoid results depending on initial starting conditions. We plotted results of log-likelihood scores to generation times to identify the point at which log-likelihood values reached an equilibrium state (stationarity) using TRACER, version 1.4 (see Acknowledgments). We also examined convergence of clade posterior probabilities as a function of generation number, as well as the correlation of clade frequencies for pairs of different independent runs using AWTY (Wilgenbusch et al. 2004). We discarded, conservatively, the first 5 million generations and, on the basis of convergence of different runs, combined results of the 15,000 trees of each data set, for a majority-rule consensus tree of 60,000 trees.

We calculated pairwise differences and haplotype and nucleotide diversity indexes, and carried out demographic analyses described below using the program DNASP, version 4.20.2 (Rozas et al. 2003). The haplotype diversity index measures the probability that two randomly chosen haplotypes in the sample are different, whereas nucleotide diversity is a measure of pairwise nucleotide differences per site among haplotypes in the sample (Nei 1987). We derived haplotype networks from statistical parsimony using the program TCS, version 1.13 (Clement et al. 2000). Connections among haplotypes are based on pairwise differences as long as there is a high probability ($P > 0.95$) that the difference in a site between two haplotypes is the product of just one mutation (parsimonious state). We determined mismatch distributions to investigate historical changes (demographic history) in populations (Rogers and Harpending 1992, Rogers 1995). Mismatch distributions are the distribution of pairwise nucleotide differences among haplotypes (Rogers and Harpending 1992) and, when compared to a model of population expansion, can provide insight on the demographic

history of species. The shape of these distributions indicates whether populations have gone through episodes of decline or expansion in population growth (unimodal distributions) or, on the contrary, have been characterized by constant population sizes (multimodal distributions; Rogers and Harpending 1992, Rogers 1995). We evaluated statistical significance for population expansion using Ramos-Onsins and Rozas's $R2$ statistic (Ramos-Onsins and Rozas 2002), which is based on the difference between the number of singleton mutations and the average number of nucleotide differences among sequences within a population sample, and Fu's F test (Fu 1997), which tests for deviation from neutrality but which is sensitive to departures from population stationarity (excess of young and rare mutations in non-recombining sequences) and can reveal histories of demographic events such as population expansions. We based significance on confidence limits generated under a coalescent model assuming constant population size and the number of segregating sites (Rozas et al. 2003). We also used the McDonald-Kreitman test to check for the effect of natural selection in the combined ND2 and ND3 data, because it has been shown that positive selection can affect mtDNA and bias interpretation of genetic patterns (Ballard and Whitlock 2004). As such, we compared the ratio of synonymous to nonsynonymous polymorphic sites to the ratio of synonymous to nonsynonymous fixed substitutions between pairs of clades obtained in our study, with the assumption that ratios should be the same in the absence of positive selection (Zink 2005, Zink et al. 2005).

The demographic population model described above estimates three parameters of interest—the effective population size, both present and before the expansion, and the time elapsed between the two (t). Consequently, we estimated the time elapsed from population expansion. The actual mutation rate per nucleotide and generation (μ) has to be known in order to estimate the time (number of generations) since population-size change (t), assuming that $\tau = 2ut$ (τ measures time in mutational events) and $u = \mu L$, where L is the length of the sequence of DNA analyzed (Rogers and Harpending 1992). We assumed mutation rates ranging from 0.02 (cytochrome-*b* data; Weir and Schluter 2008) to 0.03 (based on comparisons of genetic divergences between cytochrome-*b* and ND2/ND3 genes in *miniatus* sequences; J. L. Pérez-Emán et al. unpubl. data) substitutions per site per million years. Because 1-year-old birds regularly acquire territories and breed (R. L. Mumme unpubl. data), we assumed a generation time of one year as the age at which sexual maturity is reached. However, given that birds of this species have a long lifespan (Lentino et al. 2003, R. L. Mumme unpubl. data), we suspect that mean generation time is probably closer to 2 years than to 1 year and, as such, we also considered this age as an estimate of generation time.

RESULTS

Phylogenetic relationships among haplotypes.—We sequenced a total of 1,392 bp (complete ND2 and ND3 genes), all of them clean and unambiguous, with no instances of unusual stop codons, insertions or deletions, which indicates that the sequences were not nuclear copies (pseudogenes). All new sequences have been deposited in GenBank (accession numbers GQ335543–GQ335588). We included the complete sequences of both ND2 and ND3 genes combined in phylogenetic analyses, because a partition homogeneity test failed to find differences in phylogenetic signal between genes ($P = 0.65$).

Phylogenetic reconstructions identified four lineages with high support (bootstrap values > 85% and posterior probabilities = 100%; Fig. 2A). Mexican populations of the subspecies *miniatus* located to the west of the Isthmus of Tehuantepec

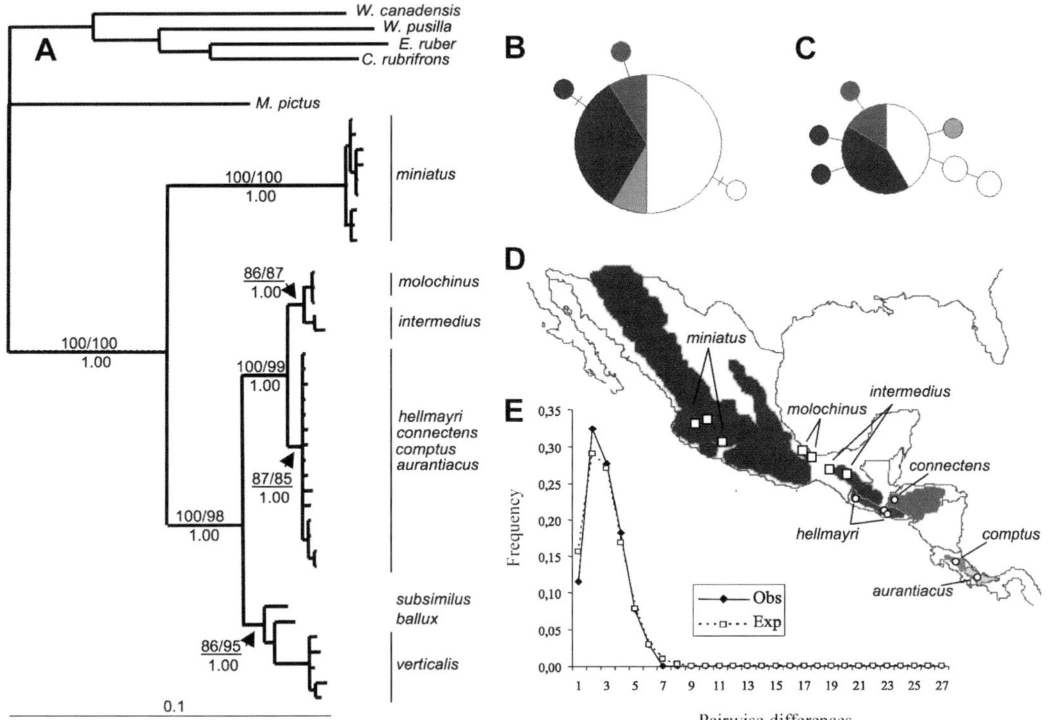

FIG. 2. Phylogenetic relationships and genetic differentiation of *Myioborus miniatus* in North America and Central America. (A) Maximum-likelihood tree topology of *miniatus* haplotypes. Numbers above the nodes represent bootstrap support values from maximum-parsimony and maximum-likelihood analyses. Numbers below the nodes represent posterior probabilities obtained from Bayesian inference. Subspecies names identify each of the major lineages. (B) Central American *miniatus* haplotype networks of both mitochondrial ND2 and (C) ND3 genes. Colors represent each of the subspecies included in this clade: black = *hellmayri*, dark gray = *connectens*, light gray = *comptus*, and white = *aurantiacus*. (D) Distributions of each subspecies of *miniatus* in North America and Central America. Circles represent voucher localities associated with each of the previous four subspecies included in haplotype networks (identified by their names). Squares represent voucher localities of other North American and Central American populations (subspecies). (E) Mismatch distribution of haplotypes included in the Central American clade, assuming a population growth model. Both observed and expected distributions are represented. See text for statistical significance of these results.

represented a basal lineage. A South American clade included populations of the subspecies *ballux* from eastern Panama, and a third lineage represented Mexican populations adjacent to the Isthmus of Tehuantepec, including the isolated subspecies *molochinus* from the Los Tuxtlas mountains of Veracruz and the subspecies *intermedius* from eastern Oaxaca and Chiapas. A fourth clade comprised individuals of four Central American subspecies (*hellmayri*, *connectens*, *comptus*, and *aurantiacus*) from western Guatemala through the Chiriqui Highlands in Panama. Populations from this last clade were very homogeneous, differing by an average of <2 pairwise nucleotide differences ($k = 1.87$) and showing no resolution in the phylogenetic tree, a result that contrasts with the morphological and plumage differentiation shown by populations distributed along this geographic range. A likelihood ratio test failed to reject a clock-like evolution of sequences, which indicates a lack of rate heterogeneity.

Genetic variation in miniatus *populations.*—We found a total of 32 haplotypes among 36 sequences of *M. miniatus* analyzed. Haplotype divergences among lineages were larger than within lineages. Average sequence divergence (*p*-distance) between Mexican populations of *miniatus* was 5.8% when compared to populations adjacent to the Isthmus of Tehuantepec and Central American clades and 5.9% when compared to the South American clade, indicating an old split between the Mexican clade and the other lineages. Sequence divergence between the Isthmus of Tehuantepec and the Central American and South American clades averaged 2.5%. Divergence within clades was low, with the largest value found in the South American clade (0.9%) and the smallest value recorded for the Central American clade including populations from Guatemala to Panama (0.1%).

These data are in agreement with haplotype and nucleotide diversities, which were highest for the South American clade and smallest for the Central American clade, highlighting the small genetic divergence in this clade (Table 1).

Demographic history of miniatus *populations.*— The genetic homogeneity found in the Central American clade is consistent with patterns of rapid diversification or rapid expansion of populations forming such a clade. Haplotype networks of both mitochondrial genes (ND2 and ND3) showed a similar star-like structure characterized by one common and frequent haplotype distributed from Guatemala to Panama, and other closely related haplotypes found in low frequencies (Fig. 2B–D). A mismatch distribution of *miniatus* from Central America (Guatemala to highlands of Panama) fit the expected pattern under population expansion (Fig. 2E; $R^2 = 0.086$, $P < 0.05$). Fu's F-test of neutrality was also significantly negative ($F_S = -4.80$, $P < 0.05$), which indicates that populations from Central America have not been stationary through time. McDonald-Kreitman tests were not significant in any of the pairwise comparisons of clades (P values ranging from 0.67 to 1.00). Estimated times from population expansion ranged from 33,500 years (67,000 years considering a generation time equal to 2 years) with a 2% calibration rate to 22,350 (44,700) years with a 3% rate, consistent with a late Pleistocene temporal framework.

DISCUSSION

Phylogeography of Myioborus miniatus *populations.*— Phylogenetic reconstructions of *M. miniatus* populations included in our study identified four well-supported clades, each of them representing monophyletic groups consistent with

TABLE 1. Genetic diversity indices for each of the four geographic clades recognized by phylogenetic analyses. Subspecies of *Myioborus miniatus* included in each clade were as follows: Mexican clade, *miniatus*; Isthmus of Tehuantepec clade, *molochinus* and *intermedius*; Central American clade, *hellmayri*, *connectens*, *comptus*, and *aurantiacus*; South American clade, *subsimilis*, *verticalis*, and *ballux* (including populations from eastern Panama).

Clade	n	Number of haplotypes	Haplotype diversity	Nucleotide diversity	Nucleotide differences
Mexican	9	8	0.97	0.0024	3.28
Isthmus of Tehuantepec	5	3	0.70	0.0026	3.60
Central American	15	9	0.89	0.0013	1.87
South American	7	7	1.00	0.0088	12.30

geography. The most distinct clade, a basal lineage, comprised populations of the nominate subspecies *miniatus* from northern and central Mexico. This clade was sister to a large group composed of a clade encompassing populations of three South American subspecies (*ballux*, *subsimilis*, and *verticalis*), which was sister to a second clade subdivided into two main lineages: one represented by populations on opposite sides of the Isthmus of Tehuantepec (including two subspecies, *molochinus* from the Sierra de Los Tuxtlas mountains of Veracruz, and *intermedius* of the interior highlands of Oaxaca and Chiapas), and another one comprising Central American populations ranging from Guatemala to the Chiriqui highlands of western Panama. This phylogeographic pattern, a Mexican basal taxon sister to a clade of Central American and South American lineages sister to each other, is largely congruent with those shown by other Neotropical taxa with similar geographic distributions—for example, *Chlorospingus ophthalmicus* (García-Moreno et al. 2004, Weir et al. 2008) and *Arremon brunneinucha* (Cadena et al. 2007, Navarro-Sigüenza et al. 2008). The phylogeographic pattern also suggests a northern origin and subsequent southern dispersal for the *Myioborus miniatus* species complex, which has also been suggested for hummingbirds in the genus *Lampornis* (García-Moreno et al. 2006) and solitaires in the genus *Myadestes* (Miller et al. 2007).

The Central American lineage, representing populations belonging to four different named subspecies (*hellmayri*, *connectens*, *comptus*, and *aurantiacus*), was characterized by a surprising lack of genetic structure. This lack of structure could be the result of either high current levels of gene flow or recent species range expansions (Zink 1997). Although extensive gene flow seems possible in populations that represent subspecies pairs that exist in parapatry (e.g., *connectens* and *hellmaryi*, *comptus* and *aurantiacus*), populations of *connectens* and *hellmaryi* are separated from *comptus* and *aurantiacus* by several hundred kilometers of Nicaraguan lowlands, which makes extensive gene flow between these two subspecies groups exceedingly unlikely, at least at present. Our results are more consistent with the hypothesis of recent population expansion. Population expansions that result from bottleneck episodes or rapid expansions from a leading front are expected to result in homogeneous populations with low levels of genetic diversity (Rogers and Harpending 1992, Hewitt 2000). The observed fit of mismatch distributions to patterns expected with population expansion, particularly when added to the starlike haplotype phylogeny found for this clade, argues for a late Pleistocene range expansion of the Central American lineage across this region.

The proposed late Pleistocene expansion of the Central American lineage now encompasses an area from Guatemala to Panama and includes regions of extensive lowland forest (i.e., Nicaragua) where *M. miniatus* is neither currently distributed nor regularly recorded. Although it is possible that the proposed range expansion may have occurred as the result of long-distance colonization across unsuitable lowland habitat, we think that the expansion was more likely facilitated by a more extensive distribution of montane forest at lower elevations during the late Pleistocene. Paleoecological analysis indicates that the late Pleistocene flora at Lake La Yeguada, Panama, at elevation 650 m, included montane forest elements that are now more characteristic of elevations above 1,500 m (Bush et al. 1992). This analysis suggests the possibility that montane forests suitable for *M. miniatus* occurred at much lower elevations during the late Pleistocene, and this could have facilitated the Central American population expansion throughout the region during this period.

Regardless of the evidence of recent population expansion, the lack of genetic structure within the Central American clade is surprising given the fairly dramatic differences in plumage, particularly belly color and tail pattern, among the four subspecies; *connectens* and *hellmayri* have salmon to red-orange bellies and relatively little white in the tail, whereas *comptus* and *aurantiacus* have yellow-orange bellies and more extensive white in the tail. One possible hypothesis to explain this pattern is that subspecific variation in plumage pattern may reflect phenotypic plasticity rather than underlying genetic variation. In other words, the subspecific variation in belly color and tail pattern within the Central American clade could be a result of geographic variation in environmental conditions during growth and development, and not of subspecific differences in genes affecting growth and development. The plasticity hypothesis is an intriguing possibility particularly for belly color, because environmental variation in dietary carotenoid intake is known to account for many cases of within- and between-population variation in the extent of yellow versus red plumage in birds (Hill 1993, 2006). However, phenotypic plasticity seems unlikely to account for variation in tail pattern. First, there is generally little evidence that variation in diet or other environmental variables can produce consistent variation in pigment deposition in melanin-based black-and-white plumage patterns (Hill 2006). Second, our preliminary examination of data from museum specimens and recaptures of individual birds between tail molts (R. L. Mumme unpubl. data) suggests that there is relatively little within-population or within-individual variation in tail pattern; if tail pattern were phenotypically highly plastic, one would predict that variation within populations would be considerable and that tail patterns of individual birds might also vary from molt to molt as social status or nutritional condition change. Neither prediction appears to be true. Finally, at least some of the subspecies that differ greatly in tail pattern (e.g, *miniatus* vs. *hellmayri*, *comptus* vs. *verticalis*) are quite different from one another in mtDNA (Fig. 2), which suggests that differences in tail patterns among those subspecies pairs may also have a genetic basis. Nonetheless, even though we believe that phenotypic plasticity is unlikely to explain subspecific variation in tail pattern in *M. miniatus*, it remains a possibility that can be eliminated only by reciprocal transplant (e.g., James 1983) or "common garden" experiments (e.g., Berthold et al. 1992) that, given the geographic distribution of these taxa across several different Central American countries, would be very difficult to conduct.

Assuming that phenotypic plasticity is not involved and that subspecific plumage variation within the Central American clade is based on underlying genetic variation, the results of our mtDNA analyses suggest that plumage divergence evolved recently (since the late Pleistocene) and in the absence of any corresponding neutral genetic divergence at the level of mtDNA. What evolutionary processes drove this rapid plumage evolution? For subspecific variation in belly color, we are unable to answer this question; we have no data that bear directly on the question of why *connectens* and *hellmayri* should have salmon or red-orange bellies whereas *comptus* and *aurantiacus* have yellow-orange bellies, or why the marked north-south gradient in belly color exists generally in the Slate-throated Redstart. In the case of tail pattern, however, we have good reason to suspect that subspecific variation within the Central American clade is a result of divergent natural selection, and we discuss the evidence in the following sections.

The black-and-white tail as foraging adaptation.—Like all redstarts in the genus *Myioborus*, the Slate-throated Redstart uses striking animated displays of its contrasting black-and-white tail to startle potential insect prey, which are then pursued and captured in flight (Jabłoński 1999, Mumme 2002). This is an innate (Jabłoński et al. 2006b) foraging tactic termed "flush-pursuit foraging" (Remsen and Robinson 1990). During flush-pursuit foraging, redstarts hop stiffly from branch to branch, erecting and fanning their contrasting black-and-white tails, drooping their wings, and pivoting and pirouetting from side to side. By experimentally darkening the white feathers of birds in the field, and by testing the response of insects to models of foraging birds, our earlier work has demonstrated that the presence of contrasting black-and-white tail feathers is critical in triggering insect escape behavior during flush-pursuit foraging, in both the Painted Redstart (*M. pictus*) in Arizona (Jabłoński 1999, 2001; Jabłoński and Strausfeld 2000, 2001) and the Slate-throated Redstart in Costa Rica (Mumme 2002, Galatowitsch and Mumme 2004, Mumme et al. 2006).

Foraging displays in *Myioborus* are effective because they exploit the simple sensory and neuromuscular pathways that govern visually evoked insect escape response. Insect escape behavior has been an important model system in insect neurobiology and is generally well understood (Bullock 1985, Holmqvist and Srinivasan 1991, Rind and Simmons 1992, Hatsopoulos et al 1995, Gabbiani et al. 1999, Jabłoński and Strausfeld 2000, Santer et al. 2006, Shin and Jabłoński 2008). Insect escape responses are visually evoked through pathways tuned to looming motion, such as an approaching predator. In particular, these pathways are sensitive to an increased rate of image expansion on the insect retina, accelerated translational movement across the visual field, and increased contrast between the object and the background. Thus, the three most striking visual aspects of foraging displays in *Myioborus*—spreading of the tail and wing to increase stimulus size, side-to-side pivoting motions to increase stimulus movement, and conspicuous display of the black-and-white tail feathers to increase stimulus contrast—exploit the visual sensitivity of escape pathways in insect prey and lead to an increase in the number of potential prey flushed by a foraging bird (Jabłoński and Strausfeld 2000, 2001). Additionally, the side-to-side pivoting promotes foraging success by affecting prey escape direction (Jabłoński 2001, Jabłoński and McInerney 2005).

Having a black-and-white tail is therefore clearly an important foraging adaptation in *Myioborus* (Jabłoński 1999, Mumme 2002). However, this does not necessarily mean that geographic (subspecific) variation within *M. miniatus* in the extent of white in the tail is adaptive and driven by diversifying selection. A legitimate nonadaptive null hypothesis is that although some white is necessary for successful flush-pursuit foraging, subspecific variation in the extent of white in the tail could be entirely inconsequential to foraging performance and simply a result of genetic drift, not natural selection (Mumme et al. 2006). The alternative adaptationist hypothesis, however, is that subspecific geographic variation in the tail pattern of *M. miniatus* has been shaped by natural selection and reflects adaptation to regional prey and habitat characteristics that maximizes flush-pursuit foraging performance.

We have tested these two alternatives with a set of field experiments conducted with the subspecies *M. miniatus comptus* in Monteverde, Costa Rica, in which we either reduced or increased the amount of white in the tail of Monteverde birds to mimic the tail patterns found in *hellmayri* of northern Central America or *verticalis* of Bolivia (Mumme et al. 2006). First, birds with reduced-white *hellmayri*-like tails, although they performed better than birds with no white, had significantly decreased flush-pursuit foraging performance compared with *comptus*-like controls. However, birds with the increased-white *verticalis*-like tail pattern had slightly but not significantly lower foraging performance than *comptus*-like controls (Fig. 3A). These results suggest that the particular black-and-white tail pattern of *comptus* represents a broad but nonetheless distinct locally adaptive peak that maximizes the flushing of insect prey under Costa Rican field conditions. This conclusion is corroborated by a series of field experiments that tested the response of seven different Costa Rican insect prey species, six species of homopterans, and an asilid robber fly, to models of foraging *Myioborus* ssp. Under Monteverde field conditions, insect prey were significantly less responsive to models of *hellmayri* versus *comptus*, and slightly but not significantly less responsive to models of *verticalis* (Fig. 3B; Galatowitsch and Mumme 2004, Mumme et al. 2006).

Collectively, the results of the bird and insect experiments indicate that the tail pattern of Costa Rican birds of the subspecies *comptus* maximizes flush-pursuit foraging performance for the specific prey and/or habitat conditions that exist within

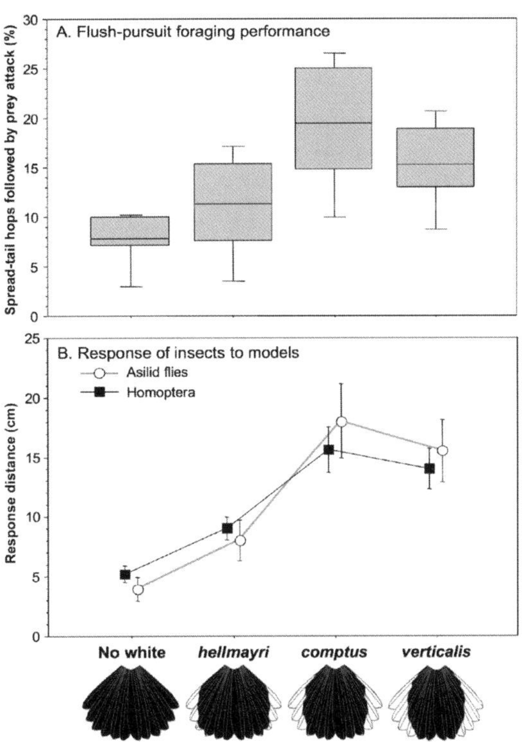

FIG. 3. (A) Flush-pursuit foraging performance of individuals of *Myioborus miniatus comptus* following experimental manipulation of the extent of white in the tail. Flush-pursuit foraging performance was measured by the percentage of hops in the characteristic spread-tail posture that were followed by an attack on a prey item. Median (horizontal line), 25th and 75th percentiles (box), and 10th and 90th percentiles (whiskers) are shown. Modified from Mumme et al. (2006). (B) Mean response distance of homopterans and asilid flies in reaction to a model of a foraging redstart representing four different plumage patterns. Mean and standard errors are shown. Based on data from Galatowisch and Mumme (2004) and Mumme et al. (2006).

its range. Because efficient foraging is known to have direct positive effects on lifetime reproductive success in birds (Lemon and Barth 1992), it is likely that the tail pattern of this subspecies has been shaped by a history of natural selection.

Evolutionary divergence of tail pattern: Prey-driven or habitat-driven?—One issue raised by our findings is the nature of the environmental variation that appears to have produced subspecific divergence in tail pattern. Two hypotheses can be proposed: a prey-specific hypothesis and a habitat-specific hypothesis (Jabłoński et al. 2006a, Mumme et al. 2006). The prey-specific hypothesis suggests that different subspecies of *M. miniatus* may exploit insect prey with different visual sensitivities or escape behavior, leading to selection for subspecific divergence in plumage pattern. However, the neural mechanisms that govern insect escape behavior are simple and highly conserved (Bacon and Strausfeld 1986; Jabłoński and Strausfeld 2000, 2001), and significant geographic variation in how those systems respond to subtly different types of visual signals may not exist (Jabłoński et al. 2006a, Mumme et al. 2006). Detailed studies of the prey base of different subspecies of *miniatus* in different geographic regions would be required to test the prey-specific hypothesis and determine whether geographic variation in insect prey could potentially produce plumage divergence.

The habitat-specific hypothesis (Jabłoński et al. 2006a) proposes that geographic variation in habitat or environmental conditions may select for subspecific divergence in plumage patterns. Under this hypothesis, geographic variation in the physical environment, not the prey base, promotes diversification of plumage pattern by selecting for tail patterns that work best under prevailing ecological conditions. This hypothesis is supported by experiments that showed that the background against which model redstarts are displayed can significantly influence the effectiveness of a particular plumage pattern in startling prey; models with relatively little white were more effective when displayed against relatively light backgrounds, whereas models with more extensive white were more effective when displayed against darker backgrounds (Jabłoński et al. 2006a). Similarly, in field experiments, muscid flies living in very different habitats responded differently to models of redstarts with a particular tail pattern (Jabłoński et al. 2006a). Although these data support the habitat-specific hypothesis, a definitive test will require detailed studies of the physical environment of *M. miniatus* throughout its broad geographic range.

Subspecific taxonomy, plumage pattern, and genetic divergence.—One of the implications of the present study is that the current subspecific taxonomy of the Central American populations of *M. miniatus*, a taxonomy based on decades-old evaluation of plumage variation, is not supported by modern evolutionary genetic analysis of sequence divergence in mtDNA. Although several of the 12 recognized subspecies of *M. miniatus* appear to represent distinct evolutionary lineages

(e.g., the nominate subspecies *M. m. miniatus* of central Mexico, as well as the subspecies *molochinus*, *intermedius*, and *verticalis*), four Central American subspecies—*hellmayri*, *connectens*, *comptus*, and *aurantiacus*—cannot be resolved by mtDNA analysis. From an mtDNA perspective, subspecific designations of these taxa do not correspond to independent evolutionary units (Zink et al. 2000).

Should we therefore conclude that the four named Central American subspecies have no genetic or evolutionary validity? This question lies at the heart of much of the recent controversy surrounding the recognition of avian subspecies (e.g., Zink 2004, Phillimore and Owens 2006, Rising 2007). We believe, however, that such a conclusion would be unwarranted. Although our genetic analysis suggests that the subspecies of the Central American clade are indeed genetically homogeneous, at least from the perspective of mtDNA, our analysis of plumage pattern and its effect on foraging performance suggests that the Central American subspecies have undergone recent evolutionary divergence that appears to have been adaptive and driven by natural selection for improved foraging performance. As pointed out by other authors (Rising 2007, Winker et al. 2007) mtDNA gene trees will not always reflect every instance of evolutionarily significant genetic change. In fact, evolutionary change in neutral characters at the intraspecific level is unlikely to result in monophyletic groups; on the contrary, paraphyletic and even polyphyletic patterns are expected as a result of incomplete lineage sorting (Funk and Omland 2003). Moreover, populations that have undergone recent historical demographic expansions will display genetic homogeneity that could confound the identification of evolutionary units. As a consequence, any adaptive evolutionary change that occurred during a short time will cause a lack of concordance among phenotypic characters that have been subject to selection and neutral molecular characters (Moritz 2002). Examples of morphological differences in clearly diagnosable taxa that are uncorrelated with genetic divergence include sparrows in the genus *Passerculus* (Remsen 2005, Zink et al. 2005, Rising 2007), warblers of the *Dendroica coronata* species complex (Milá et al. 2007), gnatcatchers in the genus *Polioptila* (Zink et al. 2000), and grouse species in the subfamily Tetraoninae (Oyler-McCance et al., this volume). In many of these examples, morphologically distinct populations deserve taxonomic status; lumping different morphological and ecological populations into one name can confound the evolutionary history of the organism and produce taxa that are not predictive of ecology or morphology (Barrowclough 1982, Remsen 2005).

An important point to consider is the diagnosibility of infraspecific categories. Most of the criticism directed toward subspecies categories focuses on the lack of unambiguous criteria to diagnose and identify geographically distinct populations (Wilson and Brown 1953). Expanding and formalizing the historical "75% rule" proposed by Amadon (1949), Patten and Unitt (1992) criticized the normal approach of considering mean differences instead of magnitude of variation. In the case of subspecies of *M. miniatus*, formal descriptions of the four subspecies included in the Central American clade focused on mean differences in some morphological characters as well as variation in plumage coloration, especially in the crown, chest, belly, and tail feathers (van Rossem 1936, Wetmore 1944). At present, we lack sufficient data to apply statistical tests of diagnosibility to these populations. However, *hellmayri–connectens* populations are clearly 100% diagnosable compared with *comptus–aurantiacus* populations, both for belly color and tail pattern, and the differences in tail pattern appear to be adaptive. Moreover, these populations are potentially isolated geographically. Thus, even in the absence of mtDNA genetic divergence, geographic variation of this sort ought to be recognized taxonomically if we want classifications to be predictive. On the other hand, the populations within the two subspecies groups (*hellmayri–connectens* and *comptus–aurantiacus*) are more similar to each other, and there is likely some geographic contact between populations within each group. In the case of *comptus* and *aurantiacus*, there seems to be evidence for a large contact zone and, possibly, intergradation in the Costa Rican central plateau (Slud 1964). Consequently, each of these populations should be studied more thoroughly if we want to recognize or reject subspecies names given to some geographic variation that may be clinal.

Our results highlight the importance of selecting a set of diverse characters when evaluating geographic variation within species and underscore the value of field experiments as tools for understanding and evaluating potential selective mechanisms responsible for evolutionary divergence within a species. We suggest, as some have previously done (Legge et al. 1996, Crandall et al. 2000, Remsen 2005), that subspecies concepts should be critically examined but, at the same time, should be sufficiently broad to include cases where subspecific adaptation has occurred in the apparent absence of overall mitochondrial genetic divergence. Recognizing both phylogenetic and adaptive changes in the evolutionary history of an organism, and correctly naming such geographic variation, will produce classifications that are consistent with evolutionary history.

ACKNOWLEDGMENTS

We thank K. Winker and S. Haig for inviting us to contribute to this volume. We also thank all institutions and individuals who provided genetic material loans: J. Bates, S. Hackett, and D. Willard (Field Museum of Natural History), F. Sheldon, D. Dittmann, and J. Babin (Louisiana State University Museum of Natural Science), E. Bermingham and M. González (Smithsonian Tropical Research Institute), J. Fjeldså and J. Bolding Kristensen (Zoological Museum, University of Copenhagen), and S. Edwards and S. Birks (University of Washington Burke Museum). I. Lovette kindly provided sequences for outgroup taxa. We thank J. Bates and S. Hackett, R. Ricklefs, and J. Klicka for providing the laboratory facilities at the Field Museum of Natural History (Pritzker Laboratory for Molecular Systematics and Evolution), University of Missouri–St. Louis, and Barrick Museum of University of Nevada Las Vegas to complete this project. For laboratory assistance we thank E. Kellog, S. Malcomber, J. Barber, J. Martinez, H. Won, B. Swanson, C. Roy, K. Halbert, M. Ramos, D. Cadena, M. Mika, J. DaCosta, and R. Carson. R. Panza of the Carnegie Museum of Natural History kindly provided access to museum specimens under her care. Funding to complete this project came from a National Science Foundation Dissertation Improvement Grant (DEB-9801524), a Frank M. Chapman Memorial Fund of the American Museum of Natural History, a Sigma Xi Grant-in-Aid of Research, and an International Center for Tropical Ecology Research Fellowship and a Parker-Gentry Tropical Research Fellowship, both from the University of Missouri-St. Louis. P.G.J. thanks the College of Natural Sciences, Seoul National University (grant no. 3344-20080067), the Korean Research Foundation (project no. 0409-20080118, grant no. KRF-2007-412-J03001), and the second stage of the Brain Korea 21 Project 2009. Thanks to J. Miranda for elaboration of the distribution maps. J.P.E. thanks S. Brugada for encouragement and invaluable assistance. M. Morrison, F. James, and an anonymous reviewer provided comments that improved the manuscript. TRACER software is available at tree.bio.ed.ac.uk/software/tracer/.

APPENDIX. List of taxa, tissue numbers, museum collections, and localities of samples used in the study.

Taxon	Tissue no.	Museum[a]	Locality
Myioborus miniatus			
miniatus	FMNH 4647	FMNH	Mexico: Michoacan, Tancitaro
miniatus	FMNH 4478	FMNH	Mexico: Jalisco, Sierra de Manantlan, Puerto Los Mazos
miniatus	FMNH 4477	FMNH	Mexico: Jalisco, Sierra de Manantlan, Puerto Los Mazos
miniatus	FMNH 5628	FMNH	Mexico: Jalisco, Sierra de Manantlan, Puerto Los Mazos
miniatus	FMNH 2185	FMNH	Mexico: Jalisco, Sierra de Manantlan, Puerto Los Mazos
miniatus	FMNH 1217	FMNH	Mexico: Jalisco, Sierra de Manantlan, Puerto Los Mazos
miniatus	FMNH 1209	FMNH	Mexico: Jalisco, Sierra de Manantlan, Puerto Los Mazos
miniatus	JK03-239	MBM	Mexico: Jalisco, Sierra de Bolaños, 5.8 km N, 9 km W of Bolaños
miniatus	GMS581	MBM	Mexico: Jalisco, Sierra de Bolaños, 5.8 km N, 9 km W of Bolaños
molochinus	PEP 2883	UNAM	Mexico: Veracruz, Sierra de Santa Marta, 700 S Santa Marta
molochinus	PEP 2884	UNAM	Mexico: Veracruz, Sierra de Santa Marta, 700 S Santa Marta
molochinus	PEP 2919	UNAM	Mexico: Veracruz, Volcan de San Martin de Tuxtla, 1 km N La Herradura
intermedius	FMNH 1693	FMNH	Mexico: Chiapas, Cerro Tzontehuitz, 7 km NE San Cristobal de las Casas
intermedius	CHIMA 530	UNAM	Mexico: Oaxaca, San Isidro La Gringa, 1 km SE San Francisco La Paz
connectens	DHB3162	MBM	Honduras: Dpto. Copan, Copan Ruinas, 15 km N
connectens	DHB3163	MBM	Honduras: Dpto. Copan, Copan Ruinas, 15 km N
hellmayri	DHB4569	MBM	Guatemala: Dpto. Quezaltenango, Santa María de Jesús 5 km SSW, Finca de Santa María
hellmayri	DHB4593	MBM	Guatemala: Dpto. Quezaltenango, Santa María de Jesús 5 km SSW, Finca de Santa María
hellmayri	KU 5917	KU	El Salvador: Ahuachapan Department
hellmayri	KU 5931	KU	El Salvador: Ahuachapan Department
hellmayri	KU 6009	KU	El Salvador: Sonsonate Department
comptus	B 27304	LSUMNS	Costa Rica: San Jose Province; 0.5 km NNE San Ramon
aurantiacus	B 26421	LSUMNS	Panama: Chiriquí Province, District Gualaca, Cordillera Central
aurantiacus	B 01561	LSUMNS	Panama: Chiriquí Province, District Gualaca, Cordillera Central
aurantiacus	B 05337	LSUMNS	Panama: Chiriquí Province, District Gualaca, Cordillera Central
aurantiacus	B 05475	LSUMNS	Panama: Chiriquí Province, District Gualaca, Cordillera Central
aurantiacus	B 29046	LSUMNS	Panama: Chiriquí Province, District Gualaca, Cordillera Central
aurantiacus	B 28167	LSUMNS	Panama: Chiriquí Province, District Gualaca, Cordillera Central
aurantiacus	B 28175	LSUMNS	Panama: Chiriquí Province, District Gualaca, Cordillera Central
ballux	B-1423	LSUMNS	Panama: Darién Province; ca. 9 km NW Cana on slopes Cerro Pirre
subsimilis	B-178	LSUMNS	Peru: Piura Dept.; km 34 on Olmos-Bagua Chica Hwy
verticalis	O 1541	ZMUC	Ecuador: Zamora-Chinchipe, near Chinapinza
verticalis	B 1694	LSUMNS	Peru: Pasco, Santa Cruz; ca. 9 km SSE Oxapampa
verticalis	B 1696	LSUMNS	Peru: Pasco, Santa Cruz; ca. 9 km SSE Oxapampa
verticalis	B 22788	LSUMNS	Bolivia: La Paz, Province B. Saavedra, 83 km by road E. Charazani
verticalis	B 22797	LSUMNS	Bolivia: La Paz, Province B. Saavedra, 83 km by road E. Charazani
Myioborus pictus	FMNH 2210	FMNH	Mexico: Sinaloa, El Batel
Wilsonia pusilla	47919	UWBM	USA: Washington; Wahkiakum, Cathlamet (JMB696)
W. canadensis	PA-WCA55	STRI	Panama: Colon Province, 2 km W Gatun locks
Ergaticus ruber	BMM-187	FMNH	Mexico: Michoacan; Pico de Tancitaro, 3 km N Zirimondiro
Cardellina rubrifrons	B 10178	LSUMNS	USA: Arizona: Santa Cruz Co.: GRNR CNYN

[a]Abbreviations: FMNH = Field Museum of Natural History; LSUMNS = Louisiana State University Museum of Natural Science; ZMUC = Zoological Museum of the University of Copenhagen; STRI = Smithsonian Tropical Research Institute; UWBM = University of Washington Burke Museum; KU = Kansas University Natural History Museum; MBM = University of Nevada Las Vegas, Barrick Museum of Natural History; and UNAM = Universidad Nacional Autónoma de Mexico.

Ornithological Monographs
Volume (2010), No. 67, 103–113
© The American Ornithologists' Union, 2010.
Printed in USA.

CHAPTER 9

THE SIGNIFICANCE OF SUBSPECIES: A CASE STUDY OF SAGE SPARROWS (EMBERIZIDAE, *AMPHISPIZA BELLI*)

CARLA CICERO[1]

*Museum of Vertebrate Zoology, 3101 Valley Life Sciences Building,
University of California, Berkeley, California 94720, USA*

ABSTRACT.—Subspecies have been viewed as important biological entities that provide evidence of adaptation and early stages of speciation and that stimulate biological research on behavior, ecology, and other non-systematic questions. However, the history of subspecies and the lack of congruence with molecular data have led to questions about whether they help or hinder studies in avian biology and conservation. The Sage Sparrow (*Amphispiza belli*) provides a case study for examining the significance of subspecies. Of the five named subspecies, three breed in the continental United States (*A. b. belli*, *A. b. canescens*, *A. b. nevadensis*) and have been studied and debated for decades regarding their systematic relationships and status. I review this history and summarize our current understanding. In this particular case, subspecies have helped our understanding by alerting researchers to interesting geographic and behavioral patterns that otherwise might have been overlooked.

Key words: *Amphispiza belli*, geographic variation, intergradation, mitochondrial DNA, morphology, postbreeding movements, subspecies.

La Importancia de las Subespecies: Un Estudio de Caso sobre *Amphispiza belli* (Emberizidae)

RESUMEN.—Las subespecies han sido consideradas entidades biológicas importantes en el estudio de adaptaciones y estados tempranos de especiación. Además, su estudio ha estimulado investigaciones no sistemáticas relacionadas con la ecología o etología de los grupos estudiados. Sin embargo, la historia de las subespecies y la incongruencia que existe a veces entre datos moleculares y morfológicos, nos han llevado a preguntarnos si éstas facilitan o dificultan los estudios sobre la biología y la conservación de las aves. *Amphispiza belli* es un buen modelo para examinar la importancia de las subespecies. De las cinco subespecies conocidas, tres se reproducen en el área continental de los Estados Unidos (*A. b. belli*, *A. b. canescens*, *A. b. nevadensis*). Estudios sobre la relación filogenética entre estas subespecies han generado debates durante décadas. En este trabajo hago una revisión bibliográfica y resumo el estado actual de la información disponible. En este caso particular, las subespecies han ayudado al desarrollo de nuestro conocimiento, mostrándonos patrones geográficos y de comportamiento que de otra manera hubieran pasado desapercibidos.

THE SIGNIFICANCE OF subspecies has been hotly debated among ornithologists for decades (e.g., Wiens 1982, Zink 2004, Phillimore and Owens 2006, Rising 2007). In North America, this contentiousness can be attributed to several factors. First, most avian subspecies were described in the late 1800s to early 1900s (Fig. 1), when relatively few specimens and characters were used compared to modern standards. Second, formal subspecies names have been applied to birds that vary "from groups of populations barely discernible on the basis of weak divergence in a single

[1]E-mail: ccicero@berkeley.edu

Ornithological Monographs, Number 67, pages 103–113. ISBN: 978-0-943610-86-3. © 2010 by The American Ornithologists' Union. All rights reserved. Please direct all requests for permission to photocopy or reproduce article content through the University of California Press's Rights and Permissions website, http://www.ucpressjournals.com/reprintInfo.asp. DOI: 10.1525/om.2010.67.1.103.

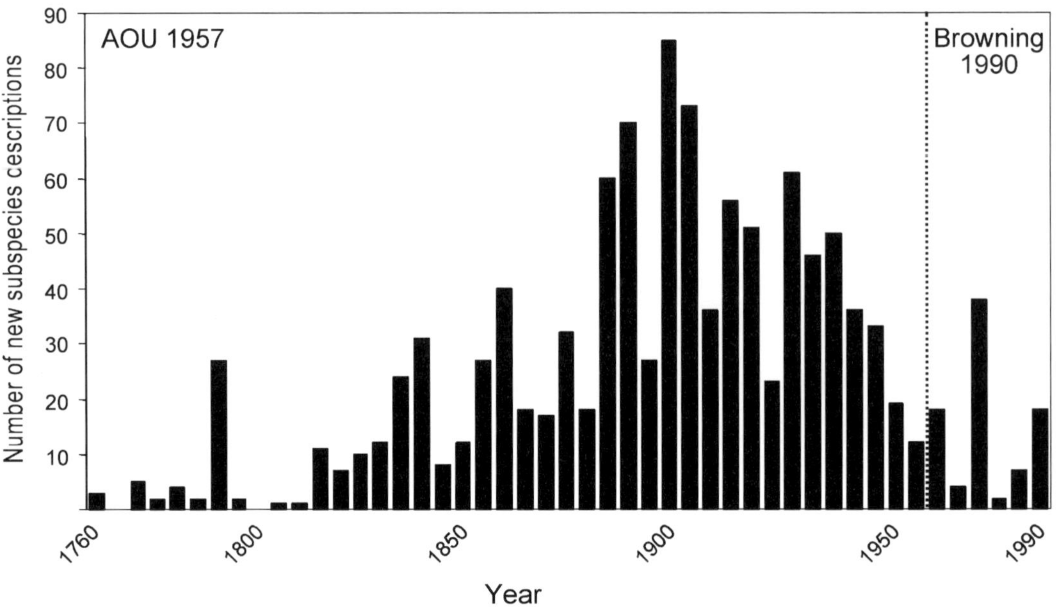

Fig. 1. Dates of descriptions of North American subspecies, in 5-year increments. Data are from American Ornithologists' Union (1957) and Browning (1990).

character to geographic forms that illustrate trenchant differences in morphology, coloration, and voice" (Johnson 1982:605). Third, molecular analyses often conflict with boundaries defined using traditional methods (Ball and Avise 1992, Greenberg et al. 1998, Zink 2004). Although incongruence between genetic data and subspecific characters based on phenotype is not surprising, such results have led to attacks on the concept of subspecies.

In a forum on the value of subspecies, Wiens (1982) posed a series of questions to several prominent American avian systematists to get their personal views on the topic. These questions focused on whether the concept of subspecies is useful and whether it should be revised, how subspecies should be defined, and whether subspecies exist as real biological units. Mayr (1982a:594–595) noted that subspecies "call attention to differences between geographically separated populations ... that might have been overlooked otherwise" and that ornithologists who study ecology or behavior "often find the subspecies designations of the taxonomist useful as to clues to problems that might be studied profitably." Johnson (1982:605) echoed this sentiment when he wrote that subspecies names "function importantly as signposts calling attention to populations of significance for

their research potential." Johnson (1982:605) also noted that "some of these 'subspecies' will turn out after careful study to be full species."

Several examples illustrate the usefulness of subspecies for guiding research. Mennill (2001) studied song variation in two Yellow Warbler subspecies, *Dendroica petechia bryanti* and *D. p. aestiva*, with a goal of determining whether these subspecies, which were characterized by Browning (1994) on the basis of visual characteristics, show similar differences in song characteristics and singing behavior. His findings showed that the two subspecies are completely separable by song and that these differences, combined with geographic and morphological evidence, indicate strong divergence between them. In another study, Valkiūnas and Iezhova (2001) compared hematozoa of three subspecies of Yellow Wagtail (*Motacilla flava*) caught during spring migration to determine whether subspecies varied in their blood parasites, and found differences in the prevalence of infection which they attributed to differences in latitudinal range and breeding habitats used by each host subspecies. Other examples involve studies of differences in migratory routes and wintering areas between subspecies of Swainson's Thrush (*Catharus ustulatus*; Ruegg and Smith 2002) and Sharp-tailed

TABLE 1. Phenotypic, ecological, and behavioral differences among the five subspecies of Sage Sparrow (*Amphispiza belli*).

Subspecies	Size	Color	Primary habitat	Migration
A. b. belli	Small	Dark	Chaparral and coastal sage scrub	Nonmigratory
A. b. cinereus	Small	Pale	Arid and semi-arid scrub	Nonmigratory
A. b. clementeae	Small	Dark	Maritime desert scrub	Nonmigratory
A. b. canescens	Medium	Pale	Saltbush, shadscale desert scrub	Short-distance migrant
A. b. nevadensis	Large	Pale	Great Basin sagebrush	Long-distance migrant

Sparrows (*Ammodramus caudacutus* and *A. nelsoni*; Greenlaw and Woolfenden 2007), which can have important conservation implications (Greenlaw and Woolfenden 2007). In Swainson's Thrush, subspecific differences in migratory pattern are congruent with genetic, ecological, and acoustic divergences, and sharp concordant clines across a narrow hybrid zone provide evidence of barriers to gene flow that may justify recognition as sister species (Ruegg 2007).

Subspecies can be both a driving force and a challenge in evolutionary biology and conservation (Haig and D'Elia, this volume; Winker, this volume). Not surprisingly, close investigation using quantitative criteria may result in the elimination of some, and perhaps many, currently named subspecies. However, if subspecies are defined as phenotypically diagnosable breeding populations (Patten and Unitt 2002, Cicero and Johnson 2006), they can be useful taxonomic units that (1) provide evidence of early stages of allopatric speciation; (2) illustrate local adaptation in spite of ongoing gene flow; (3) alert researchers to differences other than traits originally considered, leading to recognition of some subspecies as full species; and (4) inform researchers about non-breeding movements of distinct portions of species' breeding ranges (Johnson 1982, Mayr 1982a, Rising 2007). Here, I use a case study of the Sage Sparrow (*Amphispiza belli*) to illustrate the value of subspecies in ornithology. Specifically, I review the history and current knowledge of Sage Sparrow systematics, and ask whether subspecies have been useful to researchers studying its biology and evolutionary relationships.

DEBATE OVER SAGE SPARROW SUBSPECIES

The Sage Sparrow provides a suitable case study on the significance of subspecies in ornithology because it shows strong geographic differentiation, has a long history (110 years) of differing interpretations and debate about taxonomic relationships, and is of conservation concern as a result of habitat loss and degradation (Martin and Carlson 1998). Five subspecies are currently recognized (Table 1), with names dating back more than a hundred years: *A. b. belli* (Cassin 1850), *A. b. nevadensis* (Ridgway 1874), *A. b. cinerea* (Townsend 1890), *A. b. clementeae* (Ridgway 1898), and *A. b. canescens* (Grinnell 1905). Two subspecies are listed as federally threatened (*A. b. clementeae*) or of special concern in California (*A. b. belli*), and the species itself is listed as a species of special concern in several western states. I focus on the three subspecies that breed primarily in the continental United States (*A. b. belli*, *A. b. canescens*, and *A. b. nevadensis*; Figs. 2 and 3) because they have received the most systematic study. The two other subspecies (*A. b. cinerea* and *A. b. clementeae*) are resident in west-central Baja California, Mexico, and on San Clemente Island in the California Channel Islands, respectively.

Grinnell (1898b) collected *A. b. belli* and *A. b. "nevadensis"* (currently *A. b. canescens*) together in July 1897 in the mountains of central Los Angeles County, California (1,219–1,829 m elevation). He was surprised to find the two forms breeding in the same locality, and he collected adults and fully fledged young of both forms that showed no evidence of intergradation. On the basis of these observations, he argued that *A. b. belli* and "*A. b. nevadensis*" should be considered specifically distinct. Following this, Fisher (1898) countered that intermediates were collected on the east slope of the Sierra Nevada during the Death Valley Expedition in 1891 and that the birds Grinnell (1898b) collected were fully fledged and "had evidently wandered from their desert home." Thus, he concluded that the two forms were no more than subspecifically distinct.

Subsequent to these early reports, Grinnell (1905) described a new subspecies of Sage Sparrow (*A. b. canescens*) from the higher-elevation

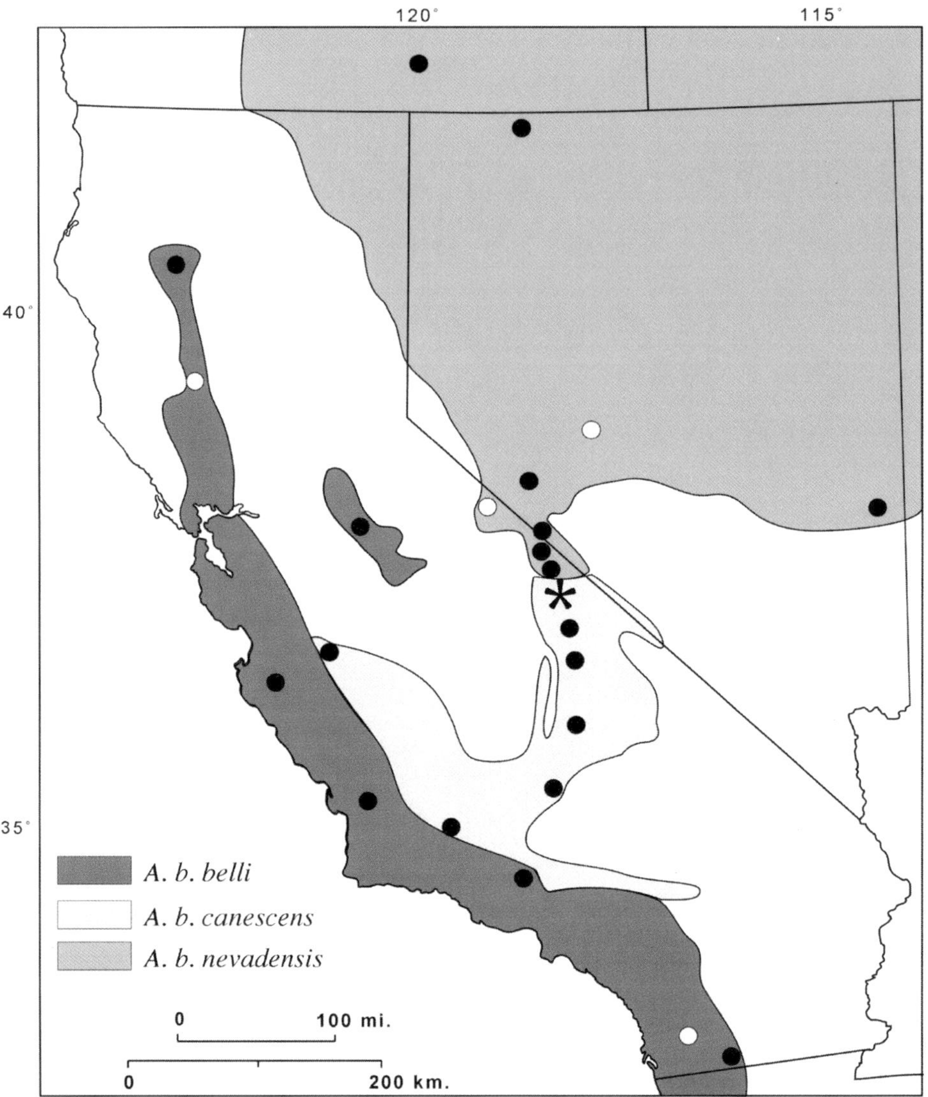

Fig. 2. Breeding distribution of three subspecies of *Amphispiza belli*. Closed circles show sites that have been studied for both allozymes (Johnson and Marten 1992) and mtDNA (Cicero and Johnson 2007, C. Cicero et al. unpubl. data); open circles show sites that were analyzed for allozymes only. Additional sampling and analyses have been done where *A. b. canescens* and *A. b. nevadensis* are in contact in northern Owens Valley, eastern California (shown by asterisk; see Cicero and Johnson 2007).

sage valleys of the southern Sierra Nevada and adjacent mountains of California and referred his earlier specimens from Los Angeles County to this subspecies. Although he described *A. b. canescens* as having characters intermediate between *A. b. belli* and *A. b. nevadensis*, he argued that all three forms were distinctive and that there were no intermediates between *canescens* and *belli* or between *canescens* and *nevadensis*. In his examination of specimens from the 1891 Death Valley Expedition, he definitively assigned them to *canescens* and thus expanded the range of this form farther northward. Although Grinnell (1905:18–19) was convinced that *A. b. canescens–A. b. nevadensis* and *A. b. belli* should be considered species because of their distinctiveness, the apparent lack of

FIG. 3. Dorsal and ventral views of specimens of three of the five subspecies of *Amphispiza belli*. (Top) *A. b. nevadensis*. (Middle) *A. b. canescens*. (Bottom) *A. b. belli*. These forms have received the most attention because of confusion and debate over their relationships. Study skins are in the Museum of Vertebrate Zoology, University of California, Berkeley. (Photographs by Anand Varma.)

intermediates, and the thought that they bred in close proximity in the mountains of Los Angeles County, he stated: "current rulings being overwhelmingly against it . . . it is therefore only under protest that I use the combination *Amphispiza belli canescens* instead of *Amphispiza nevadensis canescens.*"

The debate over how to treat phenotypically diagnosable forms of *A. belli* over a century ago, and their assignment as subspecies rather than species, catalyzed decades of interest and study aimed at understanding the evolutionary history and biology of this species. The extent to which populations move after breeding, as suggested by Fisher (1898), and whether or not different forms intergrade have received particular attention. In their analysis of the distribution of the birds of California, Grinnell and Miller (1944) countered earlier views on the lack of intergradation by reporting that *A. b. belli* and *A. b. canescens* intergrade on or near San Benito Mountain in the interior coastal range of San Benito County, California. They based this on specimens of *A. b. canescens* collected during June–July 1936 and August 1944 (housed in the Museum of Vertebrate Zoology, MVZ 69886–69904, 89797–89840), which they assumed were from breeding grounds. Their assertion that *A. b. belli* and *A. b. canescens* intergrade held for approximately half a century, until Johnson and Marten (1992) reexamined the specimens and emphasized that the birds were in non-breeding condition (i.e., small gonads, molt) and in flocks. Analysis of the series of specimens showed that they are

> typical *A. b. canescens* in fresh, post-breeding plumage acquired after an uphill migration from nesting localities in the adjacent lowlands of the San Joaquin Valley. . . . Because these specimens are in fresh plumage and are therefore more richly colored and slightly darker than worn nesting individuals of typical *A. b. canescens*, Miller evidently viewed the increased pigmentation as evidence of intermediacy with *A. b. belli*. (Johnson and Marten 1992:17)

Thus, Johnson and Marten (1992) provided the first definitive evidence that individual *A. b. canescens* wander from their hot breeding grounds into the range of *A. b. belli* while the latter is still breeding. Their finding uncovered an interesting behavioral pattern that probably would not have emerged if these forms had been treated as species per Grinnell's (1905) original inclination.

Grinnell and Miller (1944) also reported intergradation between *A. b. canescens* and *A. b. nevadensis* in the vicinity of Benton, Mono County, California. In a study of variation in allozymes and morphology, Johnson and Marten (1992) reported that populations at the northern end of the White Mountains, including those at Benton, are typical of *A. b. nevadensis*. Thus, they surmised that if contact and intergradation occurs during the breeding season, it must be somewhere in Owens Valley, eastern California, between what they reported to be the southernmost known *A. b. nevadensis* (Chalfant Valley, Mono County, ~33 km south of Benton) and the northernmost known *A. b. canescens* (Coso Junction, Inyo County). Subsequent study of populations from Benton to Coso Junction (Cicero and Johnson 2007) provided evidence that *A. b. nevadensis* and *A. b. canescens* meet in a narrow zone near Bishop, California, at the northern end of Owens Valley (about 15–20 km south of Chalfant Valley). This contact zone occurs in an area of ecological and bioclimatic transition between the Great Basin (*A. b. nevadensis*) and the Mojave Desert (*A. b. canescens*).

Johnson and Marten's (1992) study provided the first detailed analysis of population structuring in *A. b. belli*, *A. b. canescens*, and *A. b. nevadensis*. Their results showed strong morphological and genetic variation, especially between *A. b. canescens* and *A. b. nevadensis*. Reanalysis of size data combined from Johnson and Marten (1992) and Cicero and Johnson (2007) using discriminant function analysis (present study) supported the strong morphological separation between subspecies (Table 2 and Fig. 4), with the greatest overlap between *A. b. belli* and *A. b. canescens*. Although *A. b. belli* and *A. b. canescens* differ strongly in plumage, the allozyme data showed them to be genetically closely related, with some populations of *A. b. canescens* genetically closer to *A. b. belli* than to other populations of *A. b. canescens* (Johnson and Marten 1992: fig. 8). Because Johnson and Marten (1992) suggested that *A. b. canescens* and *A. b. nevadensis* putatively make contact in Owens Valley, they refrained from recommending taxonomic action that would split the subspecies into different species. Nonetheless, Rising (1996) used this study to treat *A. b. nevadensis* and *A. b. belli* as separate species, although he mistakenly placed *A. b. canescens* in "*A. nevadensis*." The American Ornithologists'

TABLE 2. Percent classification of three subspecies of *Amphispiza belli* based on specimens housed in the Museum of Vertebrate Zoology, Berkeley (*n* = 275 males). Analysis was based on seven linear measurements (wing length, tail length, bill length, bill depth, bill width, length of tarsus plus toe, and cube root of mass). Specimens included those studied by Johnson and Marten (1992) and Cicero and Johnson (2007), with the exception of those in the area of contact between *A. b. canescens* and *A. b. nevadensis* in northern Owens Valley, eastern California.

	A. b. belli	*A. b. canescens*	*A. b. nevadensis*	Percent correct classification
A. b. belli	59	3	0	95.2
A. b. canescens	10	76	3	85.4
A. b. nevadensis	0	2	122	98.4
Total	69	81	125	93.5

Union ([AOU] 1998) currently recognizes them as two groups within *A. belli*: "*A. nevadensis*" and "*A. belli*," with the latter including *A. b. belli*, *A. b. canescens*, *A. b. clementeae*, and *A. b. cinerea*. Both treatments correctly reflect the distinctiveness of *A. b. nevadensis* from the other forms in genotype, phenotype, and ecology, and future revision by the AOU is possible pending publication of additional molecular data (see below).

BREEDING VERSUS NON-BREEDING POPULATIONS IN DELINEATION OF SUBSPECIES

Debate over the taxonomic status of subspecies of *A. belli* has focused largely on *A. b. canescens*, which is geographically and phenotypically intermediate between *A. b. belli* and *A. b. nevadensis* (Tables 1 and 2 and Figs. 2–4). This subspecies is most similar in size to *A. b. belli* (both forms are smaller than *A. b. nevadensis*), but it is most similar in color to *A. b. nevadensis* (both forms are paler than *A. b. belli*). The breeding distribution of *A. b. canescens* lies between that of *A. b. belli* and that of *A. b. nevadensis*, with its center in the western and southern San Joaquin Valley and in the western and northern Mojave Desert. As noted above, *A. b. canescens* has been reported to contact or intergrade with *A. b. belli* in Los Angeles County and in San Benito County (Grinnell 1898b, 1905; Grinnell and Miller 1944), although intergradation has not been established conclusively. Likewise, *A. b. canescens* and *A. b. nevadensis* contact one another

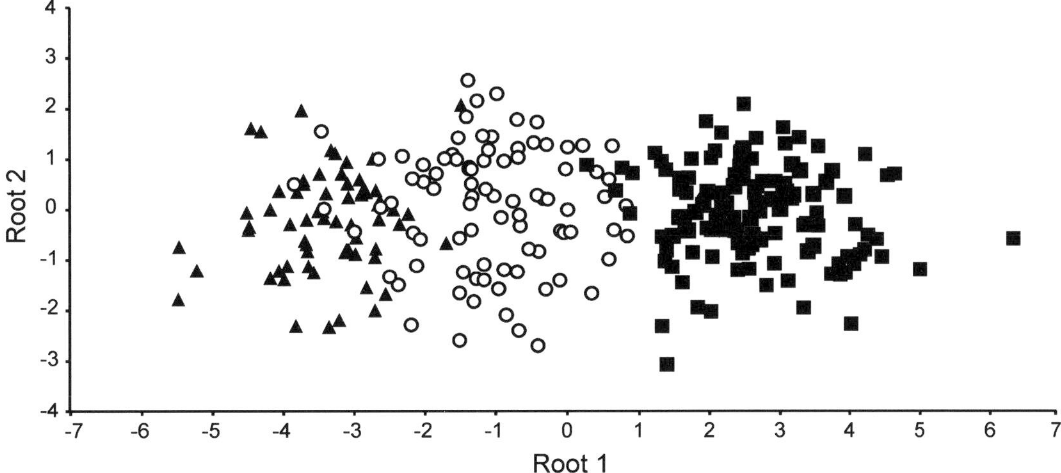

FIG. 4. Discriminant function scores for three subspecies of *Amphispiza belli*. Black triangles = *A. b. belli*, open circles = *A. b. canescens*, and black squares = *A. b. nevadensis*.

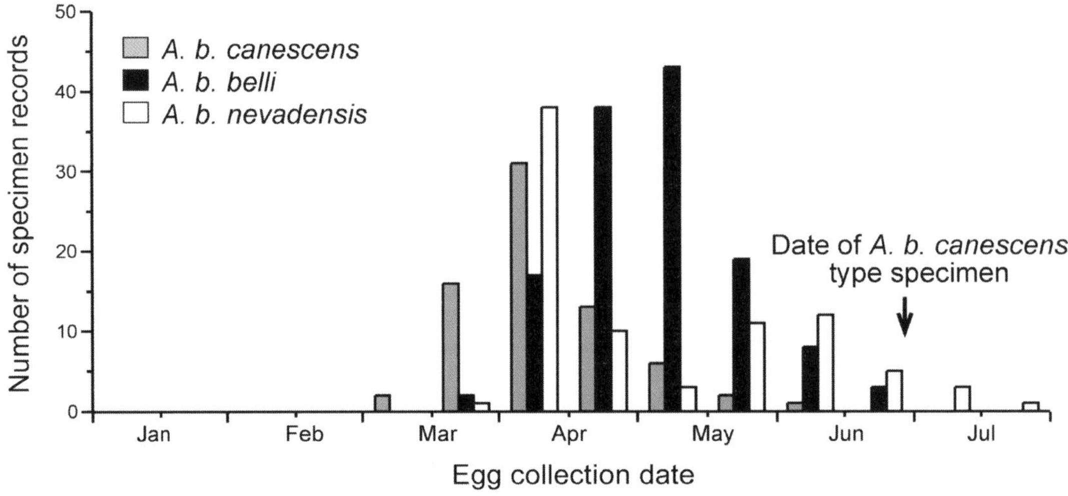

Fig. 5. Egg collection dates for three subspecies of *Amphispiza belli*, from specimen records in the Western Foundation of Vertebrate Zoology and the Museum of Vertebrate Zoology. Data were obtained from ORNIS (ornisnet.org, 1 March 2008) and from copies of egg data slips. Egg data were plotted as Julian dates in 15-day increments. Note the late (postbreeding) date of collection of the holotype of *A. b. canescens*.

narrowly in eastern California (Grinnell and Miller 1944, Johnson and Marten 1992, Cicero and Johnson 2007). The putative intergradation between *A. b. canescens* and *A. b. belli* is confounded by upslope, postbreeding movements of *canescens* into the range of *belli* while the latter is still breeding (Johnson and Marten 1992).

In reviewing the original description of *A. b. canescens* (Grinnell 1905), it is noteworthy that the type specimen (MVZ 35756) was collected at an elevation of 1,676 m on Mount Pinos, Ventura County, California, on 27 June 1904. Grinnell mistakenly assumed that this specimen was on its breeding grounds. Analysis of egg data (present study; Fig. 5) clearly showed that this was a postbreeding bird. Of the three subspecies, *A. b. canescens* breeds the earliest, with egg dates ranging from 14 March to 7 June ($n = 71$, median date = 12 April). By comparison, egg dates for *A. b. nevadensis* ranged from 25 March to 17 July ($n = 84$, median date = 19 April), and those for *A. b. belli* ranged from 25 March to 22 June ($n = 130$, median date = 2 May). Furthermore, the series of birds seen and collected by Grinnell at this locality were "moulting adults and fully fledged young . . . latter in companies in brush on summits of both Pinos and Sawmill peaks" (unpublished field notes in the archives of the Museum of Vertebrate Zoology). Thus, Fisher (1898) was correct in surmising that individuals collected in July

in the mountains of Los Angeles County, where Grinnell (1898b) postulated that *A. b. belli* and *A. b. canescens* bred sympatrically, were wandering (postbreeding) birds. Likewise, the individual *A. b. canescens* that Grinnell and Miller (1944) thought were intergrading with *A. b. belli* on the slopes of San Benito Mountain, San Benito County (see above), were clearly non-breeding individuals, given their dates of collection (13 June–7 August) and the fact that they had small gonads and were molting and in flocks (Johnson and Marten 1992).

The tendency of *A. b. canescens* to move upslope after breeding—not only into the coastal ranges but also into the Sierra Nevada and into the Inyo Mountains of Inyo County, eastern California (Squaw Flat sample of Johnson and Marten 1992:3)—has complicated interpretations and confounded prior studies of geographic variation. In a study of the morphological diagnosability of *A. b. canescens* versus *A. b. nevadensis*, Cicero and Johnson (2006) showed that incorporation of nonbreeding individuals into samples when analyzing morphologic variation (Patten and Unitt 2002) distorted the results and led to incorrect conclusions regarding diagnosability. This behavior, in which birds show regional movements into other habitats after breeding, often during molt, appears to be especially common in western North America (Rohwer et al. 2005). Recent studies that have highlighted this phenomenon (e.g., Rohwer et al. 2008) underscore the importance of paying close attention to breeding individuals when studying geographic variation (Zink and Dittman 1992), particularly when relying on museum specimens. Nevertheless, non-breeding birds may be relevant for subspecies studies in some cases (e.g., *Baeolophus inornatus* and *B. ridgwayi* [Cicero 1996], and *Branta canadensis* and *B. hutchinsii* [Anderson 2007]).

Mitochondrial DNA as a Tool for Subspecies Studies

Mitochondrial DNA (mtDNA) has proved to be an extremely useful marker for delineating evolutionary lineages within species as well as at higher levels (Avise 2004), and thus it has been the tool of choice for phylogeographic studies (Avise et al. 1987, Avise 2000). In a recent and controversial application, mtDNA has been used to develop short barcodes for species and to identify cryptic variation that might signal new species (Hebert et al. 2004, Moritz and Cicero 2004, Clare et al. 2007). Although mtDNA has many advantages, problems with gene trees have caused researchers to question whether mtDNA alone is sufficient for understanding the evolutionary history of species (Edwards and Beerli 2000, Funk and Omland 2003, Zink and Barrowclough 2008). Likewise, lack of congruence between mtDNA patterns and the boundaries of named subspecies has led some researchers to conclude that current subspecies do not reflect biological diversity (Zink 2004). On the other hand, phenotypic variation in the absence of concordant mtDNA variation can provide evidence of strong selection and local adaptation to different ecological conditions, leading to rapid phenotypic evolution (Greenberg et al. 1998, Hoekstra et al. 2005).

Analyses of mtDNA variation in *Amphispiza belli* (Cicero and Johnson 2007, C. Cicero unpubl. data) supported previous genetic data based on allozymes (Johnson and Marten 1992). As with the allozyme results, mtDNA clearly separated *A. b. canescens* from *A. b. nevadensis* and showed that populations of *A. b. canescens* in the San Joaquin Valley (e.g., Panoche Hills and Carrizo Plains; Johnson and Marten 1992: figs. 1 and 8) are genetically closer to coastal *A. b. belli* than to populations of *A. b. canescens* in the Mojave Desert (Fig. 6). Importantly, *A. b. canescens* from the San Joaquin Valley breeds in proximity to *A. b. belli*

but at an earlier date, and moves upslope after breeding into the range of *belli* without intergradation (see above). Thus, both the allozyme and mtDNA data suggest that these populations of *A. b. canescens* share an evolutionary history with *A. b. belli* that is distinct from other *A. b. canescens* and also from *A. b. nevadensis*. If *canescens* had been described originally as a subspecies of "*A. nevadensis*," there is a good chance that this unexpected relationship would still remain hidden. Nonetheless, phenotypic differences readily distinguish *A. b. belli* from *A. b. canescens*, even where their breeding ranges meet parapatrically and where they mix during the breeding and postbreeding seasons, respectively (Johnson and Marten 1992). Variation in plumage color between *A. b. belli* and *A. b. canescens* likely reflects adaptation to local ecological conditions.

A study of the contact between *A. b. canescens* and *A. b. nevadensis* (Cicero and Johnson 2007) also supported previous results from morphology and allozymes (Johnson and Marten 1992). In general, mtDNA showed congruence with morphology in the contact zone and in adjacent populations of both forms. Ecological niche models revealed that the sharp cline in mtDNA and morphology in this region is closely associated with bioclimatic changes that favor one form over the other (Cicero and Johnson 2007).

Have Subspecies Been Useful for Understanding Relationships in *Amphispiza belli*?

Over a century ago, Grinnell (1898b, 1905) argued that coastal and interior populations of Sage Sparrow (*A. b. belli* and *A. b. canescens–A. b. nevadensis*) should be recognized as full species. He based this argument on the (incorrect) perception that they breed sympatrically—or nearly so—in southwestern California and that phenotypically divergent forms do not intergrade where they supposedly make contact. If these taxa had been recognized as species over the past 100 years, it is unlikely that modern studies at the population level would have been undertaken. Because of their subspecies status, a series of detailed studies that began in the mid-1970s with collection of specimens for genetic and morphological analyses (Johnson and Marten 1992; Cicero and Johnson 2006, 2007; C. Cicero unpubl. data) are ongoing. These studies have yielded several important findings to date: (1) the common postbreeding

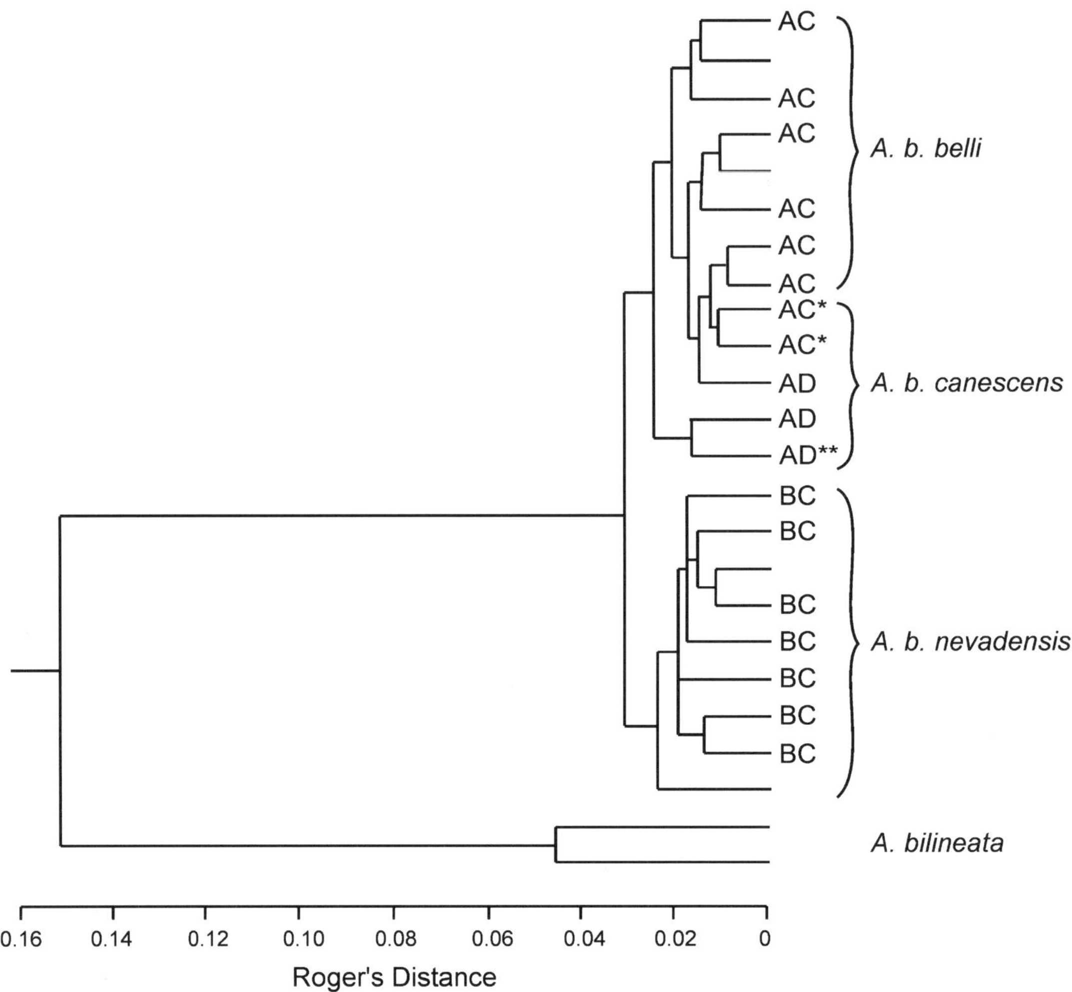

FIG. 6. UPGMA (unweighted pair group method with arithmetic mean) cluster analysis of Rogers's (1972) genetic distances among 22 population samples of *A. belli*, using *A. bilineata* as an outgroup (from Johnson and Marten 1992). The letters at the tips of branches (AC, AD, BC) show major mtDNA haplotype groupings superimposed onto the tree (Cicero and Johnson 2007, C. Cicero et al. unpubl. data) for 18 of the same populations. All are breeding populations except for one (AD**), which is from Squaw Flat, Inyo County, and contains post-breeding individuals of *A. b. canescens* that have moved upslope. The two populations of *A. b. canescens* that are genetically closer to *A. b. belli* than to other *A. b. canescens* (AC*) are also geographically closest (Panoche Hills and Carrizo Plains; see Johnson and Marten 1992: fig. 1).

upslope movement by *A. b. canescens* into the range of breeding *A. b. belli*; (2) the closer genetic relationship of *A. b. canescens* to *A. b. belli* than to *A. b. nevadensis*, contrary to treatments based on similarity in plumage color that placed *A. b. canescens* with *A. b. nevadensis* (Grinnell 1905, Rising 1996); and (3) the geographic position and extent of contact between *A. b. canescens* and *A. b. nevadensis* in Owens Valley, eastern California. Thus, as predicted by Mayr (1982a) and Johnson

(1982), the recognition of *A. b. belli, A. b. canescens,* and *A. b. nevadensis* as subspecies called attention to interesting geographic and behavioral patterns that otherwise might have been overlooked.

Cicero et al. (unpubl. data) have focused on variation in mtDNA sequences, microsatellite loci, and bioclimatic niches across populations of the three subspecies, and on song divergence between *A. b. canescens* and *A. b. nevadensis* where they come into contact in Owens Valley. Vocal differences between *A. b. belli* and *A. b. canescens* also would be worth pursuing, given that the two forms mix in different stages of their annual cycle (breeding and postbreeding, respectively) and that some populations of *A. b. canescens* are more similar genetically to *A. b. belli* than to other *A. b. canescens*. In addition, a comparison between the three continental subspecies studied here and the other two forms—especially *A. b. clementeae*, which is endemic to San Clemente Island and federally threatened—would add valuable information to the overall picture of diversification in the Sage Sparrow complex.

Are Sage Sparrow subspecies worthy of species recognition? Evidence to date suggests that *A. b. canescens* and *A. b. nevadensis* represent different evolutionary units with limited gene flow between them (Cicero and Johnson 2007). These forms are morphologically and genetically diagnosable, and secondary contact is limited to a narrow zone in northern Owens Valley where their major bioclimatic and ecological associations—the Mojave Desert and Great Basin, respectively—meet (Cicero and Johnson 2006, 2007). Although microsatellite and bioacoustic analyses (C. Cicero et al. unpubl. data) within the contact zone should shed additional light on patterns and processes of divergence, *A. b. nevadensis* clearly deserves species status. Whether *A. b. canescens* should be recognized as a species distinct from *A. b. belli* remains to be determined. Further molecular work will hopefully answer this question.

ACKNOWLEDGMENTS

I thank K. Winker and S. Haig for organizing the coordinated session on subspecies and inviting me to participate. F. James, M. Morrison, K. Winker, and an anonymous reviewer provided many useful comments on drafts of the manuscript. M. J. Fernández translated the abstract into Spanish. This paper is dedicated to Joseph Grinnell, Alden H. Miller, and Ned K. Johnson (my academic lineage), who made significant contributions to understanding geographic variation and subspecies in birds and who had a long history of interest in Sage Sparrows.

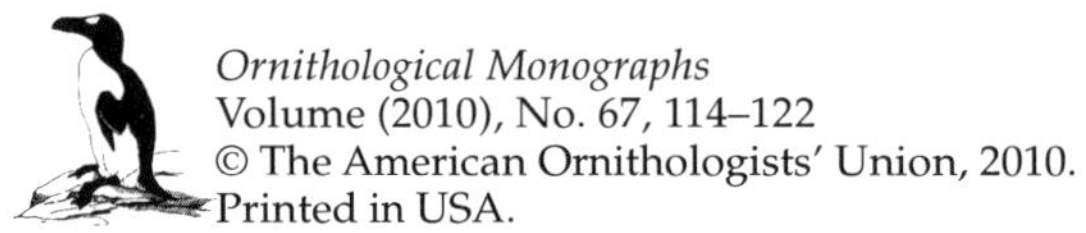

Ornithological Monographs
Volume (2010), No. 67, 114–122
© The American Ornithologists' Union, 2010.
Printed in USA.

CHAPTER 10

RAPID EVOLUTION IN LEKKING GROUSE: IMPLICATIONS FOR TAXONOMIC DEFINITIONS

SARA J. OYLER-MCCANCE,[1,2,3] JUDY ST. JOHN,[2] AND THOMAS W. QUINN[2]

[1]*U.S. Geological Survey, Fort Collins Science Center, 2150 Centre Avenue, Building C, Fort Collins, Colorado 80526, USA; and*
[2]*Rocky Mountain Center for Conservation Genetics and Systematics, Department of Biological Sciences, University of Denver, Denver, Colorado 80208, USA*

ABSTRACT.—Species and subspecies delineations were traditionally defined by morphological and behavioral traits, as well as by plumage characteristics. Molecular genetic data have more recently been used to assess these classifications and, in many cases, to redefine them. The recent practice of utilizing molecular genetic data to examine taxonomic questions has led some to suggest that molecular genetic methods are more appropriate than traditional methods for addressing taxonomic uncertainty and management units. We compared the North American Tetraoninae—which have been defined using plumage, morphology, and behavior—and considered the effects of redefinition using only neutral molecular genetic data (mitochondrial control region and cytochrome oxidase subunit 1). Using the criterion of reciprocal monophyly, we failed to recognize the five species whose mating system is highly polygynous, with males displaying on leks. In lek-breeding species, sexual selection can act to influence morphological and behavioral traits at a rate much faster than can be tracked genetically. Thus, we suggest that at least for lek-breeding species, it is important to recognize the possibility that morphological and behavioral changes may occur at an accelerated rate compared with the processes that led to reciprocal monophyly of putatively neutral genetic markers. Therefore, it is particularly important to consider the possible disconnect between such lines of evidence when making taxonomic revisions and definitions of management units.

Key words: grouse, sexual selection, speciation, species concepts.

Evolución Rápida en los Tetraoninae con Asambleas de Cortejo: Implicaciones para las Definiciones Taxonómicas

RESUMEN.—Las delimitaciones de especies y subespecies han sido tradicionalmente definidas con base en caracteres morfológicos y de comportamiento, como también por características del plumaje. Recientemente también se han usado datos moleculares genéticos para evaluar estas clasificaciones, y en muchos casos, para redefinirlas. La práctica reciente de usar datos moleculares genéticos para responder preguntas taxonómicas ha llevado a algunos a sugerir que estos métodos son más apropiados que los métodos tradicionales para abordar la incertidumbre taxonómica y definir unidades de manejo. Comparamos las especies norteamericanas de Tetraoninae—las cuales han sido definidas utilizando caracteres del plumaje, morfológicos y de comportamiento–y consideramos los efectos de redefinir estas especies usando sólo datos moleculares genéticos neutrales (región control mitocondrial y subunidad 1 de la citocromo oxidasa). Usando el criterio de monofilia recíproca, no fuimos capaces de reconocer las cinco especies que tienen un sistema de apareamiento altamente poligínico, con machos que se exhiben en asambleas de cortejo. En las especies que se reproducen en asambleas de cortejo, la acción de la selección sexual puede

---

[3]E-mail: sara_oyler-mccance@usgs.gov

Ornithological Monographs, Number 67, pages 114–122. ISBN: 978-0-943610-86-3. © 2010 by The American Ornithologists' Union.
All rights reserved. Please direct all requests for permission to photocopy or reproduce article content through the University of California Press's Rights and Permissions website, http://www.ucpressjournals.com/reprintInfo.asp. DOI: 10.1525/om.2010.67.1.114.

influenciar a los caracteres morfológicos y de comportamiento a una tasa mucho más rápida de la que se puede detectar genéticamente. Por esto, sugerimos que, al menos en especies con asambleas de cortejo, es importante reconocer la posibilidad de que los cambios morfológicos y de comportamiento pueden ocurrir a una tasa acelerada en comparación con los procesos que llevan a la monofilia recíproca de los marcadores genéticos presumiblemente neutrales. Por lo tanto, es particularmente importante considerar la posible desconexión entre esas líneas de evidencia al hacer revisiones taxonómicas y definir unidades de manejo.

THE DEBATE AS to how to classify organisms into species has been ongoing for over 150 years (Darwin 1859, Mayr 1942b, Wiley 1978, Cracraft 1983, de Queiroz 1998, Wheeler and Meier 2000). New species concepts are added almost continually (Hey 2001) to address perceived failures of the ones in use, and the debate continues as biologists attempt to place discrete boundaries on a continuous process (Winker et al. 2007). Consensus on how to define units below the species level is even more difficult to achieve, because subspecific boundaries are necessarily even less discrete and more changeable through time. As a result, the utility of the subspecies as a taxonomic rank has been debated widely (e.g., Wilson and Brown 1953, Gill 1982, Mayr 1982a, Storer 1982, Cracraft 1983, Haig et al. 2006, Phillimore and Owens 2006). If correctly delineated, however, intraspecific taxonomic units can be important for conservation efforts because they represent evolutionary capability within a species and likely represent incipient species in some cases (Moritz 1999, Haig et al. 2006). Additionally, such units provide an avenue for protection, at least within North America, where legislation recognizes a range of designations below the species level (Haig et al. 2006; Haig and D'Elia, this volume).

Although subspecies (and species) have traditionally been defined using traits related to plumage, morphology, and behavior, advances in molecular biology have led to the availability of relatively simple genetic markers that measure patterns of genetic variation contained in discrete loci that are presumed to be selectively neutral. In many cases, molecular data sets are not congruent with subspecies defined by traditional methods (Zink 1989, O'Brien and Mayr 1991, Ball and Avise 1992, Burbrink et al. 2000, Zink 2004). Further, Zink (2004) argued that subspecies defined with traditional methods may actually misinform conservation efforts, through misrepresentation of underlying patterns of intraspecific variation. This lack of concordance among approaches has led some to suggest that molecular methods should be used as the primary approach to defining such units for conservation (Moritz 1994, Zink 2004).

More specifically, it has been suggested that the criterion of reciprocal monophyly among mitochondrial sequences (i.e., all members of a group share a more recent common ancestor with one another than with other such monophyletic groups on a phylogenetic tree) should be used to define such units (Moritz 1994, Zink 2004). Others have advocated more inclusive approaches that combine data from plumage, morphology, and behavior with neutral molecular markers (Dizon et al. 1992, Vogler and DeSalle 1994, Haig et al. 2006).

Here, we highlight a situation that illustrates the continuing importance of considering both molecular genetic and more traditional types of data when making inferences about species and subspecies delineations. Specifically, in taxa with highly skewed mating systems that are subject to sexual selection, patterns of variation in neutral molecular genetic markers may not appropriately reflect patterns of genetic variation that underlie traditional characteristics such as plumage, morphology, or behavior that may be subject to strong selection. Within these taxa, using data from neutral molecular markers alone or elevating their significance in relation to other forms of evidence may also misinform conservation efforts.

Many instances of accelerated evolutionary change resulting from natural or sexual selection have been examined (Meyer 1993, Nagel and Schluter 1998, Uy and Borgia 2000, Panhuis et al. 2001, Genner and Turner 2005, Spaulding 2007). Organisms that are subject to strong sexual selection because of highly skewed reproductive success among males can undergo rapid changes in morphology and behavior that can be the driving force in speciation (Ellsworth et al. 1994, Uy and Borgia 2000, Panhuis et al. 2001, Spaulding 2007). Among three lekking species of prairie grouse, Ellsworth et al. (1994) noted a disconnect between strong morphological and behavioral differences and relatively low levels of mitochondrial and nuclear differentiation. They suggested that changes in morphology and behavior in these species occurred more rapidly than usual, compared with rates of change in mitochondrial and nuclear markers (Ellsworth et al. 1994). Thus, taxa with skewed mating systems and strong sexual selection may, as a general rule, accumulate differences in morphology and behavior at a greater rate, in relation to the amount of differentiation of neutral molecular markers, than is typical in species with more balanced mating systems. Consequently, if predetermined amounts or patterns of differentiation in neutral genetic markers are used as a criterion in species or subpecies definitions (such as a requirement for reciprocal monophyly), the magnitude of morphological and behavioral differences separating recently diverged species will differ depending on the natural history—particularly the mating systems—of the organisms under consideration. Therefore, examining only neutral genetic data or elevating the importance of such data over morphological and behavioral characteristics may mislead the conservation of real evolutionary units in these organisms.

The molecular phylogeny of grouse (Tetraoninae) and other galliforms has been studied previously using various mitochondrial and nuclear markers (Gutiérrez et al. 2000, Lucchini et al. 2001, Dimcheff et al. 2002, Drovetski 2002). These studies examined the historical relationship among all Tetraoninae and, in some cases, their placement within Galliformes. In the present study, we used North American Tetraoninae to reexamine phylogenetic relationships with a focus on the role of mating systems. Building on the work of Ellsworth et al. (1994), we investigated the relationship between mating systems and speciation by examining the group of grouse (family Phasianidae, subfamily Tetraoninae) found in North America, which includes a range of morphologically distinct species, widely accepted by taxonomists, with mating systems that vary from monogamous to highly skewed (Wittenberger 1978). Our objective was to overlay taxonomic delineations determined using traditional morphological, behavioral, and geographic methods with molecular genetic data. We examined the level of concordance between data types and determined whether discontinuities were consistent with different mating systems. We hypothesized that in species subjected to strong sexual selection either now or in the recent past, there would be less concordance between traditional and molecular methods than in species without such strong sexual selection.

METHODS

Most previous molecular studies of grouse characterized each species using a single exemplar

for phylogenetic reconstruction (Gutiérrez et al. 2000, Lucchini et al. 2001, Dimcheff et al. 2002). Drovetksi (2002), however, used multiple individuals from each species to reconstruct phylogenies using different genes. In the present study, we chose mitochondrial genes for which multiple exemplars from each taxon could be included. We obtained all published complete mitochondrial control-region sequence for North American grouse species that were available through GenBank, including species with three types of mating systems: monogamous, promiscuous with males dispersed, and highly promiscuous with lekking males (Wittenberger 1978). These three groups of species included Willow, Rock, and White-tailed ptarmigan, all considered monogamous; Ruffed Grouse, "Blue Grouse" (see below), and Spruce Grouse, all considered promiscuous, with males dispersed; and Greater Sage-Grouse, Gunnison Sage-Grouse, Sharp-tailed Grouse, Greater Prairie-Chicken, and Lesser Prairie-Chicken, all considered highly promiscuous, with lekking males (Wittenberger 1978; scientific names of species are given in Table 1). For most of these species, there were only a few complete control-region sequences, and these were used in our analysis. There were 59 sequences for Blue Grouse, so we chose 13 of those sequences loosely representing different geographic locations and spanning the two subspecies that are now recognized as full species, Dusky Grouse (*Dendragapus obscurus*) and Sooty Grouse (*D. fuliginosus*) (Barrowclough et al. 2004). We refer to both these species as "Blue Grouse" (*D. obscurus*) because this is how they were defined originally using morphological characters. Within each Blue Grouse location, we randomly chose one sequence. Additionally, we sequenced the entire control region in five Gunnison Sage-Grouse and an additional seven Greater Sage-Grouse known to represent both clades described by Kahn et al. (1999), because the complete control-region sequences for Greater Sage-Grouse available in GenBank represented only one of two deeply divergent clades.

To amplify the complete mitochondrial control region in Greater and Gunnison sage-grouse, a 25-µL polymerase chain reaction (PCR) was performed with primers 16775L (Quinn 1992) and H595 (Oyler-McCance et al. 2007) using the following thermal profile: preheat at 94°C for 2 min followed by 35 cycles of denature at 94°C for 40 s, anneal at 55°C for 1 min, and extend at 72°C for 4 min. The reactions concluded with a 10-min post-heat at 72°C. The PCR products were prepared for

TABLE 1. Species included in the study, their mating system as defined by Wittenberger (1978), and the GenBank accession numbers of the sequences used in the study.

Latin name	Common name	Mating system	Control-region accession numbers	COI accession numbers
Lagopus muta	Rock Ptarmigan	Monogamous	AF184299, AF532445, AF532447, AF532449, AF532446, AF184294	DQ433739, DQ433738, DQ433737, DQ433736, DQ433734
L. lagopus	Willow Ptarmigan	Monogamous	AF532444, AF532440, AJ297169, AF532443, AF532442, AF532441	DQ433713, DQ433712, DQ433711, DQ433710
L. leucura	White-tailed Ptarmigan	Monogamous	AF532437, AF532439, AJ297167, AJ297168, AF532438	DQ433714, DQ433717, DQ433718, DQ433715, DQ433716
Bonasa umbellus	Ruffed Grouse	Promiscuous, males dispersed	AF532415, AF532416, AF532417, AJ297157	DQ432768, DQ434343, AY666563, AY666214
Dendragapus obscurus	Dusky Grouse ("Blue Grouse" herein; see text)	Promiscuous, males dispersed	AY570309, AY570302, AY570308, AY570318, AY570310, AY570331, AY570356, AY570347, AY570346, AY570352, AY570354, AY570343, AY570332	DQ433565, DQ433564, DQ433563, DQ433562, DQ433561, DQ432884
Falcipennis canadensis	Spruce Grouse	Promiscuous, males dispersed	AF532454, AF532453	DQ432923, DQ433635, DQ433636, DQ433637
Centrocercus urophasianus	Greater Sage-Grouse	Promiscuous, lek breeding	AY569303, AF532424, AJ297158, AJ297159, AF532423, GQ902779, GQ902780, GQ902781, GQ902782, GQ902783, GQ902784, GQ902785	DQ433466, DQ433465, DQ433464, DQ433463, DQ432834, GQ902786, GQ902787, GQ902788, GQ902789
C. minimus	Gunnison Sage-Grouse	Promiscuous, lek breeding	AF532425, GQ902774, GQ902775, GQ902776, GQ902777, GQ90278	DQ432833, DQ432832
Tympanuchus phasianellus	Sharp-tailed Grouse	Promiscuous, lek breeding	AJ297176, AF532436, AF532435, AJ297177, AY569304	DQ434206, DQ434205, DQ434204
T. cupido	Greater Prairie-Chicken	Promiscuous, lek breeding	AY569305, AJ297171, AJ297172, AF532432, AF532431, AF532435	AY666333
T. pallidicinctus	Lesser Prairie-Chicken	Promiscuous, lek breeding	AF532434, AJ297174, AJ297175, AF532433	DQ434203, DQ434202, DQ434201, DQ434200, DQ434199

sequencing by adding 5 U exonuclease I (10 U μL⁻¹, USB, Cleveland, Ohio) and 0.5 U shrimp alkaline phosphatase (1 U μL⁻¹, USB) and incubating at 37°C for 30–45 min. The enzymes were denatured by a 15-min 80°C incubation. Sequencing was performed using 2 μL prepared template and a Quick Start Kit (Beckman Coulter, Fullerton, California) following the manufacturer's protocol except using half reaction volumes (10 μL). Each product was sequenced using five primers to increase accuracy: 521H (Quinn and Wilson 1993), 16775L,

H595, grouse internal CR A (AGTGTCAAGAT-GATTCCCCATAC), and grouse internal CR B (CTCTGGTTCCTCGGTCAG). Sequences were visualized on a CEQ8000 XL DNA Analysis System (Beckman Coulter).

For the aforementioned grouse taxa, we also obtained all published sequences of a portion of the mitochondrial cytochrome-*c* oxidase I gene (COI), also known as the barcoding gene (Hebert et al. 2003). There were fewer published sequences in this region, and all the sequences were published in two studies (Hebert et al. 2003, Kerr et al. 2007). To ensure that the COI sequences for Greater Sage-Grouse represented individuals from both of the deeply divergent control-region clades (Kahn et al. 1999), we sequenced an additional four Greater Sage-Grouse known to be representative of both clades.

To sequence the Greater Sage-Grouse COI gene, 25-µL PCRs were performed using primers Bird F1 and Bird R1 (Kerr et al. 2007) with the following touch-down thermal profile: denature at 94°C for 30 s, anneal at 60°C for 1 min, and extend for 2 min at 72°C; subtract 1°C from the annealing temperature per cycle for 12 cycles; continue for 23 cycles with a 30-s denature at 94°C anneal for 1 min at 45°C and extend at 72°C for 2 min. The reactions concluded with a 20-min post-heat at 72°C. The products were prepared for sequencing and sequenced as above using both Bird F1 and Bird R1 primers.

Sequences from both mitochondrial regions were aligned in SEQUENCHER, version 4.5 (Gene Codes, Ann Arbor, Michigan). Wild Turkey (*Meleagris gallopavo*) control region and COI sequences (AF532414, DQ433016) were used as outgroups. Phylogenetic analyses were performed on both data sets using Bayesian inference within MR-BAYES, version 3.12 (Huelsenbeck and Ronquist 2001, Ronquist and Huelsenbeck 2003). Analysis of aligned sequences from each mitochondrial region was done by running four chains in each of the two independent analyses that MRBAYES executes as a default. The chain heating temperature was set to 0.2. Tree and parameter values were recorded every 100 generations. At the end of the analysis, the first 25% of stored trees were eliminated and the remaining trees were compiled into a consensus tree by the program. For both data sets, 1 million generations were completed, at which time the final convergence diagnostic (average standard deviation of split frequencies) was 0.0073 for the COI data set and 0.0119 for the control-region data set.

To determine whether different analytical methods gave congruent results, phylogenetic analyses of these data were also done with maximum-parsimony analysis using the heuristic search algorithm of PAUP*, version 4.0b10 (Swofford 2003). Maximum trees saved (MaxTrees) was set at 10,000 and random branch swapping was done using tree bisection–reconnection (TBR). Gaps were scored as a fifth base. A consensus of 1 million bootstrap replicates was used for the final tree.

RESULTS

DNA sequence alignments were straightforward across COI, with 73% of sites (414 of 566) completely conserved. For the control-region sequences, 67% of sites (788 of 1,177) were completely conserved. This is consistent with the relative ease of alignments using coding regions (i.e., COI) as compared with those using non-coding regions (i.e., control region).

Phylogenetic analyses (Bayesian and maximum parsimony) of the control-region sequence were concordant with taxonomic delineations defined using traditional methods in all non-lekking grouse. All formed well-supported reciprocally monophyletic groups, each with a posterior probability of 100% (Bayesian) and a bootstrap value of 100 (maximum parsimony). Blue Grouse formed two reciprocally monophyletic clades corresponding to the split described by Barrowclough et al. (2004) that ultimately led to the recent elevation of these two groups to full species status. The five taxa that exhibit lekking behavior did not form reciprocally monophyletic groups with either Bayesian analysis (Fig. 1) or maximum-parsimony analysis (not shown). Kahn et al. (1999) reported that sequence data from the control region yield two deeply divergent clades within Greater Sage-Grouse. Gunnison Sage-Grouse fell into one of those two distinct Greater Sage-Grouse clades; thus, some Greater Sage-Grouse are more closely related in mitochondrial DNA (mtDNA) to Gunnison Sage-Grouse than they are to members of their own species. The remaining three lekking grouse (Sharp-tailed Grouse, Lesser Prairie-Chicken, and Greater Prairie-Chicken) were even less well resolved and did not form reciprocally monophyletic groups (Fig. 1). Maximum-parsimony analysis of the same data yielded a bootstrap consensus tree with the same key features described above. However, among deeper topological features, there was no support (bootstrap < 50%) for placing the *D. obscurus* and *Tympanuchus* complex as sister clades, nor was there support for the deeper clade that includes those two plus the *Centrocercus* group.

Analysis of the COI region revealed that all non-lekking grouse formed well-supported reciprocally monophyletic clades that match prior species designations with posterior probabilities of 100 (Fig. 2, Bayesian) and bootstrap values ≥97. Among the lekking grouse, none formed reciprocally monophyletic groups despite there

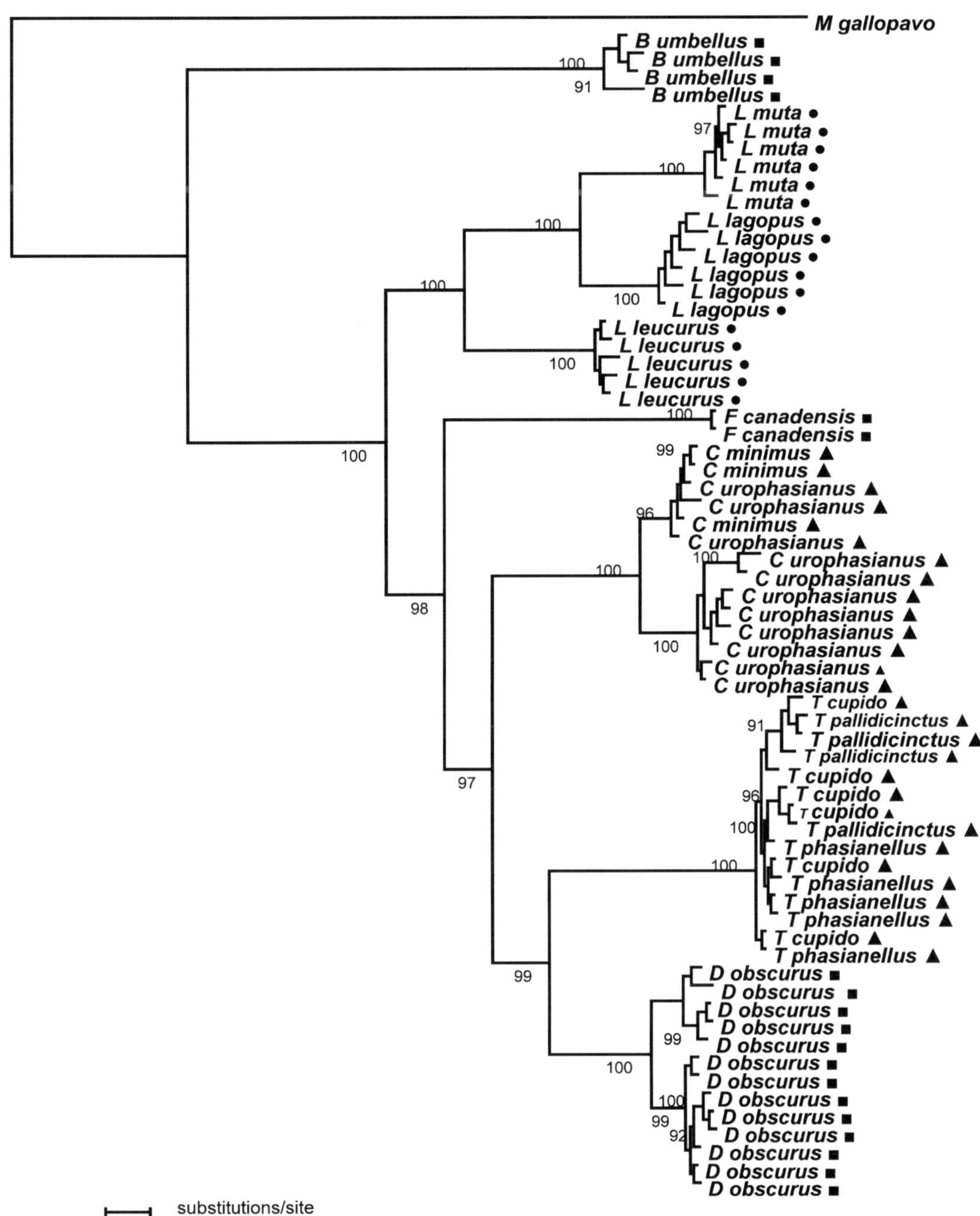

FIG. 1. Phylogenetic tree based on Bayesian analysis of the control region. Circles represent species with monogamous mating systems, squares represent promiscuous species with dispersed males, and triangles represent promiscuous species with a lek mating system. The numbers are posterior probabilities as calculated by Bayesian analysis.

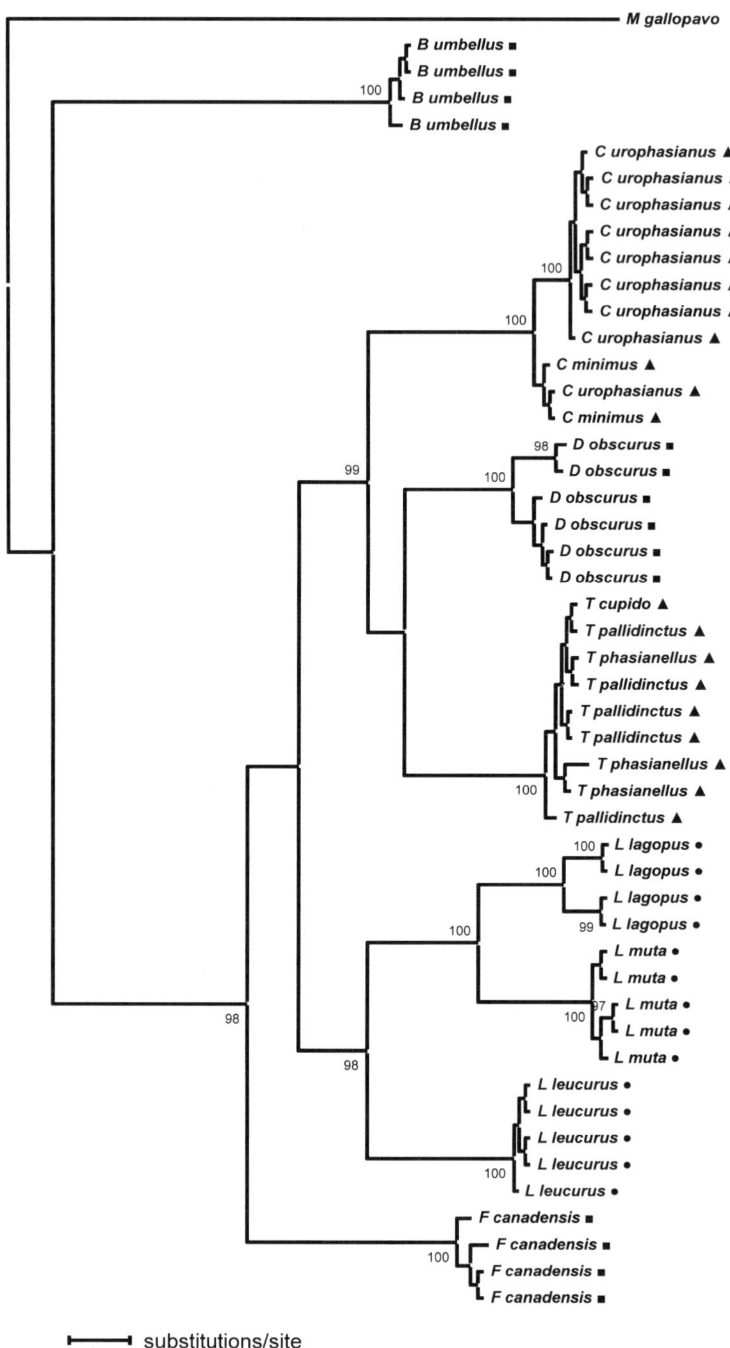

Fig. 2. Phylogenetic tree based on Bayesian analysis of the mitochondrial cytochrome-*c* oxidase I gene (COI). Circles represent species with monogamous mating systems, squares represent promiscuous species with dispersed males, and triangles represent promiscuous species with a lek mating system. The numbers are posterior probabilities as calculated by Bayesian analysis.

being more than one defined species in each case. Gunnison Sage-Grouse haplotypes were allied with one of the two deep clades of Greater Sage-Grouse (Fig. 2). Similar to the results for the control-region, Lesser Prairie-Chicken, Greater Prairie-Chicken, and Sharp-tailed Grouse were all intermixed within a single clade. Maximum-parsimony analysis of the same data revealed a similar topology but did not place *L. leucurus* as a sister group to the *L. muta* + *L. lagopus* clade.

Discussion

Our findings support the hypothesis that in lekking species subjected to strong sexual selection either now or in the recent past, there was less concordance between traditional and molecular methods than in species without such strong sexual selection. Like Drovetski (2002), we found that all non-lekking taxa formed monophyletic groups. And like Ellsworth et al. (1994) and Drovetski (2002), we found that none of the three taxa within the *Tympanuchus* group were reciprocally monophyletic. Ellsworth et al. (1994) and Drovetski (2002) suggested that speciation within this group is very recent. Johnson (2008) proposed that this group experienced rapid diversification in the late Pleistocene (10,000–18,000 years ago), which resulted in species with little or no interchange since divergence. Our data expand upon those from Drovetski (2002) by including additional samples from Greater Sage-Grouse and multiple samples from Gunnison Sage-Grouse. Drovetski (2002) showed Gunnison Sage-Grouse and Greater Sage-Grouse as sister groups exhibiting reciprocal monophyly. By including additional samples in the present study, we detected both of the deep clades present in the Greater Sage-Grouse (Kahn et al. 1999), rather than the single clade represented in previous phylogenetic studies (Drovetski 2002). Gunnison Sage-Grouse fell within one of those clades (Figs. 1 and 2), as previously recognized. Thus, Greater Sage-Grouse and Gunnison Sage-Grouse lack reciprocal monophyly, which is consistent with the pattern that we see in the other lekking species in North America. We suggest that speciation within this *Centrocercus* group is probably recent, like that within the *Tympanuchus* group.

Within North American grouse, non-lekking species are reciprocally monophyletic for neutral molecular markers. By contrast, among lekking species, taxonomic boundaries based on observations of plumage, morphology, and behavior are not reflected by similar diagnostically consistent characters at the molecular level. In the lek mating system, in which a few males do most of the mating, sexual selection can act to influence morphological and behavioral traits (Spaulding 2007) at a rate much faster than can be tracked using neutral genetic markers (Ellsworth et al. 1994). In some cases, reciprocal monophyly may appear long after complete and irreversible isolating mechanisms are in place. Further, the time that it takes to reach reciprocal monophyly in mitochondria depends on multiple factors, such as the effective population size of females (Avise and Wollenberg 1997).

Although most of the analysis presented here has focused on the species level, there are obvious implications for subspecies delineations as well. Species are more reasonably expected to be reciprocally monophyletic than subspecies. Here, we have shown that even at the species level, lekking grouse are not reciprocally monophyletic, and, thus, we should not expect such a relationship at the subspecies level. Lesser and Greater Prairie-Chickens exhibit distinct differences in behavior, plumage, morphology, habitat affiliation, and social aggregation that led to their recognition as distinct species (Grange 1940, Jones 1964, Sharpe 1968, Johnsgard 2002), although some have considered them subspecies in the past (Short 1967, Johnsgard 1983). There are currently no defined subspecies of Lesser Prairie-Chicken, and Greater Prairie-Chickens are divided into two subspecies (one being the endangered Attwater's Prairie-Chicken, *T. c. attwateri*). Compared to Lesser and Greater prairie-chickens, Sharp-tailed Grouse are even more distinct, particularly in morphology (Johnsgard 2002), and were at one time considered a distinct monotypic genus (Ellsworth et al. 1994). Seven subspecies of Sharp-tailed Grouse have been identified, primarily on the basis of subtle morphological differences and geographic distribution (Connelly et al. 1998). Like Ellsworth et al. (1994), Drovetski (2002), and Johnson (2008), we suspect that speciation in *Tympanuchus* is recent. Additionally, Greater Sage-Grouse and Gunnison Sage-Grouse exhibit distinct morphological, plumage, and behavioral characteristics (Hupp and Braun 1991; Young 1994; Young et al. 1994, 2000) and appear to be reproductively isolated (Young 1994, Young et al. 1994, Kahn et al. 1999, Oyler-McCance et al. 1999), which suggests that speciation within this group is recent as well. There are no recognized subspecies of Gunnison Sage-Grouse, whereas Greater Sage-Grouse were previously split into two subspecies (Eastern and Western), although the validity of this division has been questioned (Benedict et al. 2003).

Our data are consistent with the hypothesis that the strong force of sexual selection driving rapid changes in morphology, plumage, and behavior has led to rapid reproductive isolation and speciation within these lekking taxa. As such, there may not have been sufficient time to reach reciprocal monophyly even at the species level in these groups. Thus, if one were to examine only data from neutral genetic markers among these taxa, important evolutionary processes would be overlooked. Identification of subspecies within these lekking taxa is likely to require an especially astute analysis of plumage, morphology, and behavior and may be misled in cases where genetic data alone are considered. Further, the conservation of these subspecies is vital because they may ultimately represent incipient species in a time-frame shorter than that experienced by non-lekking taxa. Because such processes are imperative for conservation efforts, we think that a pluralistic approach involving morphological, behavioral, and genetic data should be used, particularly when assessing taxa with highly skewed mating systems, such as lekking grouse.

Acknowledgments

The use of any trade, product, or firm names is for descriptive purposes only and does not imply endorsement by the U.S. Government. We thank C. E. Braun, C. L. Aldridge, B. C. Fedy, and three anonymous reviewers for helpful comments on the manuscript.

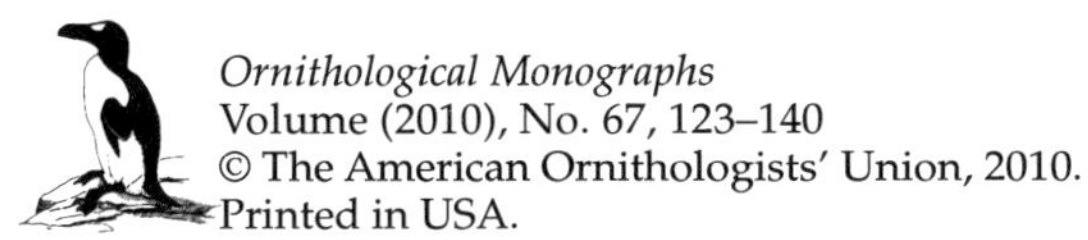

Ornithological Monographs
Volume (2010), No. 67, 123–140
© The American Ornithologists' Union, 2010.
Printed in USA.

CHAPTER 11

QUANTIFYING SUBSPECIES ANALYSIS: A CASE STUDY OF MORPHOMETRIC VARIATION AND SUBSPECIES IN THE WOODCREEPER GENUS *DENDROCOLAPTES*

CURTIS A. MARANTZ[1,4] AND MICHAEL A. PATTEN[2,3]

[1]*Museum of Natural Science, Louisiana State University, Baton Rouge, Louisiana 70803, USA;*
[2]*Oklahoma Biological Survey and Department of Zoology, University of Oklahoma, Norman, Oklahoma 73019, USA; and*
[3]*San Diego Natural History Museum, San Diego, California 92112, USA*

ABSTRACT.—Many authors have criticized the use of subspecies, but most of this criticism has been directed at the inconsistent treatment of subspecies rather than the inutility of diagnosable populations. To assess the validity of a taxon, one must include in the analysis those characters used in the original diagnosis and remember that different character sets may lack geographic concordance. We examined morphometric variation using 3,027 specimens representing all five species and 30 subspecies in the woodcreeper genus *Dendrocolaptes* (Dendrocolaptinae). Most subspecies in the genus differ in plumage patterns and coloration, but a few taxa were described using characters of size and structure. We sought to assess quantitatively, on the basis of the *D*-statistic (Patten and Unitt 2002), those subspecies described using mensural characters, to quantify morphological variation, and to examine the influence of ecological correlates in the genus. Males average slightly larger than females but have a less massive bill. *Dendrocolaptes certhia* has a wider bill than *D. sanctithomae*, and *D. picumnus puncticollis* has a slimmer bill than other *D. picumnus*. Although previously considered a subspecies-group within *D. certhia*, *D. sanctithomae* has a shorter and slimmer bill that appears to reflect a greater dependence on foraging over army ants; these species also differ vocally and in plumage. Bill length in Middle American populations of *D. sanctithomae* varies as a smooth cline and, when combined with weak and potentially clinal variation in plumage and bill coloration documented in an earlier study of plumage variation, our data failed to support the recognition of *D. s. nigrirostris* and *D. s. colombianus*. Amazonian, montane, and Chaco representatives of *D. picumnus* differ structurally, but subspecies differ little morphometrically within each region. *Dendrocolaptes hoffmannsi* broadly overlaps Amazonian *D. picumnus* morphometrically, but its plumage patterns are distinctive. The only subspecies in this complex described exclusively on the basis of mensural characters, *D. p. casaresi*, appears to be slightly larger than *D. p. pallescens*, but we had too few specimens to assess diagnosability. We conclude that reanalysis of described subspecies using quantitative, statistical methods will provide a clearer starting point for studies of biogeography, migration, and other aspects of evolutionary biology than will subspecies based solely on qualitative judgment.

Key words: *Dendrocolaptes*, *D*-statistic, ecological correlates, morphometrics, quantitative analysis, subspecies diagnosis, woodcreepers.

[4]Present address: Macaulay Library, Cornell Laboratory of Ornithology, 159 Sapsucker Woods Road, Ithaca, New York 14850, USA. E-mail: cam233@cornell.edu

Ornithological Monographs, Number 67, pages 123–140. ISBN: 978-0-943610-86-3. © 2010 by The American Ornithologists' Union.

Cuantificación en los Análisis de Subespecies: un Estudio de Caso de la Variación Morfométrica y las Subespecies en el Género *Dendrocolaptes*

RESUMEN.—Muchos autores han criticado el uso de subepecies, pero la mayor parte de esas críticas se ha dirigido más al tratamiento poco consistente de las subespecies que a la inutilidad de las poblaciones diagnosticables. Para determinar la validez de un taxón, uno debe incluir en el análisis aquellos caracteres utilizados en el diagnóstico original y recordar que diferentes conjuntos de caracteres pueden carecer de concordancia geográfica. Examinamos la variación morfométrica en 3027 especímenes representativos de las cinco especies y 30 subespecies del género *Dendrocolaptes* (Dendrocolaptinae). La mayoría de las subespecies difieren en sus patrones de plumaje y coloración, aunque unos pocos taxones fueron descritos utilizando caracteres de tamaño y estructura. Nuestros objetivos fueron evaluar cuantitativamente usando el estadístico *D* (Patten y Unitt 2002) a aquellas subespecies que fueron descritas usando caracteres mensurativos, cuantificar la variación morfológica y examinar la influencia de correlatos ecológicos en el género. Los machos fueron en promedio ligeramente más grandes que las hembras, pero presentaron picos más pequeños. *Dendrocolaptes certhia* tiene un pico más ancho que *D. sanctithomae* y *D. picumnus puncticollis* tiene un pico más delgado que el de otras formas de *D. picumnus*. A pesar de que anteriormente se consideraba un grupo de subespecies dentro de *D. certhia*, *D. sanctithomae* tiene un pico más corto y delgado que parece reflejar su gran dependencia en seguir ejércitos de hormigas para forrajear. Estas especies también se diferencian en sus vocalizaciones y plumaje. La longitud del pico en las poblaciones mesoamericanas de *D. sanctithomae* varía como una clina suave. Combinando esa variación con la variación débil y posiblemente clinal del plumaje y la coloración del pico documentada en un estudio anterior, nuestros datos no apoyan el reconocimiento de las subespecies *D. s. nigrirostris* y *D. s. colombianus*. Los representantes de *D. picumnus* de las regiones amazónicas, montanas y del Chaco difieren estructuralmente pero las subespecies de una misma región difieren poco morfométricamente. La única subespecie en este complejo que fue descrita exclusivamente con base en caracteres mensurativos, *D. p. casaresi*, parece ser levemente más grande que *D. p. pallescens*, pero contamos con muy pocos especímenes como para hacer un diagnóstico. Concluimos que el re-análisis de las subespecies descritas usando métodos estadísticos cuantitativos brindará puntos de partida más claros para los estudios de biogeografía, migración y otros aspectos de biología evolutiva que las subespecies basadas sólo en juicios cualitativos.

A SUBSPECIES IS a collection of one or more populations that occupy a distinct breeding range and that are diagnosable from other such populations (Mayr and Ashlock 1991). Well-defined subspecies play a key role in studies of geographic variation, and they are useful for studying migration, adaptation, and speciation. Indeed, in evolutionary studies it is important to distinguish primary differentiation from secondary contact with hybridization, something that may be best accomplished using a biological species concept with subspecies (Johnson et al. 1999). Still, the use of subspecies has been criticized by many authors (e.g., Wilson and Brown 1953, Selander 1971, McKitrick and Zink 1988, Zink 2004). Most such criticism appears to stem from the uneven application of the concept, which has led to the recognition of many invalid taxa, rather than from the potential importance of well-defined and readily diagnosable subspecies (Parkes 1982; Patten and Unitt 2002; Patten, this volume; Remsen, this volume). Moreover, many critics of subspecies fail to acknowledge that most such taxa were described before the advent of modern statistics and have never been examined using the numerous sophisticated techniques now applied in studies of geographic variation (Remsen 2005, this volume). Patten and Unitt (2002) emphasized that the diagnosability of populations, as opposed to mean differences between them, is the key to effectively defining subspecies. The key to diagnosability is the degree of overlap allowed between populations and not the statistical difference between the means of populations that may overlap extensively (Patten and Unitt 2002; Patten, this volume; Remsen, this volume), although authors have varied on the percentage of individuals in a given population that must differ from those in all other such populations (usually varying in the range of 75–100%; Amadon 1949, Marshall 1967, Mayr 1969, Amadon and Short 1992).

There is some disagreement about the degree of concordance expected among multiple character sets at various taxonomic levels. Wilson and

Brown (1953) argued that empirical studies contradicted the assumption of a coadapted system in which multiple character sets vary geographically in a coordinated manner within species, and they went on to say that populations that show a high degree of geographic concordance in multiple character sets are probably specifically distinct. More recently, however, Avise and Ball (1990) stressed that subspecies should be recognized only when there are concordant patterns of geographic variation in multiple, independent, genetically based characters. Even these authors noted, however, that natural selection can result in some character sets showing geographic patterns contrary to the main phylogenetic signal. In either case, the only logical place to begin the assessment of a subspecies is to examine characters originally used to describe the taxon in question, because even though new character sets may support the recognition of a particular taxon, one cannot discount taxa without examining the characters used in their original diagnosis.

Many avian subspecies have been described on the basis of plumage characters; however, there are also many taxa based on size or shape differences that are best assessed using mensural characters. Plumage coloration and patterns used to define subspecies may have arisen from adaptations for crypsis or from sexual selection; however, morphometric characters likely reflect adaptations for locomotion or foraging (Bock 1966, Karr and James 1975, Fitzpatrick 1985, Leisler and Winkler 1985) or they result from adaptations of body size to climatic variation (James 1970). Although the results of morphometric studies may be dominated by relationships of size, with the first principal component (PC1) of such analyses sometimes treated as a general size factor (see Zink and Remsen 1986, Rising and Somers 1989), shape differences are likely to provide better clues to differences in locomotory ability or foraging behavior (Karr and James 1975, Fitzpatrick 1985, Leisler and Winkler 1985). In birds, wing structure is adapted primarily for flight and the legs for walking or perching. The tail seems to be somewhat less tied to flight than the wings, and in some species this has allowed some modifications of the tail through sexual selection in addition to, or even instead of, adaptation for various forms of locomotion. Bill shape, among the most variable of structural features in birds, reflects adaptation to foraging and consuming a variety of food items (Bock

1966, Fitzpatrick 1985). One may therefore expect closely related species that differ in locomotory behavior (including migration) or foraging behavior to have concomitant differences in body size and shape.

The woodcreepers are a monophyletic group of ~50 species (Marantz et al. 2003) presently recognized as a subfamily within a large Neotropical radiation of the Furnariidae (Irestedt et al. 2002, Remsen et al. 2009). Woodcreepers represent a remarkably cohesive group with respect to both morphology and behavior, but the species differ conspicuously in size, bill shape, and foraging behavior (Marantz et al. 2003). Although most woodcreepers forage by gleaning and probing, some species, especially those that take prey primarily in association with swarms of army ants, forage more by sallying than by probing (Willis 1972, 1982, 1992; Marantz et al. 2003). The five members of the genus *Dendrocolaptes* are relatively large, slim woodcreepers with short bills that are dorsoventrally compressed to varying degrees (Marantz 1992, Marantz et al. 2003). Observations of foraging by these birds have further revealed the importance of sallying, and this is especially true of those species that forage primarily in association with army ants (especially *Eciton burchelli*; Willis 1982, 1992).

Species in the genus *Dendrocolaptes* are best treated as representatives of two groups defined on the basis of plumage patterns (see Marantz 1997). Taxa in the *D. certhia* complex (including *D. sanctithomae*) have the head and body barred extensively (the barred woodcreepers), but those in the *D. picumnus* complex (including *D. platyrostris* and *D. hoffmannsi*) have the head, neck, and breast streaked to a varying degree (the streaked birds). All *Dendrocolaptes* show some indication of barring on the belly. Two poorly marked forms were assigned to their respective groups on the basis of faint barring on the head and body (*D. certhia concolor*) or fine streaking on the breast (*D. hoffmannsi*). Within *D. picumnus*, subspecies-groups were recognized on the basis of geography (Amazonian, Chaco, and montane), and the montane taxa were further subdivided on the basis of plumage similarity. Montane subspecies that occur in the central portion of the range (group A) have the underparts less extensively streaked on a weakly barred background, but peripheral populations (group B) have more extensive streaks on an essentially unmarked background (see Marantz 1997).

Our study had two primary goals. First, using a sample combining the holdings of most museums in the New World that house large numbers of woodcreeper skins, we sought to assess the validity of taxa that were described on the basis of morphometric characters. Although most taxa in this complex were defined on the basis of plumage characters (see Marantz 1997), a few subspecies in the genus were described largely or exclusively on the basis of mensural characters. In particular, two Middle American subspecies of *D. sanctithomae* (*D. s. nigrirostris* and *D. s. colombianus*) were defined primarily on the basis of differences in bill length (Todd 1950). One additional subspecies in *Dendrocolaptes* was described exclusively on the basis of mensural differences: *D. picumnus casaresi* (Steullet and Deautier 1950), but sample sizes were small. Taken together with an earlier study of plumage variation in this complex (Marantz 1997), our morphometric analysis provides an assessment of all taxa presently recognized in this genus. Our secondary goal was to examine overall patterns of morphometric variation in all five species in the genus *Dendrocolaptes* and to determine whether differences in foraging behavior noted by Willis (1982, 1992) correspond to discrete structural differences, particularly those in bill shape. We set out to demonstrate that quantitative analysis of patterns of geographic variation provides a sounder baseline for studies of biogeography, migration, and other aspects of evolutionary biology than will studies based on qualitative fractionation of that variation.

METHODS

MORPHOMETRIC DATA

Marantz measured 3,027 specimens of the five species in the genus *Dendrocolaptes* (Table 1 and Fig. 1; see Acknowledgments). Our sample included specimens representing all 30 subspecies either listed in Peters (1951) or described since that publication, and all 23 subspecies recognized by Marantz et al. (2003). Species-level taxonomy follows Marantz et al. (2003).

Using either dial or digital calipers, we measured, to the nearest 0.1 mm, nine characters on each specimen. In addition to standard measurements of wing chord, tail length, and tarsus (tarsometatarsus) length (Baldwin et al. 1931), we

TABLE 1. Species of *Dendrocolaptes* used in the present study and, for each, the plumage group, sample size used in morphometric analyses, and tendency to forage in association with army ants (the latter based on Willis 1982, 1992). *Dendrocolaptes picumnus* was further divided into four subspecies-groups on the basis of geography and plumage patterns.

Species	Plumage group	n	Foraging behavior
D. sanctithomae	Barred	463	Army ants
D. certhia	Barred	755	Generalized
D. picumnus[a]	Streaked	560	Army ants[b]
D. hoffmannsi	Finely streaked	26	Army ants
D. platyrostris	Streaked	738	Generalized

[a] *D. picumnus* subspecies-groups (and their sample sizes) are as follows. Amazonian (*n* = 207): *D. p. picumnus*, *D. p. validus*, *D. p. transfasciatus*. Chaco (*n* = 73): *D. p. pallescens*, *D. p. extimus*, *D. p. casaresi*. Montane (group A; *n* = 168): *D. p. costaricensis*, *D. p. veraguensis*, *D. p. multistrigatus*. Montane (group B; *n* = 122): *D. p. puncticollis*, *D. p. seilerni*, *D. p. olivaceus*.
[b] Amazonian *D. picumnus* rely extensively on foraging in association with army ants, but it is likely that army ants are less important to montane and Chaco populations, which occur at elevations or in a region, respectively, in which army ants are probably an unreliable resource.

obtained six bill measurements. Visual inspection from above revealed that the bills of most woodcreeper species are compressed laterally, and therefore are quite narrow, whereas those of *Dendrocolaptes* are generally broad and bowed outward along their edges. To quantify this expansion, we measured three sets of bill characters. The width and depth of the bill were measured both at the anterior edge of the nares (bill width and depth at the nares) and at the anterior feathering along the culmen (bill width and depth at the base). Bill length was measured both from the anterior edge of the nares (bill length) and from the insertion point at the skull (total culmen). Body mass was recorded when reported on the label, and we noted molt, wear, or damage that could have affected our measurements.

MEASUREMENT ERROR

The variability among repeated measures of the same character on a given individual should be considered whenever mensural characters are studied, because measurement error can affect subsequent morphometric analyses (Francis

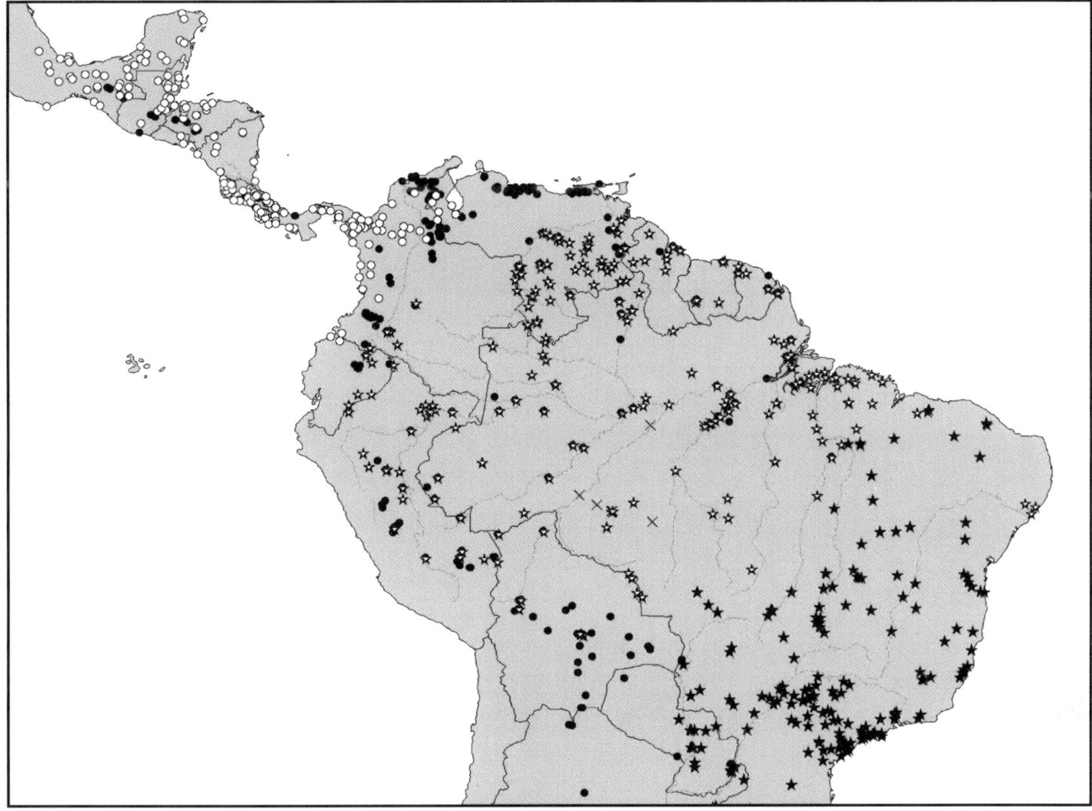

Fig. 1. Geographic sampling of *Dendrocolaptes* specimens used in this study (white circles are *D. sanctithomae*, $n = 553$; black circles are *D. picumnus*, $n = 714$; open stars are *D. certhia*, $n = 881$; black stars are *D. platyrostris*, $n = 851$; multiplication symbols [×] are *D. hoffmannsi*, $n = 28$).

and Mattlin 1986, Lougheed et al. 1991). Accordingly, the extent of measurement error associated with each morphometric character under study ought to be assessed and reported. If measurement error is too high for a particular character, that character should be excluded from subsequent analyses. We used repeated measurements and a model II analysis of variance (ANOVA) to estimate measurement error for a subset of our sample (see Bailey and Byrnes 1990). Most specimens were measured only once, but to assess the extent of measurement error, we took three sets of measurements from 25 individuals each from eight taxa (*D. sanctithomae sanctithomae, D. certhia radiolatus, D. c. concolor, D. hoffmannsi, D. picumnus validus, D. p. puncticollis, D. platyrostris platyrostris,* and *D. p. intermedius*). All specimens measured for this analysis were males, with the exception of 8 female *D. hoffmannsi* that were included because too few males were available.

Because tarsus length had extremely high measurement error (Table 2)—that is, it was not repeated with high precision—we excluded it from all analyses. Measurements of bill length, width, and depth taken at the nares and at the base were highly correlated. Of these, length measurements were the most highly correlated ($r = 0.87, n = 2{,}928$; Sokal and Rohlf 1995), but corresponding measurements of bill width and depth also had a strong overall correlation (width: $r = 0.74, n = 3{,}013$; depth: $r = 0.74, n = 3{,}005$). Because length, width, and depth measurements taken at the base of the bill had higher measurement error than comparable measurements taken at the nares (Table 2), we discarded from multivariate analyses (described below) all three measurements taken at the base of the bill. The remaining five characters—wing chord, tail length, and bill length, depth, and width measured at the anterior edge of the nares—had relatively low measurement error ($<5\%$).

TABLE 2. Measurement error (% ME), estimated from a model II ANOVA (Bailey and Byrnes 1990) of nine mensural characters of *Dendrocolaptes*.

Character	MS_{total}	MS_{among}	MS_{within}	s^2 among	s^2 within	% ME
Bill length at nares	3.58	10.62	0.075	0.075	3.52	2.09
Total culmen	4.44	13.00	0.178	0.178	4.27	3.99
Bill width at nares	0.50	1.47	0.013	0.013	0.49	2.65
Bill width at base	0.81	2.24	0.097	0.097	0.71	12.00
Bill depth at nares	0.30	0.88	0.016	0.016	0.29	5.22
Bill depth at base	0.56	1.61	0.042	0.042	0.52	7.52
Tarsus length	1.12	2.64	0.362	0.362	0.76	32.35
Wing chord	38.99	116.89	0.139	0.139	38.92	0.36
Tail length	29.98	88.71	0.684	0.684	29.34	2.28

STATISTICAL ANALYSES

Depending on the analysis performed, males and females were either pooled or examined separately (the male bill averaged ~1.5% longer, but ~4% narrower and ~1.5% shallower; male wing chord and tail length averaged 1–1.5% longer). To reduce the effects of age-related variation (see Marantz et al. 2003), we excluded birds in juvenal plumage and those in which bill dimensions had not yet reached full size. We included in multivariate analyses only those individuals for which all characters could be measured.

We conducted canonical discriminant analysis, by sex, to determine whether morphology alone could distinguish among the five *Dendrocolaptes* spp., and we examined concordance between mensural differences among species and subspecies-groups and expectations based on ecomorphology. Prior probabilities were set equal among groups (i.e., any given specimen had a one-in-five chance of being classified as a particular species). We conducted similar analyses for subspecies within species. Because discriminant analysis may result in inflated Type I error rates, especially when variation is clinal in nature (Skalski et al. 2008), we used this method in a heuristic manner as a first attempt to determine whether subspecies and subspecies-groups are separable. Classification to a group was considered correct only if the posterior probability of assignment to that group was ≥0.90. We also used discriminant analysis to generate synthetic multivariate scores to be used in subsequent pairwise analyses. All such statistics were performed using SAS, version 9.1.3 (SAS, Cary, North Carolina). We used the *D*-statistic (Patten and Unitt 2002) to test diagnosability for all pairwise comparisons among taxa for which there was a mensural component in the original diagnosis.

RESULTS

Our bill measurements attempted to quantify lateral and dorsoventral compression of the bill. Of the three sets of bill characters, only the two width measurements differed conspicuously among species: the bill was consistently wider at the nares in relation to the base in *D. certhia* than in *D. sanctithomae* (Fig. 2) and markedly narrower at the nares in relation to the base in *D. p. puncticollis* than in all other *D. picumnus* (not shown). All remaining discussion of the bill characters (length, width, and depth) reflects exclusively those sets of measurements taken from the anterior edge of the nares.

Canonical discriminant analysis distinguished *Dendrocolaptes* spp. on the basis of mensural characters (Fig. 3). Although *D. certhia* and *D. sanctithomae* were treated as conspecific for many years, *D. certhia* differed markedly in structure from all other species as a result of its generally larger bill (Fig. 3 and Table 3), and *D. certhia* and *D. sanctithomae* were seldom misclassified as one another (Table 4). Still, *D. sanctithomae*, especially, and *D. picumnus* exhibit considerable morphometric variation, so specimens of these species were misclassified more frequently than those of other taxa (Table 4). Belying appearances in two-dimensional space (Fig. 3), *D. hoffmannsi* differs structurally from other species (Table 4), but our small sample size for this species did not allow for its finer discrimination.

When the barred and streaked groups were examined separately, discrimination improved. Discriminant analysis of the two barred woodcreepers

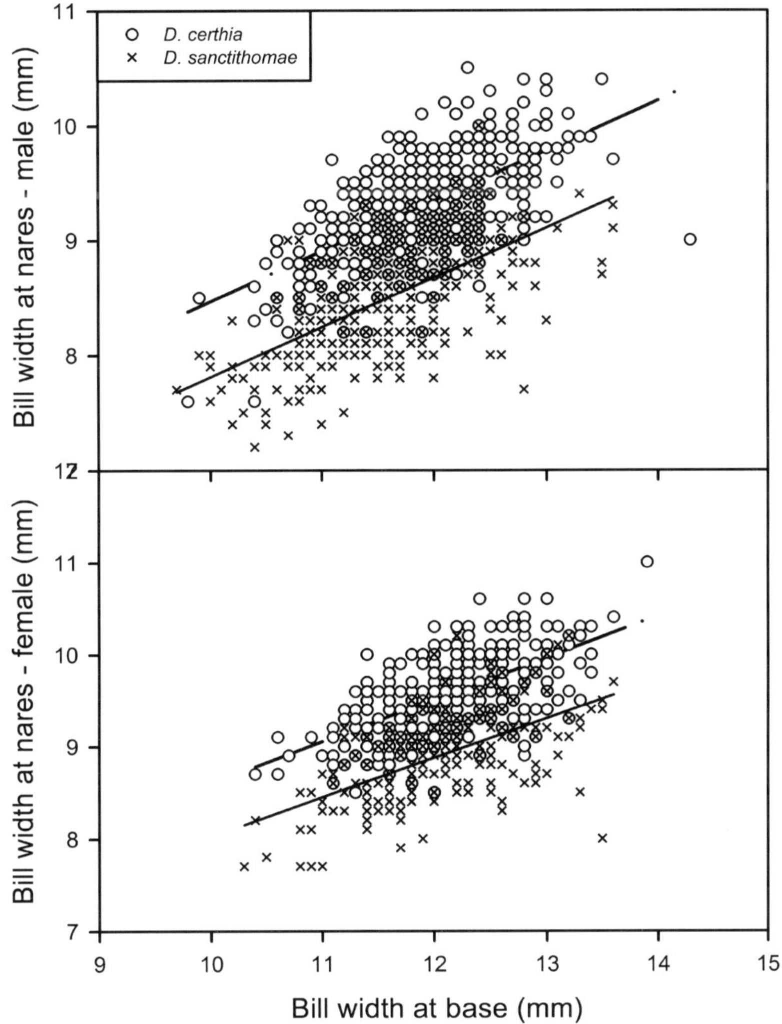

FIG. 2. Larger values for bill width at the nares in relation to those at the base in *Dendrocolaptes certhia* reflect greater lateral expansion of the bill than in *D. sanctithomae* (above males, below females).

resulted in the correct classification of ~85% of individuals for both species (Fig. 4; 85.3% for *D. certhia*, $n = 755$; 84.3% for *D. sanctithomae*, $n = 463$). Bivariate plots further revealed that the longer and wider bill of *D. certhia* separated most males and many females from *D. sanctithomae* (Fig. 5). Similar analyses of the streaked taxa (*D. picumnus*, *D. platyrostris*, and the weakly streaked *D. hoffmannsi*) resulted in effective discrimination of the three species in this complex and of biogeographically defined subspecies-groups within *D. picumnus* (Table 5 and Fig. 6); however, two groups of montane subspecies in this complex

that differ in plumage patterns were not separated morphometrically.

Contrary to the effective separation of species, most subspecies were not readily diagnosed on the basis of mensural characters. With one exception (*D. c. polyzonus*), diagnosability was low (<50% correct classification) among subspecies of *D. certhia*, none of which were described on the basis of morphometric differences. Despite a small sample size ($n = 9$), *D. c. polyzonus* of Bolivia was the most distinct subspecies: 77.8% (7 of 9) were classified correctly because of the taxon's shallower bill and longer wings and tail

TABLE 3. Loadings (correlations between raw variables and synthetic variates) for canonical discriminant analyses of morphometric data for *Dendrocolaptes*. Males and females were analyzed separately. The first two axes are shown, except in the analysis for the *D. certhia* complex, for which the two species were separated along a single axis.

| | All taxa combined | | | | *D. certhia* complex | | *D. picumnus* complex | | | |
| | Male (n = 1,523) | | Female (n = 1,008) | | Male (n = 740) | Female (n = 478) | Male (n = 792) | | Female (n = 532) | |
Character	DFI	DFII	DFI	DFII	DFI	DFI	DFI	DFII	DFI	DFII
Bill length	0.283	0.338	0.129	0.439	0.455	0.511	0.341	0.198	0.255	0.231
Bill width	0.415	0.257	0.449	0.104	0.876	0.849	0.261	0.732	0.151	0.803
Bill depth	0.570	0.042	0.522	0.243	−0.097	−0.146	0.169	0.936	0.033	0.944
Wing chord	0.111	0.352	0.260	0.508	0.079	0.162	0.992	0.057	0.988	0.107
Tail length	−0.032	0.213	0.354	0.290	0.499	0.556	0.750	0.045	0.734	0.133

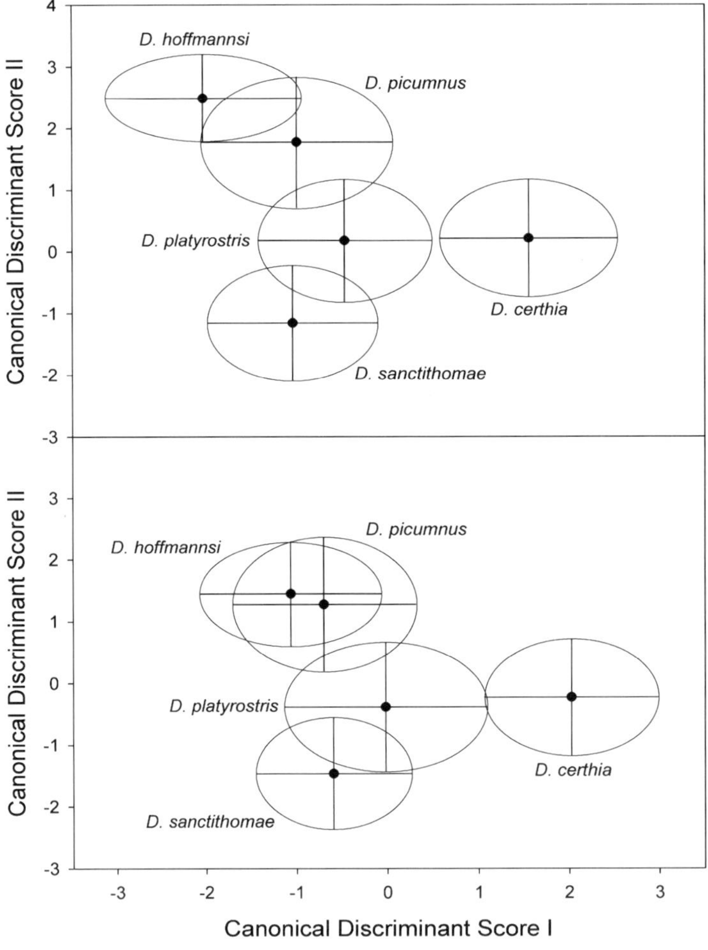

FIG. 3. Mean scores (ellipses = SD) on the first two canonical axes of discriminant analyses of *Dendrocolaptes* spp. using morphometric data (male top, female bottom). The first axis represents primarily bill depth and width in both sexes, and less so tail length in females; the second axis has its highest correlations with wing chord and bill length in both sexes (Table 3).

TABLE 4. Classification matrix from a canonical discriminant analysis of morphometric data from 2,542 adults representing all species of *Dendrocolaptes*. Classification percentages were averaged across the results from each sex ($n = 1,532$ males, $n = 1,010$ females).

Species	Percent classified as species					
	sanctithomae	*certhia*	*hoffmannsi*	*picumnus*	*platyrostris*	*n*
D. sanctithomae	**42.1**	14.3	4.5	13.4	25.7	463
D. certhia	12.1	**82.8**	0.3	2.1	2.8	755
D. hoffmannsi	3.9	0.0	**80.8**	15.4	0.0	26
D. picumnus	11.7	3.3	21.4	**60.0**	3.7	560
D. platyrostris	16.5	3.3	0.7	2.0	**77.5**	738

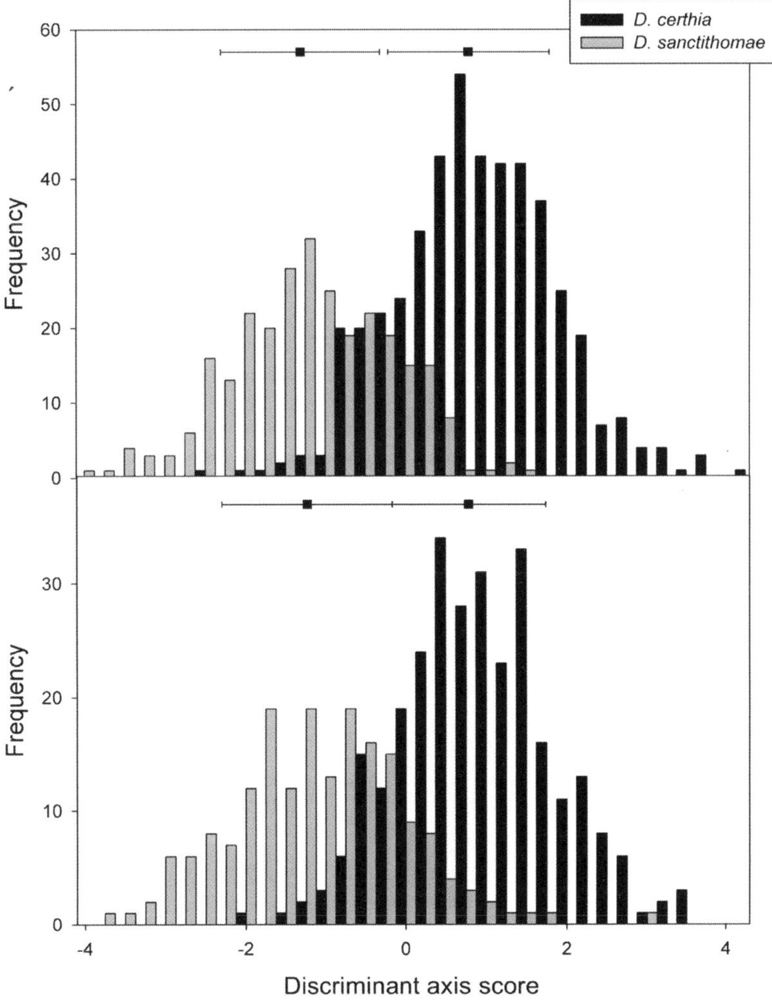

FIG. 4. Histograms of canonical discriminant scores from a morphometric analysis of *Dendrocolaptes certhia* and *D. sanctithomae* (male top, female bottom).

TABLE 5. Classification matrix from a canonical discriminant analysis of morphometric data from the streaked species of *Dendrocolaptes* and, within *D. picumnus*, subspecies-groups (see Table 1). Classification percentages were averaged across the results from each sex ($n = 792$ males, $n = 532$ females).

Group	Amazonian	Chaco	Montane A	Montane B	*hoffmannsi*	*platyrostris*	n
			Percent classified as group				
D. picumnus							
Amazonian	**65.3**	6.3	2.8	3.6	18.8	3.1	207
Chaco	8.4	**58.7**	8.3	9.5	5.5	9.5	73
Montane (A)	7.9	23.0	**49.0**	10.6	8.2	1.2	168
Montane (B)	4.6	8.8	6.2	**76.6**	0.8	2.8	112
D. hoffmannsi	11.4	11.1	2.9	5.6	**69.0**	0.0	26
D. platyrostris	1.1	10.0	0.6	2.4	0.4	**85.5**	738

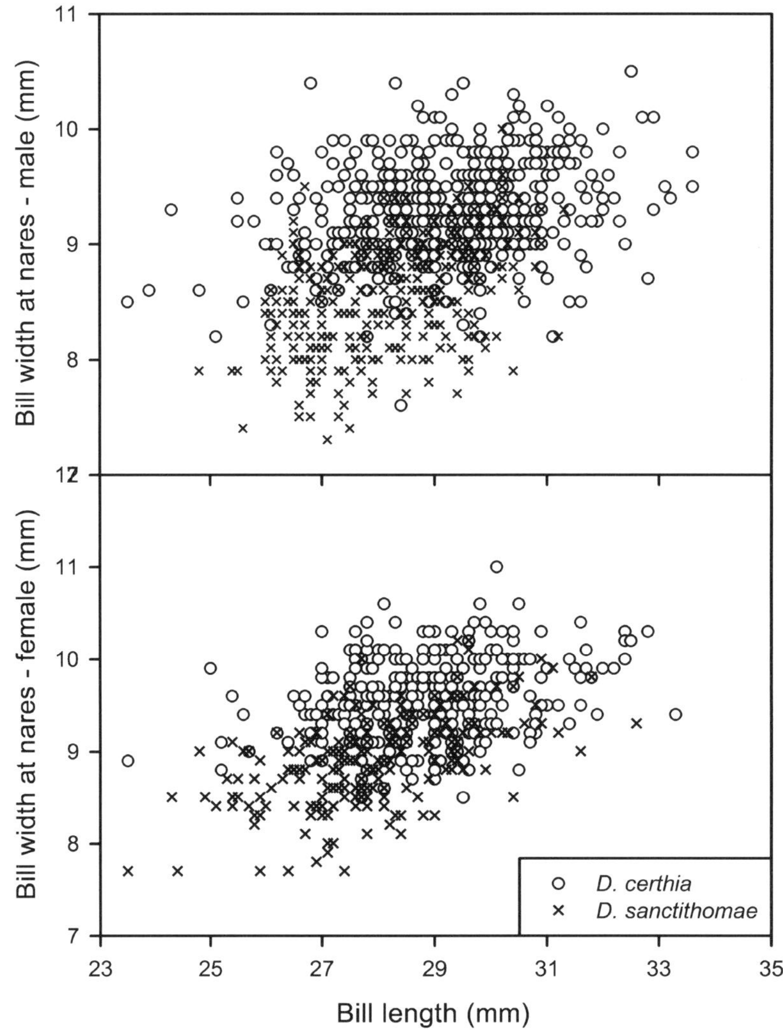

FIG. 5. Bill length vs. bill width of *Dendrocolaptes certhia* and *D. sanctithomae* (male top, female bottom).

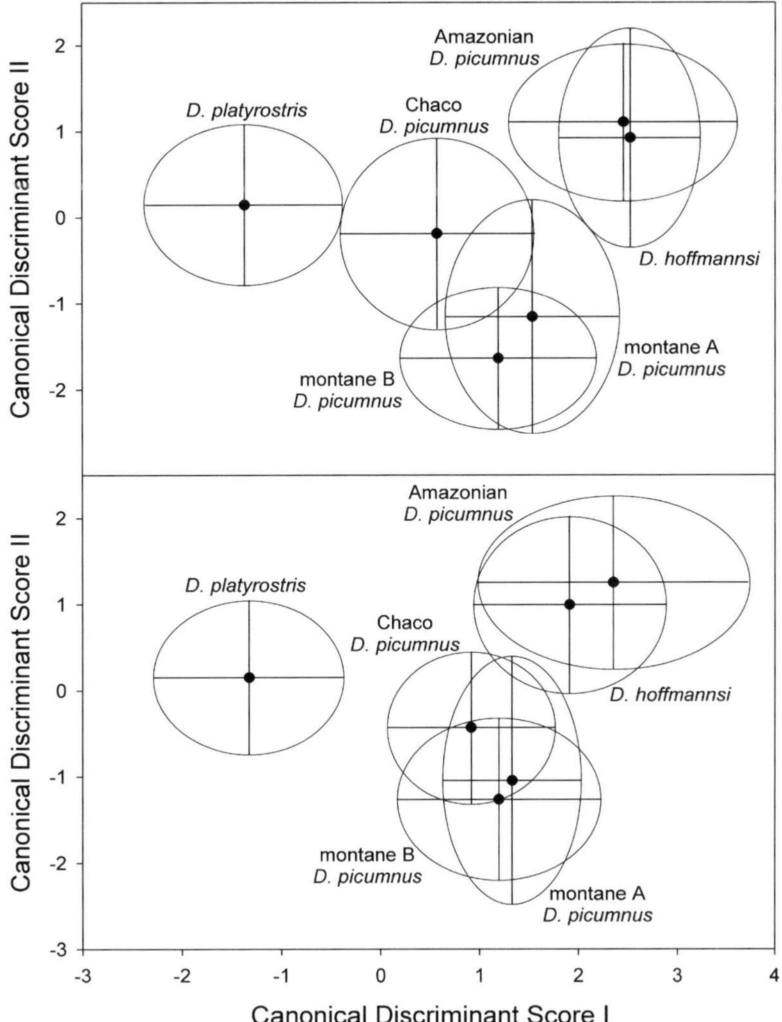

Fig. 6. Mean scores (ellipses = SD) on the first two canonical axes of discriminant analyses using morphometric data for species and subspecies in the *Dendrocolaptes picumnus* complex (Table 6). In both sexes (male top, female bottom), the first axis reflects size, with high positive scores for wing chord and tail length, and the second reflects bill mass, with high positive scores for both width and depth (Table 3).

(not shown). Despite weighing less on average (Marantz et al. 2003), *D. c. concolor* differed little in structure from other subspecies of *D. certhia* (not shown).

Subspecies of *D. sanctithomae* likewise could not be diagnosed with high confidence. Sample sizes were small (<10 specimens) for four of the eight described subspecies, but for those with adequate sample sizes, only nominate *D. s. sanctithomae sensu stricto* could be diagnosed at >50%. Subspecies described from Middle America and

northwestern South America (*D. s. nigrirostris* and *D. s. colombianus*), in part on the basis of bill length, were particularly problematic because bill length appeared to vary in a smooth cline from southern Mexico south to northern Colombia (Fig. 7).

Only one subspecies of *D. picumnus* was described solely on the basis of mensural differences; even so, diagnosability on the basis of morphometrics was somewhat higher among subspecies of *D. picumnus* (Table 6) than in *D. certhia*. Diagnosability

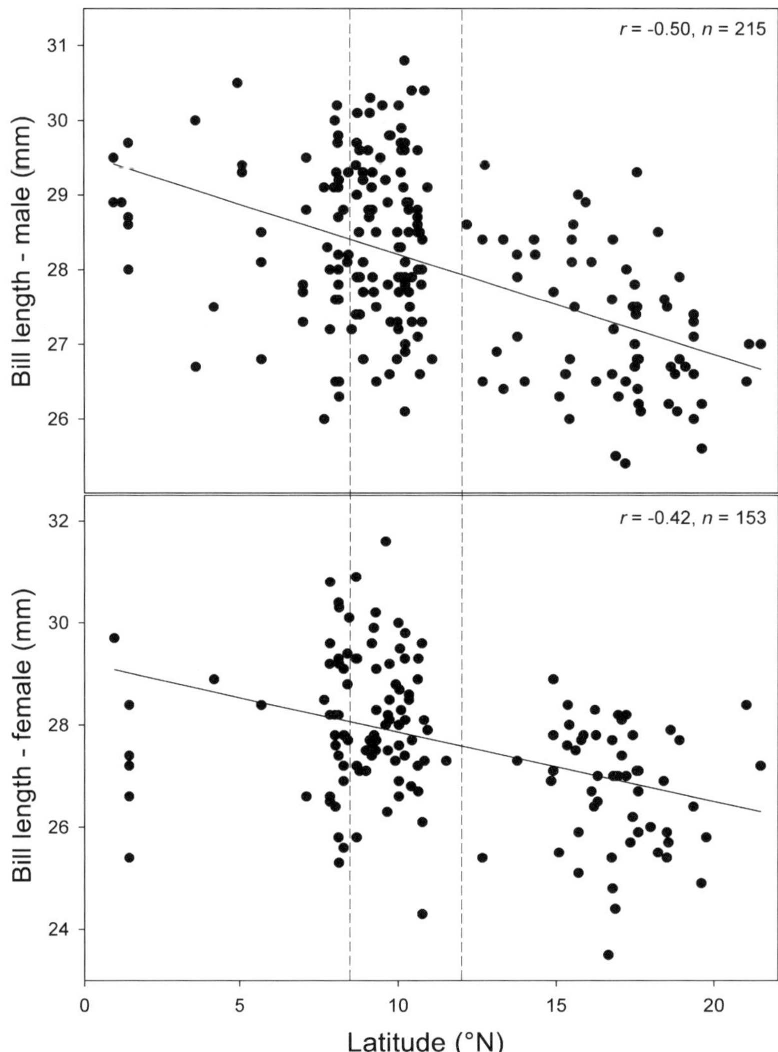

FIG. 7. Variation in bill length of *Dendrocolaptes sanctithomae* across latitude from western Ecuador to southern Mexico represents a smooth cline (adult male top, adult female bottom). Vertical lines mark geographic breaks between named subspecies: from south to north, *D. s. colombianus*, *D. s. nigrirostris*, and *D. s. sanctithomae*.

was especially high for *D. p. puncticollis*, which has an unusually long and narrow bill, but morphometric diagnosability of *D. p. extimus*, *D. p. picumnus*, *D. p. transfasciatus*, and *D. p. veraguensis* was also reasonably high (>70%; Table 6). The only subspecies described on the basis of morphometrics is *D. p. casaresi*, particularly in relation to *D. p. pallescens*. Diagnosis on bill length was perhaps possible (Fig. 8), but other characters overlap considerably. Additional specimens of *D. p. casaresi* are needed to clarify its taxonomic status.

Separation on the basis of morphometric data of the two subspecies of *D. platyrostris* was relatively high along a single axis that contrasted bill size and wing length. Averaged across sex, 77.5% of *D. p. intermedius* and 74.0% of *D. p. platyrostris* specimens were classified correctly as a result of the relatively larger bill yet shorter wing of the former.

Among those taxa for which either the original description or subsequent analyses found a morphometric component, only two (*D. p.*

TABLE 6. Classification results for subspecies of *Dendrocolaptes picumnus* from a canonical discriminant analysis using morphometric data for adults. Classification percentages were averaged across the results from each sex.

Group/subspecies	Percent classified correctly	n
Amazonian		
D. p. picumnus	75.3	85
D. p. validus	38.9	108
D. p. transfasciatus	71.4	14
Chaco		
D. p. pallescens	40.3	62
D. p. extimus	80.0	5
D. p. casaresi	66.7	6
Montane (A)		
D. p. costaricensis	57.1	21
D. p. veraguensis	80.0	5
D. p. multistrigatus	46.5	86
Montane (B)		
D. p. puncticollis	95.0	60
D. p. seilerni	18.2	88
D. p. olivaceus	40.0	20

multistrigatus and *D. p. puncticollis*) were statistically diagnosable on the basis of mensural characters (Table 7). Two additional taxa may be diagnosable (*D. s. sheffleri* and *D. p. casaresi*), but our sample sizes were too small.

MEASUREMENT ERROR

Most characters were measured with an error of <5%, but measurement error accounted for >30% of the variation in tarsus length and ~12% in measurements of bill width at its base. Using a similar method, Lougheed et al. (1991) found error to be minimal in wing chord, high in tarsus length, and intermediate in a measure of tail length for the American Coot (*Fulica americana*), but they did not include external measurements comparable to our bill characters. In most birds, both wing chord and tail length are large values that can be measured with minimal error, so measurement error is expected to be small. That the tips of woodcreeper rectrices represent the feather shafts further results in a low error compared with corresponding measurements in other passerines, in which worn feather tips become frayed and difficult to measure. That is, although the tail tip wears over time, resulting in seasonal variation, repeated measurements of the same specimen can be made with a high degree of accuracy. Tarsus length is challenging to measure repeatedly because it is often difficult to determine the last undivided scute. This character is therefore measured either with high precision, if the correct scute is chosen, or with large error (1–2 mm) if the incorrect scute is chosen. We suspect that the "glaring error" noted

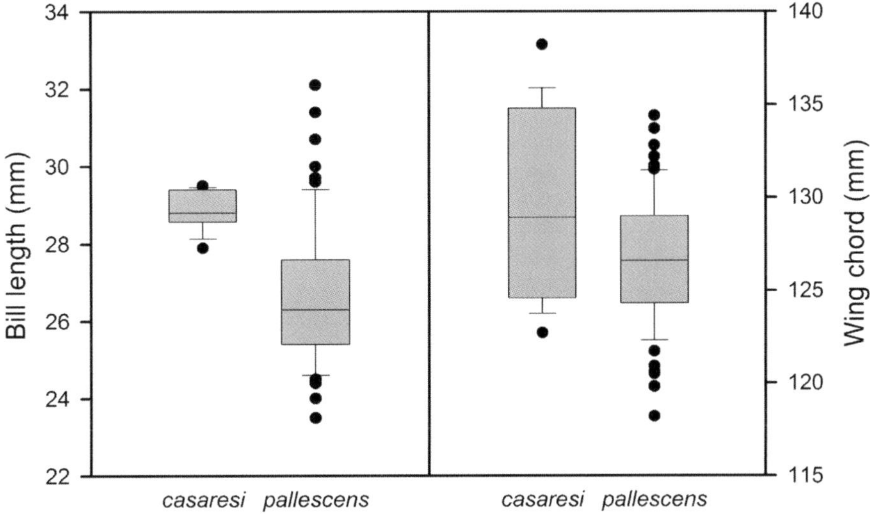

FIG. 8. Box plots of bill length and wing chord in *Dendrocolaptes picumnus casaresi* ($n = 9$ for bill length, $n = 12$ for wing chord) vs. *D. p. pallescens* ($n = 70$ for both characters). Sexes were combined for this analysis.

TABLE 7. Summary of primary characters used in the original description (or mentioned by Cory and Hellmayr [1925] for older taxa) for all recently recognized (Peters 1951, Marantz et al. 2003) subspecies in the genus *Dendrocolaptes*. Our assessment of morphometric diagnosability and our recommendations for subspecies validity are based on plumage and morphometrics; for invalid subspecies, we list the subspecies name that should be considered the senior synonym. Pairwise comparisons were made between taxa with a mensural component in their diagnosis or those we found to be relatively diagnosable using discriminant analysis. The sexes were combined for all comparisons except that between *D. p. multistrigatus* and *D. p. picumnus*, for which we averaged values for the two sexes because of apparent sexual differences.

Subspecies	Characters used in description	Morphometrically diagnosable?	Taxonomic status
D. sanctithomae			
D. s. sheffleri	Plumage, bill color, size	Unclear[a]	Valid
D. s. legtersi	Plumage	No	*D. s. sanctithomae*
D. s. nigrirostris	Plumage, bill length	No[b]	*D. s. sanctithomae*
D. s. colombianus	Plumage, bill length	No[b]	*D. s. sanctithomae*
D. s. hesperius	Plumage	No	Valid
D. s. hyleorus	Plumage, bill shape	No[b]	*D. s. punctipectus*
D. s. punctipectus	Plumage	No	Valid
D. certhia			
D. c. radiolatus	Plumage	No	Valid
D. c. juruanus	Plumage	No	Valid
D. c. polyzonus	Plumage, size	No[b]	Valid
D. c. concolor	Plumage	No	Valid
D. c. ridgwayi[c]	Plumage	No	*D. c. concolor*
D. c. medius	Plumage	No	Valid
D. picumnus			
D. p. validus	Plumage, bill color	No	Valid
D. p. transfasciatus	Plumage	No[b]	Valid
D. p. pallescens	Plumage	No	Valid
D. p. extimus	Plumage	No[b]	*D. p. pallescens*
D. p. casaresi	Morphometric	Unclear[d]	Insufficient data
D. p. costaricensis	Plumage	No	Valid
D. p. veraguensis	Plumage	No[b]	*D. p. costaricensis*
D. p. multistrigatus	Plumage, bill shape	Yes[b]	Valid
D. p. puncticollis	Plumage, bill shape	Yes[b]	Valid
D. p. seilerni	Plumage, size, bill shape	No[b]	Valid
D. p. olivaceus	Plumage	No	Valid
D. platyrostris			
D. p. intermedius	Plumage	No[b]	Valid

[a] Sample size for *D. s. sheffleri* was too small to make a morphometric comparison, but its plumage and bill coloration differ strikingly from those of other subspecies.

[b] D_{ij} values are from the pairwise index of Patten and Unitt (2002); $D_{ij} > 0$ indicates that the taxon is diagnosable morphometrically, whereas $D_{ij} < 0$ indicates that it is not. Comparisons and the sample sizes for each are as follows: *D. s. nigrirostris* ($n = 215$) vs. *D. s. sanctithomae* ($n = 156$) (bill length: $D_{ns} = -2.4$, $D_{sn} = -2.1$); *D. s. columbianus* ($n = 82$) vs. *D. s. nigrirostris* ($n = 215$) (bill length: $D_{cn} = -4.0$, $D_{nc} = -3.7$); *D. s. hyleorus* ($n = 14$) vs. *D. s. punctipectus* ($n = 13$) (bill depth: $D_{hp} = -0.9$, $D_{ph} = -0.6$); *D. c. polyzonus* ($n = 11$) vs. *D. c. juruanus* ($n = 69$) (wing chord: $D_{pj} = -10.3$, $D_{jp} = -15.6$); *D. p. transfasciatus* ($n = 14$) vs. *D. p. picumnus* ($n = 88$) (discriminant function scores of all characters: $D_{tp} = -0.75$, $D_{pt} = -0.81$); *D. p. extimus* ($n = 5$) vs. *D. p. pallescens* ($n = 64$) (discriminant function scores of all characters: $D_{ep} = -1.73$, $D_{pe} = -0.93$); *D. p. casaresi* ($n = 9$) vs. *D. p. pallescens* ($n = 70$) (bill length: $D_{cp} = -2.4$, $D_{pc} = -0.5$); *D. p. veraguensis* ($n = 5$) vs. *D. p. costaricensis* ($n = 25$) (discriminant function scores of all characters: $D_{vc} = -2.4$, $D_{cv} = -1.7$); *D. p. multistrigatus* ($n = 91$) vs. *D. p. picumnus* ($n = 90$) (discriminant function scores of bill characters averaged across the sexes: $D_{mp} = 0.58$, $D_{pm} = 0.78$); *D. p. puncticollis* ($n = 73$) vs. all other subspecies of *D. picumnus* ($n = 546$) (discriminant function scores of bill characters: $D_{po} = 0.9$, $D_{op} = 1.0$); *D. p. seilerni* ($n = 119$) vs. *D. p. multistrigatus* ($n = 91$) (bill length: $D_{sm} = -4.8$, $D_{ms} = -4.1$); *D. platyrostris intermedius* ($n = 286$) vs. *D. p. platyrostris* ($n = 486$) (discriminant function scores of all characters: $D_{ip} = -1.7$, $D_{pi} = -1.9$).

[c] Previous analysis of plumage variation suggested that *D. c. ridgwayi* was described on the basis of a population showing characters of hybridization between *D. c. medius* and *D. c. concolor* (Marantz 1997).

[d] We did not find our small sample of *D. c. casaresi* to be diagnosable.

by Lougheed et al. (1991) was one such error. Bill characters are usually measured with high precision, but the total value for width and depth measurements is small, so even a small error is large in relation to the total. Measurements of bill characters are more repeatable from the anterior edge of the nares than from the anterior feathering along the culmen because precise determination of the latter is more difficult (see also Winker 1998). Similarly, bill depth measurements are less repeatable than those for width, probably because to take these measurements one often needs to compress the partially open bill on a study skin.

SEXUAL DIMORPHISM

Woodcreepers are not sexually dimorphic in plumage pattern or coloration, but the sexes of some species differ in size or shape. Size dimorphism, as measured by wing length, is marked in both species of *Deconychura* (Bierregaard 1988) and, to a lesser degree, in two species of *Dendrocincla* (Willis 1972, 1979), with males slightly larger than females in both cases. In *D. picumnus*, *D. certhia*, and *D. sanctithomae*, males have slightly longer wings than females, yet in small samples of birds with body-mass data, females weighed slightly more (Willis 1982, 1992); in all three species, differences in both characters were small. Willis (1992) did not find sexual differences in bill length. Using a larger number of measurements in a sample that included many of the same birds examined by Willis, we found that males averaged slightly (1–1.5%) larger than females in all length characters (bill length, wing chord, and tail length). Although behavioral observations suggest that female *Dendrocolaptes* are dominant over their mates (Willis 1982, 1992), the contradiction between the slightly larger size of males suggested by our mensural data and the apparently greater body mass of females found by Willis (1982, 1992) supports Winker's (1998) caution regarding the comparison of body mass in males and females taken during the breeding season.

In spite of their greater length, the bills of male *Dendrocolaptes* were both narrower and shallower than those of females, which hints at the possibility of differences in prey taken or in the methods used to obtain prey. Extensive foraging observations by Willis (1982, 1992) apparently did not reveal any sexual differences in foraging behavior; however, such determination would be challenging given the difficulty of sexing birds in the field.

Apparent differences in head plumage, with birds appearing to have either "ruffed" (*sensu* Willis 1982:272, Willis and Oniki 2001:232) or smooth heads (representing, respectively, either males or females), may be the best way to sex *Dendrocolaptes* in the field (Willis 1982, 1992; Willis and Oniki 2001), but such differences were not easily assessed from museum skins.

MORPHOMETRIC VARIATION AND ECOMORPHOLOGY IN *DENDROCOLAPTES*

Woodcreeper species differ most conspicuously in their overall size and bill shape (Marantz et al. 2003), and in large part these two characters define the genera. Bills in members of the genus *Dendrocolaptes* differ from those of most other woodcreepers in being relatively short and compressed dorsoventrally rather than laterally, yielding a relatively wide but shallow bill that resembles those of many flycatchers (Tyrannidae); *Dendrocolaptes* also have relatively long tails (Marantz 1992, Marantz et al. 2003). Within the genus, two groups of species are distinguished on the basis of plumage patterns. Two species are barred throughout the body (*D. sanctithomae* and *D. certhia*), and two other species are streaked boldly on the head, neck, and breast (*D. picumnus* and *D. platyrostris*); two taxa (*D. hoffmannsi* and *D. c. concolor*) are weakly marked and thus difficult to assign to a group on this basis alone (Marantz 1997). Morphologically and behaviorally, *D. concolor* is a weakly marked, southern Amazonian representative of *D. certhia* (Willis 1992, Marantz 1997), with which it was merged recently (Marantz et al. 2003). *Dendrocolaptes hoffmannsi* is more difficult to place, having been treated in the past as a member of either the barred group or the streaked group (see Willis 1982, Raikow 1994, Marantz 1997). Vocal and behavioral data suggest that it is best recognized as a plain member of the *D. picumnus* complex (Marantz et al. 2003), and Willis (1982) even suggested that it may represent a subspecies of *D. picumnus*, a treatment that makes sense both biogeographically and morphometrically (Figs. 1 and 6).

We found a moderate degree of separation among the five presently recognized species of *Dendrocolaptes*; however, the barred and streaked groups were not supported by structural characters. Because discriminant analysis using mensural characters tends to superimpose the two groups atop one another, we also examined structural patterns of each group on its own. Morphometric analysis of the barred species resulted in separation of most *D. sanctithomae* and *D. certhia*, primarily on the basis of the longer and consistently wider bill of *D. certhia*. Willis (1992) documented marked differences in foraging behavior between *D. sanctithomae* and *D. certhia*. *Dendrocolaptes sanctithomae* foraged almost exclusively in association with swarming army ants (primarily *Eciton burchelli*), whereas *D. certhia* foraged extensively away from army ants and often with mixed-species flocks. *Dendrocolaptes sanctithomae* was observed foraging closer to the ground on average, by dropping down to take prey flushed by the ants, but *D. certhia* took prey primarily at higher levels by sallying, often to the undersides of leaves. It would therefore seem that *D. certhia* uses a foraging technique similar to that of an "upward-striker" (*sensu* Fitzpatrick 1980:45), a guild that Fitzpatrick (1985: table 1) found to posses a relatively long, wide bill in Tyrannidae. By contrast, *D. sanctithomae*, observed by Willis (1992) to more often drop down onto prey over ant swarms, which presumably requires less specialization (see Fitzpatrick 1985), possesses a bill structure more typical of other *Dendrocolaptes*. The recent split of *D. sanctithomae* from *D. certhia* is supported not only by our morphometric data and by foraging behavior, but also by vocalizations and well-defined differences in plumage (Marantz 1992, 1997; Willis 1992; American Ornithologists' Union 1998; Marantz et al. 2003).

Morphometric patterns among the streaked *Dendrocolaptes* appear to be dominated by size, with *D. platyrostris* small and the Amazonian *D. picumnus* and *D. hoffmannsi* large. In fact, we found less distinction morphometrically between *D. hoffmannsi* and Amazonian *D. picumnus* than between the other subspecies-groups of *D. picumnus*, thus providing some support for Willis's (1982) suggestion that these two may be conspecific. The songs of these two species are also remarkably similar (Ridgely and Tudor 1994, Marantz et al. 2003, C. A. Marantz field recordings), and its geographic distribution is almost identical to that of the equally poorly marked *D. c. concolor* discussed above (Peters 1951, Marantz et al. 2003). The peripheral populations of *D. picumnus* of the dry Chaco and humid montane regions are intermediate in size. The Amazonian taxa forage primarily over swarming army ants, where they are subordinate only to the even larger woodcreepers in the genus *Hylexetastes* (Willis 1982). Their larger size compared with syntopic *D. certhia* is presumably advantageous in these encounters. Willis

(1992) suggested that aggression from the larger woodcreepers may shape foraging behavior in *D. certhia*, but it remains unclear to what degree the small size of peripheral populations of streaked *Dendrocolaptes* in the Atlantic Forest, Chaco, and montane regions reflects an absence of *D. certhia* (i.e., character release). The less massive bill of the montane subspecies of *D. picumnus* presumably reflects a tendency toward more gleaning and probing by birds that occur at elevations at which Willis and Oniki (1978) indicated that army ants are an unreliable source of flushed prey. It is also possible that they take smaller prey.

MORPHOMETRIC DIAGNOSIS OF SUBSPECIES

Most woodcreeper subspecies have been described on the basis of plumage patterns, including those in *Dendrocolaptes* (Marantz 1997). Within presently recognized species in this group, taxa are similar structurally, yet they have subtly different plumage coloration and patterns that are usually manifest in the degree of richness in the coloration and in the width and definition of the streaking or barring. Given that few taxa in this complex were described largely or exclusively on the basis of morphometric characters, it was not surprising that morphometric analyses found few taxa to be diagnosable (Table 7).

Dendrocolaptes sanctithomae.—Three subspecies of *D. sanctithomae* from Middle America and northwestern South America (*D. s. sanctithomae*, *D. s. nigrirostris*, and *D. s. colombianus*) were distinguished primarily on the basis of subtle differences in plumage, bill length, and the extent of pale coloration at the base of the lower mandible (Todd 1950). Shortly after Todd (1950) described *D. s. nigrirostris* and *D. s. colombianus*, Binford (1965) questioned the validity of the former, indicating that he could not detect any differences in plumage between it and *D. s. sanctithomae*, and that subtle differences between these forms in bill size and coloration were insufficient to merit recognition. Marantz (1997) likewise found variation in plumage and the extent of pale coloration at the base of the bill to be subtle, inconstant, and potentially clinal in nature. Our morphometric analyses further revealed that bill length varies as a smooth cline, with shorter-billed birds in the north and longer-billed birds in the south (Fig. 7), thus supporting the recommendations by Binford (1965) and Marantz (1997) that all three populations be recognized under the name *D. s. sanctithomae*. Our sample of two *D. s. sheffleri*, apparently the only two specimens of this taxon in existence, was too small to analyze morphometrically. Samples were larger for *D. s. hesperius*, yet we were unable to differentiate it morphometrically, again supporting Binford's (1965) finding on the basis of a smaller sample. Still, the conspicuously narrower barring and more grayish-brown cast to its underparts compared with *D. s. sanctithomae* supports its recognition (Binford 1965, Marantz 1997).

Dendrocolaptes certhia.—No taxa within *D. certhia* were described on the basis of mensural characters, so it was not surprising that our analyses were unable to distinguish subspecies. The longer wings and tail of *D. c. polyzonus* suggest that this southernmost form is larger than other taxa, but a larger sample than we had available would be necessary to examine the possible influence of Bergmann's Rule.

Dendrocolaptes picumnus.—We examined morphometric variation in the *D. picumnus* complex by looking at variation among species and subspecies-groups as a whole as well as among the subspecies of *D. picumnus* alone. Discrimination by subspecies was generally poor for *D. picumnus*, even when the function was based exclusively on this species. Although subspecies in *D. picumnus* were not strongly separated morphometrically, with few taxa distinguished at >75% confidence, biogeographically defined subspecies-groups are relatively well defined. Amazonian taxa are not only larger than the other subspecies (Fig. 6), they also have a relatively more massive bill that may facilitate capture of large prey available in association with swarming army ants. By contrast, as a potential result of their occurring at elevations above those in which army ants occur reliably (see Willis and Oniki 1978), the smaller taxa in montane regions have slimmer and more shallow bills that are probably better adapted for gleaning and probing and, perhaps, for taking smaller prey. Despite some variation in plumage patterns among montane subspecies of *D. picumnus* (see Marantz 1997), with one exception the montane subspecies were found to be similar structurally. Only *D. p. puncticollis* stands out from other taxa as a result of its particularly long, slim bill. This northernmost population of *D. picumnus* is restricted to highlands from southern Mexico to western Honduras (Morales-Pérez et al. 2000, Marantz et al. 2003), occurring at elevations largely above those frequented by army ants. Its long, slim bill, unusual in *Dendrocolaptes*, suggests that it forages more by gleaning and probing like a *Xiphorhynchus* than by sallying like other *Dendrocolaptes*.

The only subspecies in this complex described on the basis of mensural characters is *D. p. casaresi*. Our data suggest that *D. p. casaresi* may indeed have a longer bill and possibly longer wings than the similarly plumaged *D. p. pallescens*, but our sample was too small even to determine with confidence that birds from northwestern Argentina are larger than those from Bolivia and Paraguay, much less to show that there is no overlap between measurements of these taxa (Table 7 and Fig. 8) or to dismiss the possibility that such variation is clinal in nature. A definitive answer regarding this subspecies must await examination of specimens of *D. p. casaresi* housed in Argentine collections.

Dendrocolaptes platyrostris.—We found moderate separation between *D. p. platyrostris* and *D. p. intermedius* (not shown), largely reflecting the smaller size and relatively larger bill of the latter, which is most easily distinguished on the basis of its duller crown and overall paler coloration (Cory and Hellmayr 1925, Marantz 1997). We did not find an obvious relationship between plumage and morphometric intermediacy in the apparent contact zone for these forms in central Paraguay and São Paulo, Brazil (see Marantz 1997, Willis and Oniki 2001), but the complex geographic nature of this zone may complicate such a determination.

In sum, only three subspecies in the genus *Dendrocolaptes* were described largely or exclusively on the basis of morphometric characters. Two of these taxa, *D. s. nigrirostris* and *D. s. colombianus*, appear to refer to arbitrary breaks in clinal variation compounded by a transcription error (see Binford 1965) and should thus be synonymized with the nominate form. The third subspecies, *D. p. casaresi*, may be valid, but our sample was too small to support its recognition unequivocally.

THE UTILITY OF SUBSPECIES IN *DENDROCOLAPTES*

We began by suggesting that most criticism of the use of subspecies stems not from the importance of well-defined populations but instead from uneven application of the subspecies concept (see also Patten, this volume; Remsen, this volume). Only subspecies that are readily diagnosable using repeatable techniques will be useful in studies of geographic variation and other aspects of evolutionary biology. We therefore sought to assess taxa in the woodcreeper genus *Dendrocolaptes* using a large sample that represented all presently recognized taxa. The present morphometric analysis, together with an earlier study of plumage patterns (Marantz 1997), examined the suite of characters originally used to diagnose these taxa. Our combined data set supports the synonymy of 7 of the 30 taxa that have been described in this complex, with the diagnosis of an eighth taxon uncertain, given our sample. A more superficial review of woodcreeper subspecies during the preparation of accounts for Marantz et al. (2003) suggested that a similar percentage of taxa should be synonymized across the family.

Biogeographic patterns are clarified by the reduced subset of diagnosable taxa in *Dendrocolaptes*. The continuous nature of lowland forest along the Caribbean Slope of Central America is more conducive to the clinal variation we documented, as opposed to the steps that were suggested by the past recognition of *D. s. colombianus* and *D. s. nigrirostris*. Similarly, the tiny geographic ranges of both *D. s. legtersi* and *D. p. extimus* (both known essentially from their type localities) were not supported biogeographically, given that both of these taxa were described from geographic regions that are relatively homogeneous (the eastern Yucatan Peninsula and the Paraguayan Chaco, respectively). *Dendrocolaptes s. hyleorus* and *D. p. veraguensis* likewise have geographically small ranges, but at least these taxa were described from somewhat more heterogeneous regions in which differentiation may be more likely (lowland valleys in northern Colombia and the mountains of central Panama, respectively). Only *D. c. ridgwayi* occupies a large, geographically bounded region (in southern Amazonia between the Rio Tapajós and the Rio Tocantins), but in this case, phenotypic heterogeneity and apparent clinal variation suggest that this population represents a hybrid swarm (Marantz 1997). With the synonomy of these taxa, the remaining subspecies occur in larger and geographically better-defined regions that better match the areas of endemism outlined by Haffer (1985) and Cracraft (1985).

We agree with Patten (this volume) and Remsen (this volume) that, rather than discarding the subspecies concept, the reanalysis of previously described taxa using quantitative methods and stressing diagnosability rather than mean differences will provide a robust starting point for studies of biogeography, migration, and other aspects of evolutionary biology. Our work provides an example in which the quantitative analysis of plumage and morphometric characters supported the synonymy of numerous poorly defined subspecies, many with small and biogeographically unsupported ranges, thus resulting in a set of well-defined taxa whose distributions are concordant with recognized biogeographic characteristics.

ACKNOWLEDGMENTS

This study was an extension of a project begun as part of an M.S. thesis by C.A.M. and supervised by J. V. Remsen, Jr. (who provided thoughtful comments on an earlier version of this manuscript), with additional guidance by R. M. Zink and M. S. Hafner. We thank the following institutions and their staff for loans and access to collections: J. V. Remsen, Jr., and S. W. Cardiff (Louisiana State University Museum of Natural Science; LSUMZ); F. Vuilleumier, A. V. Andors, and M. LeCroy (American Museum of Natural History); K. L. Garrett (Los Angeles County Museum of Natural History); B. C. Livezey, J. Loughlin, R. Panza, and the late K. C. Parkes (Carnegie Museum of Natural History); C. R. Brown (Peabody Museum, Yale University); J. M. Bates, S. J. Hackett, S. M. Lanyon, D. E. Willard, and T. S. Schulenberg (Field Museum of Natural History); D. Oren (Museu Paraense Emilio Goeldi, Belém, Brazil); M. Lentino, M. Martinez, C. Rodner, and R. Restall (Colección Ornitología de Phelps, Caracas, Venezuela); G. R. Graves and P. Angle (U.S. National Museum of Natural History); L. F. Silveira and F. D'Horta (Museu de Zoologia, Universidade de São Paulo, Brazil); R. B. Payne and J. Hinshaw (Museum of Zoology, University of Michigan); the late R. A. Paynter, Jr., and A. Pirie (Museum of Comparative Zoology, Harvard University); F. B. Gill and M. B. Robbins (Academy of Natural Sciences, Philadelphia); L. E. Kiff, R. Corado, and J. C. Fisher (Western Foundation of Vertebrate Zoology); the late Ned K. Johnson (Museum of Vertebrate Zoology, University of California, Berkeley); J. C. Hafner (Moore Laboratory of Zoology, Occidental College); J. Sanchez (Estación Biológica de Rancho Grande, Maracay, Venezuela); M. Salcedo (Museo de Historia Natural La Salle, Caracas, Venezuela); K. J. McGowan (Cornell University Museum of Vertebrates); M. K. Peck (Royal Ontario Museum, Toronto, Canada); J. R. Northern (University of California, Los Angeles); R. O. Prum (Museum of Natural History, University of Kansas); the late D. J. Klingener (Department of Biology, University of Massachusetts, Amherst); and G. K. Hess (Delaware Museum of Natural History). S. W. Cardiff helped with numerous loans of specimens to LSUMZ. J. M. Bates and S. J. Hackett facilitated work in both New York and Chicago, as did J. Minns in São Paulo and L. Bevier and C. Bevier in Philadelphia. L. Bevier also provided information on the characters used to diagnose subspecies of *Dendrocolaptes*. We thank K. Winker and S. Haig for the opportunity to contribute to this monograph, and K. Winker, R. C. Banks, F. James, M. Morrison, and an anonymous reviewer for comments on earlier versions of the manuscript.

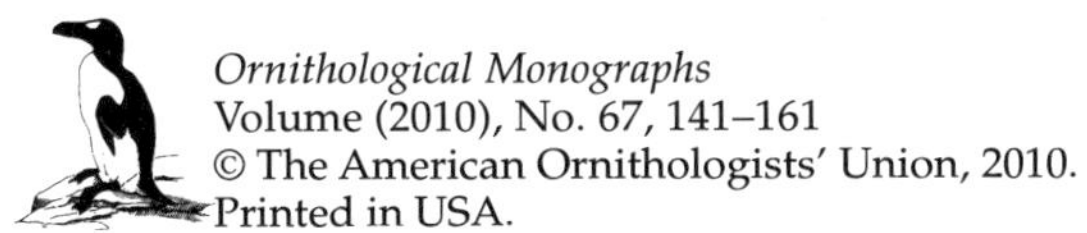
Ornithological Monographs
Volume (2010), No. 67, 141–161
© The American Ornithologists' Union, 2010.
Printed in USA.

CHAPTER 12

ECOGEOGRAPHIC VARIATION IN CINNAMON TEAL (*ANAS CYANOPTERA*) ALONG ELEVATIONAL AND LATITUDINAL GRADIENTS

ROBERT E. WILSON,[1,3] THOMAS H. VALQUI,[2] AND KEVIN G. MCCRACKEN[1]

[1]*Institute of Arctic Biology, Department of Biology and Wildlife, and University of Alaska Museum, University of Alaska Fairbanks, Fairbanks, Alaska 99775, USA; and*
[2]*Centro de Ornitología y Biodiversidad, Santa Rita 105, Oficina 2, Urbanización Huertos de San Antonio, Surco, Lima 33, Peru*

ABSTRACT.—The Cinnamon Teal (*Anas cyanoptera*) comprises five subspecies that inhabit a variety of habitats along an elevational gradient at temperate and tropical latitudes. North American and South American subspecies differ in their migratory behavior, which may have contributed to differences in body size. We measured body size of the five recognized subspecies (*A. c. cyanoptera, A. c. orinomus, A. c. borreroi, A. c. tropica,* and *A. c. septentrionalium*) throughout their ranges and evaluated morphometric differentiation in relation to Bergmann's rule. Subspecies and geographic regions differed significantly, with the largest subspecies and the largest individuals found at high elevations in the central Andes (*A. c. orinomus*) and at high latitudes in southern Patagonia (*A. c. cyanoptera*). Smaller-bodied individuals (*A. c. cyanoptera*) were found at the northern and southern limits of the Altiplano, where intermixing between subspecies with different body sizes might occur. However, there is no direct evidence of *A. c. cyanoptera* breeding at high elevations (>3,500 m). In contrast to patterns within South America, the migratory subspecies in North America (*A. c. septentrionalium*) showed few significant correlations with elevation and no relationship between latitude and body size. Morphological diversity within Cinnamon Teal appears to have arisen from spatial and temporal heterogeneity in selection pressures resulting in adaptations to their local environments.

Key words: Andes, Bergmann's rule, body size, geographic variation, high altitude, morphology.

Variación Ecogeográfica en *Anas cyanoptera* en Gradientes de Altitud y Latitud

RESUMEN.—*Anas cyanoptera* comprende cinco subespecies que se encuentran en una variedad de hábitats a lo largo de un gradiente altitudinal en latitudes templadas y tropicales. Las subespecies norteamericanas y suramericanas difieren en su comportamiento migratorio, lo que puede haber conducido a diferencias en su tamaño. Medimos el tamaño corporal de cinco subespecies reconocidas (*A. c. cyanoptera, A. c. orinomus, A. c. borreroi, A. c. tropica* y *A. c. septentrionalium*) a través de sus áreas de distribución y evaluamos su diferenciación morfométrica con relación a la regla de Bergmann. Las subespecies y las regiones geográficas se diferenciaron significativamente de manera que las subespecies e individuos de mayor tamaño corporal se encontraron a altitudes mayores en los Andes centrales (*A. c. orinomus*) y a altas latitudes en el sur de la Patagonia (*A. c. cyanoptera*). Individuos de *A. c. cyanoptera* de menor tamaño se encontraron en los límites norte y sur del altiplano, donde podrían mezclarse subespecies con diferentes tamaños corporales. Sin embargo, no existe evidencia directa de que *A. c. cyanoptera* se reproduzca a altitudes mayores que 3500 m. En contraste con los patrones observados en Suramérica, la subespecie migratoria de Norteamérica *A. c. septentrionalium* mostró pocas correlaciones significativas con la altitud y

---

[3]E-mail: rewilson@alaska.edu

Ornithological Monographs, Number 67, pages 141–161. ISBN: 978-0-943610-86-3. © 2010 by The American Ornithologists' Union. All rights reserved. Please direct all requests for permission to photocopy or reproduce article content through the University of California Press's Rights and Permissions website, http://www.ucpressjournals.com/reprintInfo.asp. DOI: 10.1525/om.2010.67.1.141.

ninguna relación entre la latitud y el tamaño corporal. La diversidad morfológica en *A. cyanoptera* parece haber surgido debido a adaptaciones a los ambientes locales impulsadas por la heterogeneidad espacial y temporal de las presiones selectivas.

GEOGRAPHIC VARIATION IN morphology is common, and widespread patterns are often explained within an adaptive framework (Price 2008). One of the best-known ecogeographical patterns of variation in body size among vertebrates is Bergmann's rule, which states that individuals from populations in colder climates tend to be larger than those from populations in warmer climates (Bergmann 1847; Mayr 1956, 1963). Modifications to Bergmann's rule showed that larger body size would also be expected at higher latitudes and elevations, or in cooler or drier climates (Snow 1954; James 1968, 1970, 1991). Even though birds show a strong tendency to conform to modified definitions of Bergmann's rule (Ashton 2002, Meiri and Dayan 2003), the adaptive mechanisms responsible for this pattern have been debated. Various mechanisms have been proposed, such as heat conservation, fasting endurance, and competition for resources (Bergmann 1847; McNab 1971; Calder 1974, 1984; James 1991). Thus, ecotypic variation may result from complex underlying processes involving various interrelated variables (Millien et al. 2006).

South American ducks (Anseriformes: Anatidae) are particularly good candidates for a study of ecogeographic variation. Unlike their Northern Hemisphere relatives, which are migratory and show little morphological variation, ducks in South America tend to be less migratory, more restricted in geographic range, and well differentiated into two or more subspecies that differ in plumage and other morphological characters (Phillips 1923, Johnsgard 1978, Williams 1991, Bulgarella et al. 2007). For example, the ducks that inhabit the puna grasslands and wetlands of the high Andes (3,000–5,000 m) tend to have overall larger body size and differ in conspicuous traits, such as plumage, bill color, or eye color, from those in southern Patagonia, where most breeding habitat occurs below 1,500 m (Fjeldså and Krabbe 1990). Most Andean waterfowl thus comprise one or more predominantly lowland subspecies (or species) and one or more highland subspecies (Phillips 1923).

Cinnamon Teal (*Anas cyanoptera*) are widespread throughout the Western Hemisphere, and five subspecies that inhabit distinct geographic and ecological zones are currently recognized: *A. c. cyanoptera, A. c. orinomus, A. c. borreroi, A. c. tropica,* and *A. c. septentrionalium* (Snyder and Lumsden 1951, Delacour 1956, American Ornithologists' Union 1957, Gammonley 1996). *Anas c. septentrionalium* breeds throughout western North America (Bellrose 1980, Madge and Burn 1988, Gammonley 1996), whereas the other four subspecies breed in South America. In South America, *A. c. borreroi* is endemic to the Colombian Andes and is replaced by *A. c. tropica* in the adjacent tropical lowlands; intermediate elevational habitat is unsuitable for either subspecies (Snyder and Lumsden 1951, Delacour 1956). *Anas c. orinomus* is endemic to the Altiplano and adjacent puna region of Argentina, Bolivia, Chile, and Peru. *Anas c. cyanoptera* occurs throughout the Andean lowlands of Peru, Bolivia, Chile, Paraguay, Brazil, Uruguay, and Argentina and is occasionally found sympatrically with *A. c. orinomus* in the high Andes (Evarts 2005). Each Cinnamon Teal subspecies thus has a distinct geographic distribution with little or no overlap, with the exception of *A. c. cyanoptera* and *A. c. orinomus* where they co-occur in the central high Andes.

We collected Cinnamon Teal throughout their range in North America and South America and compared differences in body size among geographic regions. Evidence for Bergmann's rule was evaluated to gain insight into factors shaping morphological divergence over elevational and latitudinal gradients.

METHODS

Specimen collection and subspecies classification.— We collected 153 Cinnamon Teal (39 females and 114 males) from Argentina (2001, 2003), Bolivia (2001), Peru (2002), and the western United States (2002–2003) during the breeding season (Fig. 1 and Appendix 1). Voucher specimens are archived at the University of Alaska Museum (Fairbanks), Museo de Historia Natural de la Universidad de San Marcos (Lima), and Colección Boliviana de Fauna (La Paz). Measurements from Colombian vouchered specimens from the Royal Ontario Museum and Smithsonian Institution National Museum of Natural History were obtained for *A. c. borreroi* (5 females and 13 males) and *A. c. tropica* (2 females and 2 males); new specimens could not be obtained because these subspecies are endangered.

We used a combination of geography and wing chord length to classify each specimen (Snyder and Lumsden 1951, Blake 1977). Despite differences in

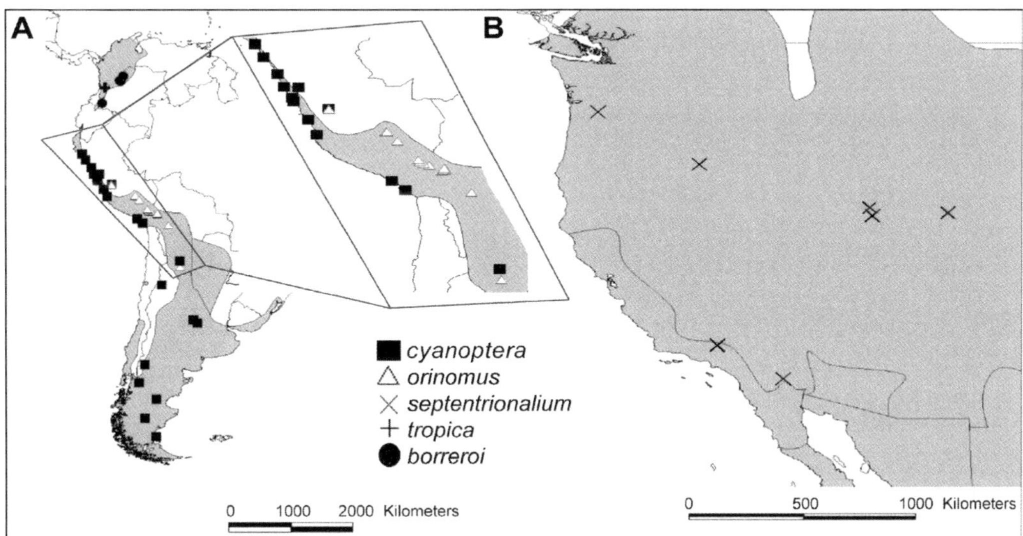

Fig. 1. Sampling localities and geographic ranges for Cinnamon Teal (Ridgely et al. 2003).

plumage (e.g., Blake 1977), coloration is variable, and *A. c. cyanoptera*, *A. c. orinomus*, and *A. c. septentrionalium* were difficult to classify to subspecies on the basis of plumage color alone (Wilson et al. 2008). We classified all individuals from North America as *A. c. septentrionalium* because it is the only subspecies known to occur there. The Colombian specimens we used were the basis of the original subspecies descriptions (Snyder and Lumsden 1951), and we followed them in classifying highland specimens as *A. c. borreroi* and lowland specimens as *A. c. tropica*. *Anas c. orinomus* is the most distinct of all the subspecies, which led some early researchers to consider it a separate species (Oberholser 1906). *Anas c. orinomus* was easily differentiated from *A. c. cyanoptera* by overall body size. To check the accuracy of classifications of *A. c. cyanoptera* and *A. c. orinomus* in areas of sympatry, we compared wing chord length to individuals of known classification. All initial classifications were confirmed.

Body measurements.—We took nine body-size measurements (±0.1 mm) from each bird: wing chord length (carpal joint to longest primary feather unflattened, ±1 mm), tail length (base of the uropygial gland on back to tip of the center tail feather, ±1 mm), exposed culmen length, bill length at nares (posterior edge of nares to tip of nail), tarsus bone length (tarsometatarsus), bill height (height of upper mandible at posterior edge of nares), bill width (width of upper mandible at posterior edge of nares), and body mass (g). Body mass was not

available from the Colombian subspecies and, therefore, was used only as a secondary character in subspecies identification. Measurements for all but 45 recently collected specimens were taken the day of collection and prior to preparation as museum specimens (wet measurements), and again several months or years after preparation (dry measurements; Appendix 2) by R.E.W. For 52 individuals from Argentina, Bolivia, and Colombia, only measurements from museum specimens were available (dry measurements). Specimen shrinkage during drying is a universal phenomenon and can cause analytical problems if not properly accounted for in studies that combine live or freshly killed birds and museum specimens (e.g., Winker 1996). Fresh and dry measurements taken by the first author differed significantly (Wilson and McCracken 2008). Therefore, dry measurements of those 52 individuals could not be directly substituted for wet measurements.

We chose to analyze wet measurements, and to use individuals missing these data we used a multiple imputation (MI) procedure implemented in the program NORM (Schafer 1999) to estimate wet measurements for the 52 individuals with only dry measurements, because we had both wet and dry measurements for most of the data set. An expectation-maximization algorithm (EM) was used to obtain starting values for the multiple imputation procedure, followed by data augmentation using Markov-chain Monte Carlo to produce multiple imputations of the missing data. We used a random number seed and 10,000 iterations, with imputation every 1,000 iterations. The resulting 10 data sets were combined following Rubin's (1987) rules for scalar estimates to provide a single set of estimates for each specimen with missing data. The combined data composed of original wet measurements obtained from 123 specimens and estimated wet measurements from 52 specimens were used for all statistical analyses.

Statistical analysis of measurements.—Statistical analyses were performed with MINITAB Statistical Software (Minitab, State College, Pennsylvania). All traits were tested for normality with Kolmogorov-Smirnov tests and were normally distributed ($P_s > 0.05$). A multivariate analysis of variance (MANOVA) was performed to evaluate overall differences among subspecies and geographic regions for each sex. Geographic regions were defined as follows (with the corresponding subspecies inhabiting each area): (1) North America (*A. c. septentrionalium*); (2) Colombian highlands (*A. c. borreroi*); (3) Colombian lowlands (*A. c. tropica*); (4) Peruvian coast (*A. c. cyanoptera*); (5) central high Andes of Argentina, Bolivia, and Peru (*A. c. orinomus* and *A. c. cyanoptera*); and (6) lowland Argentina (includes Patagonia and lowland areas of Cordoba; *A. c. cyanoptera*). Collection locations in North America (California, Oregon, and Utah) were treated as a single geographic unit, which is consistent with low levels of male breeding-site fidelity in North America (Anderson et al. 1992). Analysis of variance (ANOVA) and pairwise comparisons for each individual measurement were performed using a general linear model with Bonferroni correction for multiple comparisons. Pairwise comparisons were not made with *A. c. tropica* (lowland Colombia) because of low sample size. We used a principal component analysis to illustrate overall differences in body size among subspecies. Only those principal components with eigenvalues >1 were used for partial correlation and subspecies classification analyses.

Finally, the joint relationships between elevation and latitude and morphological variables were examined using partial correlation analysis for the following areas: all populations pooled, North America, South America, and southern South America (Altiplano and associated lowlands and Patagonia). In addition, correlations between latitude and body size were examined for *A. c. cyanoptera* (lowland and highland) separately, because it is the only subspecies with populations distributed over a large latitudinal gradient. Analyses were conducted separately for each sex, and

significance levels were corrected for multiple comparisons using Bonferroni methods.

Subspecies classification.—We used two methods to evaluate subspecies identifications. We first used linear discriminant analysis to evaluate whether the Cinnamon Teal subspecies conformed, on the basis of body-size measurements, to the 75% rule (Amadon 1949, Mayr 1969), which states that 75% of the individuals of one subspecies must be distinguishable from all other subspecies. Measurements found to be significantly different between at least two subspecies (classified based on overall body size) from the MANOVA and ANOVAs were included in this analysis. The reliability of the discriminant analysis was assessed using a cross-validation (jackknife) procedure, in which each observation was omitted one at a time and then reclassified using a classification function derived from the remaining observations (Manly 2000). Cross-validation gives a less biased error rate in classification, because it does not include observations that are used to create the classification function. We performed a discriminant analysis for each sex and locality of collection and did not include *A. c. tropica* because of low sample size.

We also tested the diagnosability of subspecies using the method of Patten and Unitt (2002), which focuses on the extent of overlap rather than detecting mean differences. Diagnosability of subspecies was determined for each measurement separately and for overall body size (PC1). An index value (D_{ij}) ≥ 0 indicates that subspecies i is diagnosable from subspecies j. Reciprocal tests were performed to determine whether subspecies i is diagnosable from subspecies j and whether subspecies j is diagnosable from subspecies i.

RESULTS

Subspecies differed significantly in overall body size (Wilks's $\lambda = 0.05$, $F = 25.38$, df = 28 and 574, $P < 0.001$), as did the sexes (Wilks's $\lambda = 0.74$, $F = 7.16$, df = 7 and 159, $P < 0.001$; Table 1). There was no significant interaction between subspecies and sex (Wilks's $\lambda = 0.78$, $F = 1.46$, df = 28 and 574, $P = 0.061$). *Anas c. orinomus* was significantly larger than *A. c. tropica* (e.g., wing chord: 32.50 mm difference; tarsus: 4.01 mm difference) and *A. c. septentrionalium* (e.g., wing chord: 31.50 mm difference; tarsus: 4.83 mm difference) in most measurements, with *A. c. borreroi* and *A. c. cyanoptera* intermediate in body size (Tables 1 and 2). In addition, when individuals were grouped on the basis of collection locality instead of

TABLE 1. Wet measurements (mm) and body mass (g) for male *Anas cyanoptera cyanoptera*, *A. c. orinomus*, and *A. c. septentrionalium* and dry measurements for *A. c. borreroi* and *A. c. tropica*. Letters after mean value correspond to subspecies (o = *orinomus*, c = *cyanoptera*, b = *borreroi*, and s = *septentrionalium*) and indicate significant pairwise differences determined using Bonferroni corrected P values ($P_{adjusted} < 0.05$). Pairwise comparisons were not performed for *A. c. tropica* because of the small sample size.

		Mass	Wing chord	Tarsus	Tail	Nare	Culmen	Bill height	Bill width
orinomus	Mean	498.8	223.4 cbs	35.52 cbs	93.28 cbs	37.28 cbs	47.85 cbs	14.67 cs	17.41 cs
(*n* = 13)	SE	10.7	2.20	0.37	1.14	0.38	0.57	0.19	0.18
	Range	425–550	211–237	33.8–37.6	86.7–99.9	35.6–40.4	44.9–52.1	13.7–15.9	16.1–18.3
cyanoptera	Mean	414.5	191.9 ob	32.95 os	81.95 ob	33.96 o	44.55 o	14.02 obs	16.92 o
(*n* = 28)	SE	7.60	1.20	0.32	1.33	0.35	0.41	0.19	0.19
	Range	340–515	181–205	29.1–35.9	69.5–97.6	30.6–37.6	40.5–48.7	12.4–16.3	15.2–19.2
septentrionalium	Mean	361.8	188.8 ob	31.01 ocb	80.47 ob	35.00 ob	45.63 o	13.39 ocs	16.76 ob
(*n* = 50)	SE	3.30	0.90	0.15	0.58	0.17	0.20	0.08	0.07
	Range	310–420	168–201	28.1–33.4	66.0–87.0	32.4–37.1	42.5–47.9	12.3–15.1	15.7–17.8
borreroi	Mean	—	196.0 ocs	31.93 os	99.54 ocs	32.91 os	43.65 o	14.39 cs	16.57 s
(*n* = 13)	SE	—	1.86	0.43	2.77	0.88	0.73	0.21	0.25
	Range	—	179–205	28.7–35.2	82–115	28.3–40.2	38.9–46.7	13.1–15.9	15.2–18.2
tropica	Mean	—	184.0	30.93	105	29.68	40.99	14.13	16.54
(*n* = 2)	SE	—	2.50	0.59	6.00	0.36	0.03	0.17	0.62
	Range	—	182–187	30.2–31.4	99–111	29.3–30.0	40.9–41.0	13.9–14.3	15.9–17.2
P[a]		<0.001	<0.001	<0.001	<0.001	<0.001	<0.001	<0.001	<0.001

[a] ANOVAs for subspecies effect based on pooled data (wet measurements and transformed dry measurements). Sample sizes: *orinomus*, *n* = 30; *cyanoptera*, *n* = 34; *septentrionalium*, *n* = 50; *borreroi*, *n* = 13; and *tropica*, *n* = 2.

TABLE 2. Wet measurements (mm) and body mass (g) for female *Anas cyanoptera cyanoptera*, *A. c. orinomus*, and *A. c. septentrionalium* and dry measurements for *A. c. borreroi* and *A. c. tropica*. Letters after mean values correspond to subspecies (o = *orinomus*, c = *cyanoptera*, b = *borreroi*, and s = *septentrionalium*) and indicate significant pairwise differences determined using Bonferroni corrected P values ($P_{adjusted} < 0.05$). Pairwise comparisons were not performed for *A. c. tropica* because of the small sample size.

		Mass	Wing chord	Tarsus	Tail	Nare	Culmen	Bill height	Bill width
orinomus	Mean	450.6	209.9 cbs	34.59 cbs	88.96 s	33.94 c	44.34 c	13.81 s	17.26 cs
(*n* = 9)	SE	11.30	1.32	0.24	1.62	0.24	0.51	0.26	0.22
	Range	390–495	204–217	33.7–36.1	77.9–95.7	33.1–35.1	41.7–46.2	12.6–14.8	16.1–18.3
cyanoptera	Mean	394.2	180.5 bo	31.14 o	77.85 b	31.85 o	42.15 o	12.29 b	16.01 o
(*n* = 13)	SE	10.60	1.10	0.37	1.97	0.26	0.41	0.22	0.22
	Range	340–470	172–185	29.2–33.1	63.1–90.0	30.1–33.6	38.8–44.6	12.1–14.9	15.1–17.6
septentrionalium	Mean	363.5	180.7 b	30.69 o	76.30 ob	32.74	43.10	12.59 ob	16.13 o
(*n* = 10)	SE	14.20	1.60	0.55	2.02	0.48	0.61	0.23	0.25
	Range	315–430	171–187	29.2–34.9	67.0–86.0	30.5–35.1	40.1–46.0	11.1–13.8	15.0–17.4
borreroi	Mean	—	190.6 ocs	31.22 o	104.8 cs	33.81	41.58	14.06 cs	15.92
(*n* = 5)	SE	—	2.68	0.99	10.4	1.19	1.01	0.46	0.50
	Range	—	182–198	30.1–35.2	86–145	29.6–36.9	38.1–43.8	12.6–14.9	14.9–17.6
tropica	Mean	—	171.5	28.94	104.00	24.82	37.84	14.28	15.39
(*n* = 2)	SE	—	4.50	0.78	5.00	0.21	0.52	0.82	0.40
	Range	—	167–176	28.2–29.7	99–109	24.6–25.0	37.3–38.4	13.5–15.1	14.9–15.8
P[a]		<0.001	<0.001	<0.001	<0.001	<0.001	<0.001	<0.001	0.003

[a] ANOVAs for subspecies effect based on pooled data (wet measurements and transformed dry measurements). Sample sizes: *orinomus*, *n* = 15; *cyanoptera*, *n* = 14; *septentrionalium*, *n* = 10; *borreroi*, *n* = 5; *tropica*, *n* = 2.

TABLE 3. Measurements (mm) and body mass (g) for populations of *Anas cyanoptera cyanoptera* in lowland Argentina, the Peruvian coast, and the central high Andes.

Sex		Mass	Wing chord	Tarsus	Tail	Nare	Culmen	Bill height	Bill width
Male									
Argentina	Mean	429.2	194.96	32.50	85.53	35.64	46.02	14.56	17.68
($n = 10$)	SE	9.67	1.20	0.55	1.08	0.46	0.54	0.26	0.28
	Range	450–540	190–200	29.6–34.8	80.9–91.0	33.0–37.6	43.3–48.7	13.5–15.9	16.7–19.2
Peruvian	Mean	394.4	190.5	33.06	80.39	33.21	43.98	13.74	16.51
coast	SE	4.88	1.61	0.37	1.52	0.36	0.43	0.16	0.14
($n = 18$)	Range	340–430	181–205	29.1–35.4	69.5–97.6	30.6–36.1	40.5–47.6	12.7–15.1	15.2–17.6
Andes	Mean	429.2	195.68	34.07	82.45	34.71	44.45	13.79	17.13
($n = 6$)	SE	14.4	3.18	0.48	2.61	0.38	0.89	0.60	0.38
	Range	380–470	186–204	32.5–35.9	75.7–92.6	33.4–35.9	42.6–47.3	12.4–16.3	15.5–18.3
Female									
Argentina	Mean	418.8	182.23	30.75	81	32.24	42.09	14.14	16.73
($n = 4$)	SE	21.4	0.65	0.87	3.32	0.53	0.84	0.34	0.34
	Range	365–470	181–184	29.3–32.3	76.0–90.0	31.2–33.6	40.3–43.6	13.3–14.9	15.9–16.9
Peruvian	Mean	387.0	179.8	31.5	76.4	31.68	41.98	12.95	15.71
coast	SE	10.4	1.37	0.40	2.15	0.26	0.48	0.16	0.18
($n = 10$)	Range	340–430	172–185	29.2–33.1	63.1–86.7	30.1–32.8	38.8–44.6	12.1–13.9	15.1–16.9

subspecies identification, geographic regions differed significantly in body size (Wilks's $\lambda = 0.06$, $F = 17.54$, df = 35 and 662, $P < 0.001$), as did the sexes (Wilks's $\lambda = 0.75$, $F = 7.57$, df = 7 and 157, $P < 0.001$). There was no significant interaction between geographic region and sex (Wilks's $\lambda = 0.78$, $F = 1.16$, df = 35 and 662, $P = 0.244$). The same basic overall pattern was observed, regardless of whether individuals were grouped by subspecies or by geographic region. Highland individuals were significantly larger and intermediate body sizes were found in Patagonia (*A. c. cyanoptera*) and the Colombian highlands (*A. c. borreroi*), except for some notable exceptions. Among females, the lowland Argentine population (*A. c. cyanoptera*) was not significantly different from the central high Andean population (*A. c. orinomus*) in either bill length measurements (bill length at nares: 2.04 mm difference; culmen length: 2.59 mm difference) or bill width (0.39 mm difference). In males, the lowland Argentine population was similar to the Andean populations (*A. c. cyanoptera* and *A. c. orinomus* combined) in tail length, bill length at nares, bill height, and bill width and was significantly larger than the Peruvian coastal population in most measurements (Table 3). Specimens of *A. c. cyanoptera* collected at high elevation in the Andes were significantly smaller than *A. c. orinomus* only in wing chord (27.7 mm difference) and tail length (10.83 mm difference). The North American and Argentine lowland populations of *A. c. cyanoptera* had similar bill lengths, but the Argentine population had significantly greater bill height (1.17 mm difference) and bill width (0.91 mm difference). North American populations had significantly larger bill length at nares (1.78 mm difference) and culmen length (1.65 mm difference) than Peruvian coastal populations of *A. c. cyanoptera*.

Principal component analysis.—The first principal component (PC1; female eigenvalue = 3.54, male eigenvalue = 3.33) accounted for 50.5% and 47.6% of the variance for females and males, respectively,

TABLE 4. Principal components (PC1 and PC2), eigenvectors, eigenvalues, and percent of variance calculated from male and female Cinnamon Teal (*Anas cyanoptera*).

	Males		Females	
	PC1	PC2	PC1	PC2
Wing chord	0.48	0.01	0.48	−0.06
Tarsus bone	0.41	−0.09	0.42	0.05
Tail	0.32	−0.48	0.14	−0.66
Bill length–nare	0.33	0.56	0.39	0.41
Culmen length	0.33	0.54	0.40	0.36
Bill height	0.36	−0.39	0.30	−0.51
Bill width	0.39	−0.09	0.42	−0.12
Eigenvalue	3.33	1.63	3.54	1.57
Variance (%)	47.6	23.2	50.5	22.5
Cumulative (%)	47.6	70.9	50.5	73.0

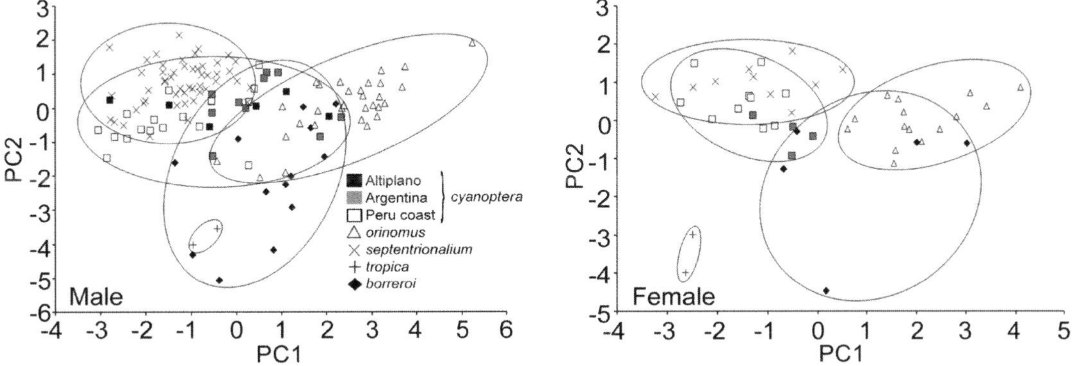

FIG. 2. Principal component analysis (PC1 vs. PC2) of nine body-size measurements for male (left) and female (right) Cinnamon Teal.

and represented an overall body-size difference (Table 4). The second principal component (PC2; female eigenvalue = 1.57, male eigenvalue = 1.63) accounted for 23.2% and 22.5% of the variance for females and males, respectively, and represented a bill-shape difference among the subspecies, as bill measurements were the most influential variables. A longer, thinner bill corresponded with a higher score. Even though plots of PC1 versus PC2 showed some overlap among subspecies, only *A. c. septentrionalium* and *A. c. cyanoptera* did not differ in PC1, and *A. c. orinomus* and *A. c. cyanoptera* did not differ in PC2 (Fig. 2). When subsets of *A. c. cyanoptera* were analyzed geographically (Argentina, Peruvian coast, and Andes), the Argentine population was significantly larger in overall body size

(PC1), whereas the Peruvian coastal population was more similar to *A. c. septentrionalium* (North America). *Anas c. orinomus* had the largest overall body size, with *A. c. borreroi* and the lowland Argentine population and individual *A. c. cyanoptera* collected in northwest Argentina showing intermediate body size. *Anas c. septentrionalium* had the longest bill (PC2) after controlling for variation in body size (Fig. 2).

Partial correlation analysis.—Several significant patterns were found after Bonferroni correction in relation to latitude and elevation (Tables 5 and 6 and Figs. 3–10). Most measurements showed a significant increase with elevation for males and females among all individuals and within South America only (elevation increase of ~4,000 m). In

TABLE 5. Partial correlation coefficients between latitude [a] and body measurements and principal components for Cinnamon Teal (*Anas cyanoptera*). Significant values determined using Bonferroni corrected P values ($P_{adjusted}$ < 0.05) are in bold.

| | Male | | | | Female | | | |
	Pooled data	*A. c. cyanoptera*	Southern South America	North America	Pooled data	*A. c. cyanoptera*	Southern South America	North America
Wing chord	−0.096	0.381	0.097	0.145	−0.092	0.044	−0.061	0.329
Tarsus bone	**−0.343**	−0.280	−0.213	0.029	−0.178	−0.219	−0.290	−0.371
Tail	**−0.425**	0.319	0.184	0.039	−0.362	−0.321	0.041	0.449
Bill length–nare	**0.476**	**0.619**	0.382	0.088	0.223	0.081	0.113	−0.101
Culmen	**0.358**	**0.544**	0.294	−0.155	0.291	0.081	0.052	−0.173
Bill height	**−0.278**	**0.557**	**0.434**	0.197	−0.236	**0.681**	**0.569**	−0.334
Bill width	−0.005	**0.662**	**0.560**	−0.33	0.022	0.407	0.438	−0.589
PC1	−0.065	**0.629**	0.383	−0.013	0.007	0.320	0.271	−0.340
PC2	**0.582**	0.259	0.133	−0.111	0.398	−0.240	−0.366	−0.261

[a] Latitude is calculated in degrees as the absolute value of distance from the equator.

TABLE 6. Partial correlation coefficients between elevation and body measurements and principal components for Cinnamon Teal (*Anas cyanoptera*). Significant values determined using Bonferroni corrected P values ($P_{adjusted} < 0.05$) are in bold.

	Male			Female		
Elevation	Pooled data	South America	North America	Pooled data	South America	North America
Wing chord	**0.724**	**0.782**	−0.038	**0.905**	**0.928**	−0.122
Tarsus bone	**0.603**	**0.619**	−0.018	**0.705**	**0.754**	0.018
Tail	**0.417**	0.354	−0.265	0.388	0.333	−0.529
Bill length–nare	**0.381**	**0.547**	0.270	**0.454**	**0.575**	−0.356
Culmen	0.262	**0.426**	0.312	**0.458**	**0.611**	−0.167
Bill height	0.292	0.275	−0.243	0.357	0.373	−0.590
Bill width	0.265	0.356	−0.188	**0.497**	**0.605**	−0.461
PC1	**0.649**	**0.710**	−0.040	**0.784**	**0.861**	−0.419
PC2	**−0.018**	0.214	**0.494**	−0.118	0.007	0.584

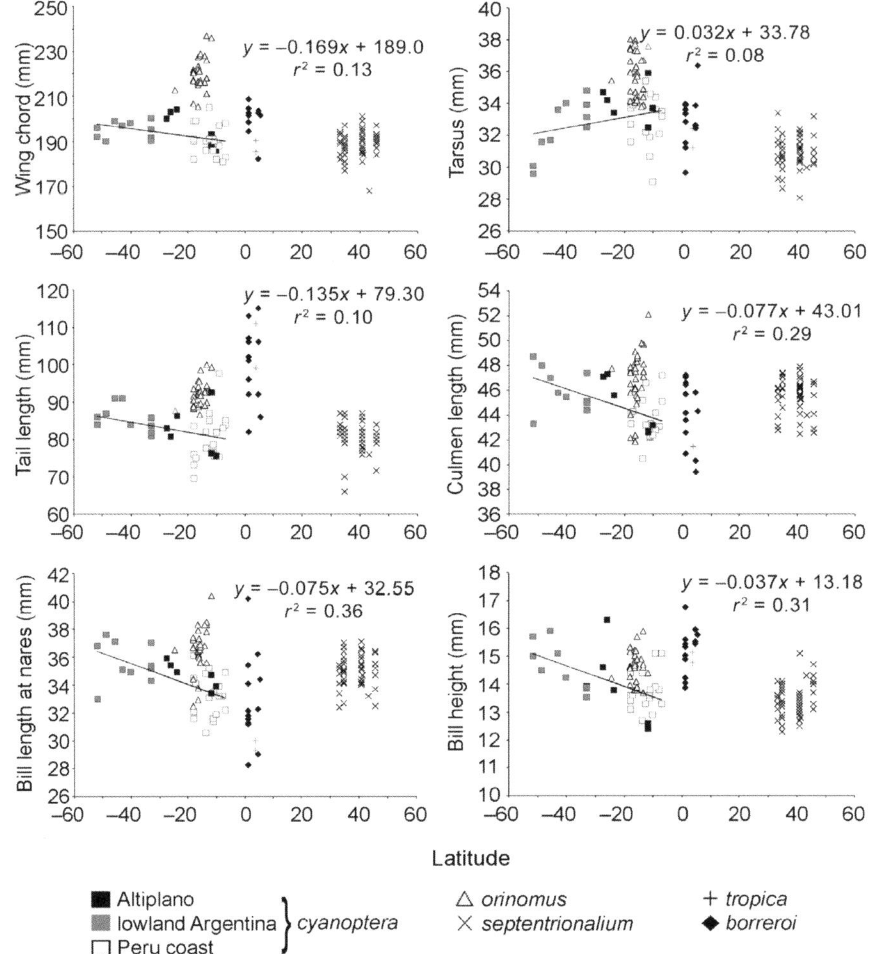

FIG. 3. Relationships between latitude and body-size measurements for male Cinnamon Teal. Regression line is for populations of *Anas cyanoptera cyanoptera* only.

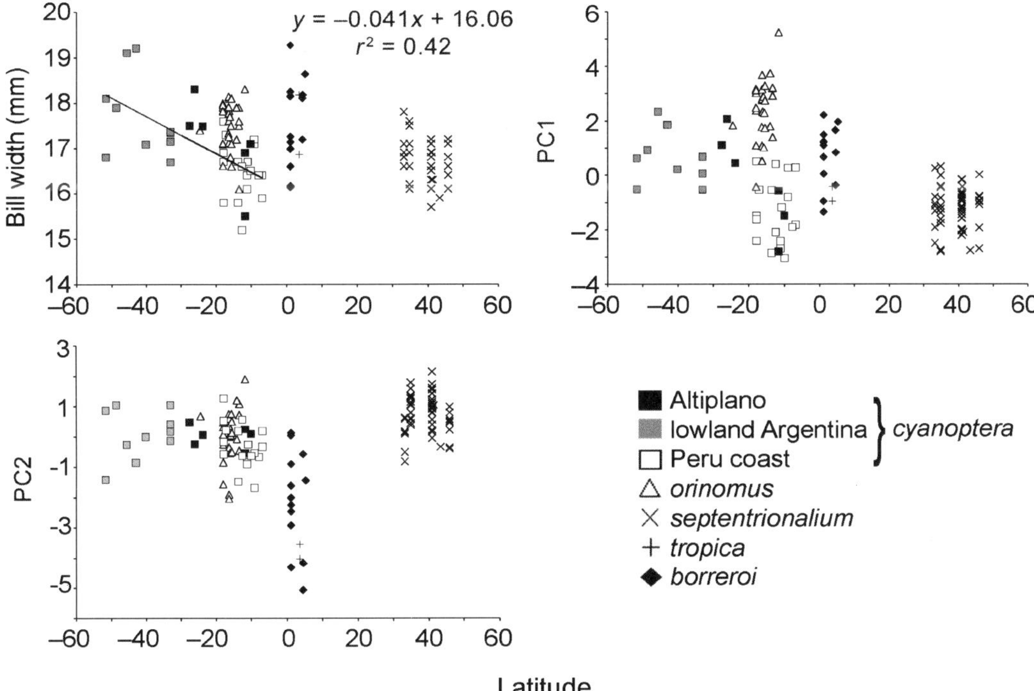

FIG. 4. Relationships between latitude and body-size measurements for male Cinnamon Teal. Regression line is for populations of *Anas cyanoptera cyanoptera* only.

males, PC2 (bill shape) decreased when all individuals were pooled and increased within North America over an elevational increase of ~1,600 m.

Significant correlations with latitude were primarily restricted to males. In females, only bill height showed a positive correlation with latitude within southern populations in South America (*A. c. cyanoptera* and *A. c. orinomus*). In males, tarsus, tail length, and bill height showed a negative correlation, and bill length at nares, culmen length, and PC2 were positively correlated with increasing distance from the equator. Within southern South America, only bill height and bill width were positively correlated with increasing latitude. When only *A. c. cyanoptera* (lowland subspecies) was considered, there was a strong positive correlation between latitude and bill length at nares, culmen length, bill height, bill width, and PC1 from the Peruvian coast to southern Patagonia.

Subspecies classification.—Discriminant analysis with cross-validation correctly classified males to originally assigned subspecies with 69–100% and females with 40–100% accuracy (Table 7).

Discriminant analysis correctly assigned 53.8–86.0% of males and 40.0–100.0% of females to their area of origin (Table 8). Six misclassified male individuals from the central high Andes were assigned to the nearest lowland population adjacent to the area where they were collected, Argentina ($n = 3$) or the Peruvian coast ($n = 3$). All of these individuals were assigned correctly as *A. c. cyanoptera* in the subspecies discriminant analysis.

Male *A. c. orinomus* were diagnosable from *A. c. cyanoptera* using wing chord, tarsus, tail length, and PC1; from *A. c. septentrionalium* using wing chord, tarsus, tail length, bill height, and PC1; and from *A. c. borreroi* using wing chord, tarsus, and bill length at nares (Table 9). The same pattern was found in female *A. c. orinomus*, except that females could not be distinguished from *A. c. borreroi* using bill length at nares (Table 10). *Anas c. septentrionalium* and *A. c. borreroi* were diagnosable using bill length at nares ($D_{sb} = 9.74$, $D_{bs} = 1.50$), culmen length ($D_{sb} = 7.63$, $D_{bs} = 0.01$), bill height ($D_{sb} = 0.02$, $D_{bs} = 2.60$), and PC1 ($D_{sb} = 4.40$, $D_{bs} = 0.91$) for males. When all three subpopulations of *A. c. cyanoptera* were pooled, *A. c.*

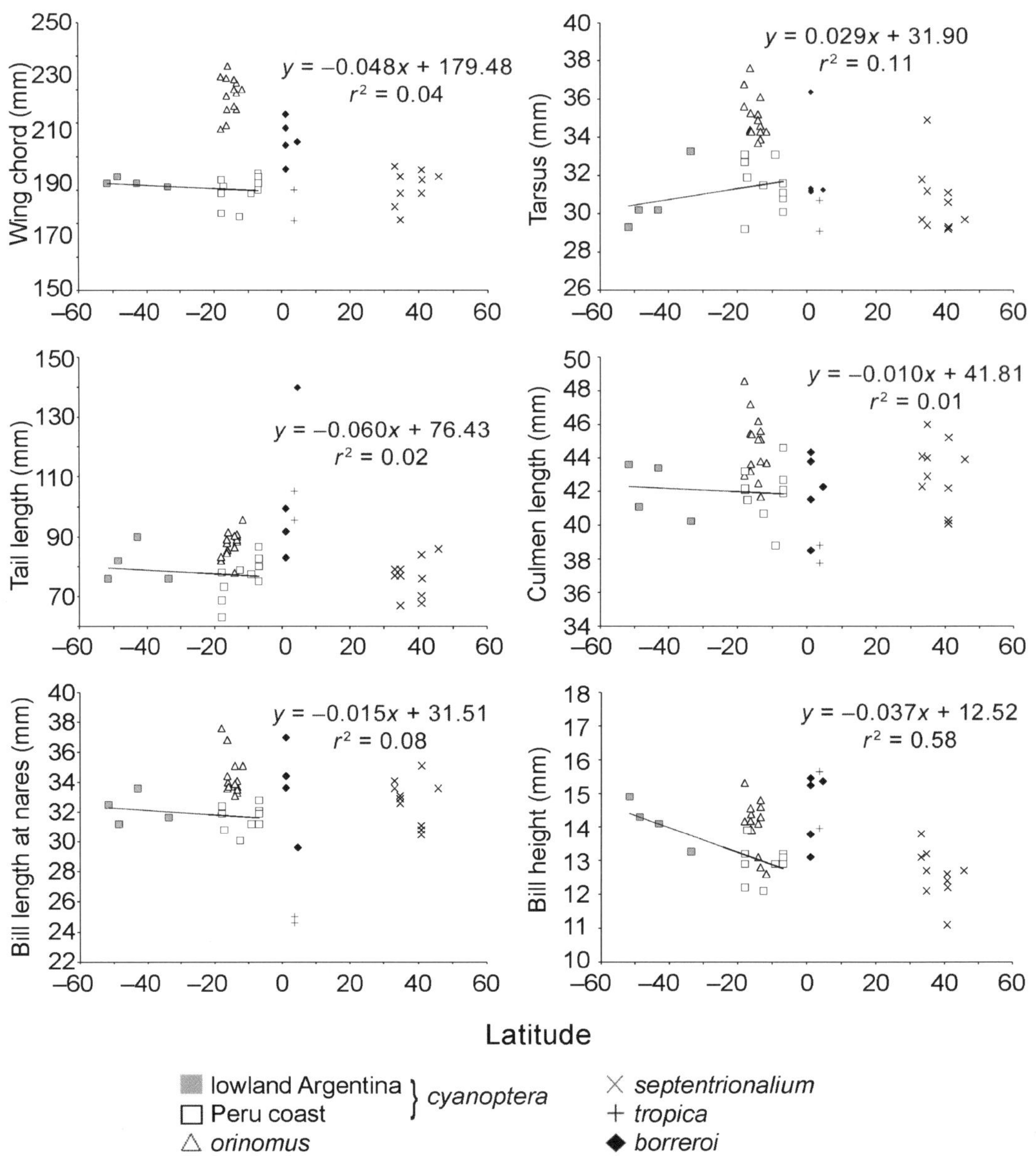

FIG. 5. Relationships between latitude and body-size measurements for female Cinnamon Teal. Regression line is for populations of *Anas cyanoptera cyanoptera* only.

cyanoptera was not diagnosable from *A. c. septentrionalium* or *A. c. borreroi* for any single measurement or PC1. However, at the individual population level, lowland Argentina (*A. c. cyanoptera*) was diagnosable from North America (*A. c. septentrionalium*) using tarsus ($D_{sa} = 5.24$, $D_{as} = 0.90$) and PC1 ($D_{sa} = 4.01$, $D_{as} = 0.61$) and from males in the Colombian highlands (*A. c. borreroi*)

using bill length at nares ($D_{ba} = 10.16$, $D_{ab} = 0.80$). The Peruvian coastal population (*A. c. cyanoptera*) was diagnosable from *A. c. septentrionalium* using bill length at nares ($D_{sp} = 4.91$, $D_{ps} = 0.06$). Female *A. c. borreroi* were diagnosable from both the Peruvian coast ($D_{pb} = 7.94$, $D_{bp} = 1.60$) and lowland Argentine ($D_{ab} = 7.12$, $D_{ba} = 0.37$) populations of *A. c. cyanoptera* using PC1.

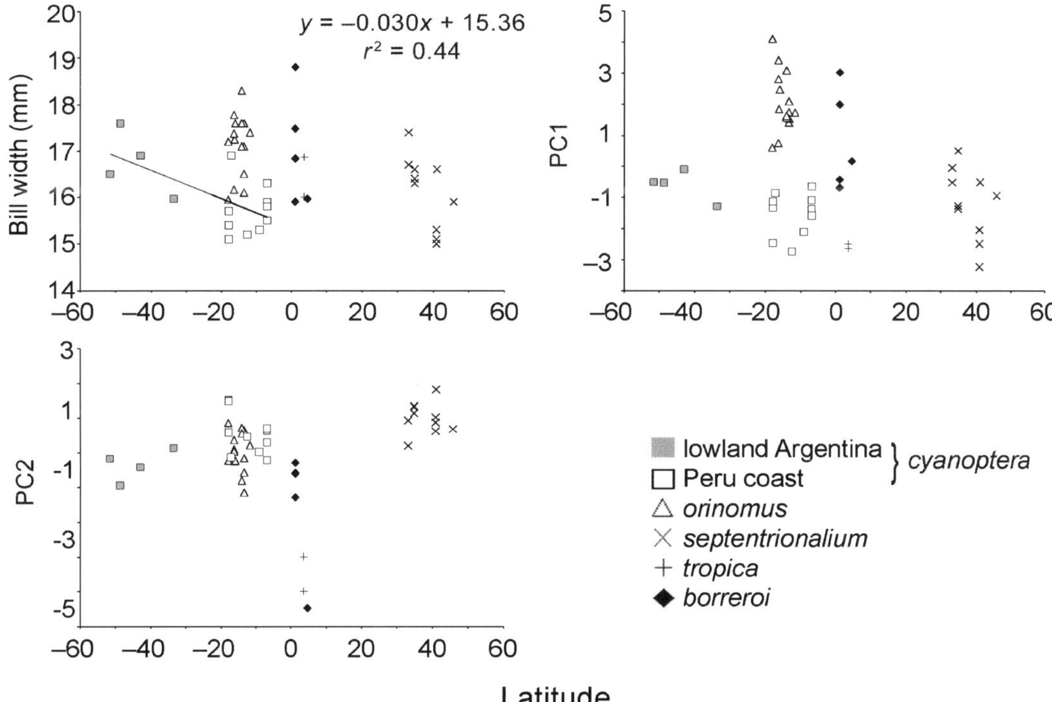

FIG. 6. Relationships between latitude and body-size measurements for female Cinnamon Teal. Regression line is for populations of *Anas cyanoptera cyanoptera* only.

DISCUSSION

Cinnamon Teal are distributed along elevational and latitudinal gradients, and within these gradients climatic and habitat variables change abruptly, placing different selection pressures on different populations (e.g., subspecies). Variances in morphological characteristics appear to conform to ecogeographic regions, given that larger individuals occupied higher elevations in the Andes (*A. c. orinomus* and *A. c. borreroi*) and occur at higher latitudes in Patagonia (*A. c. cyanoptera*), whereas smaller conspecifics resided at lower elevations in temperate regions (*A. c. cyanoptera*, *A. c. septentrionalium*, and *A. c. tropica*). Environmental variables as a function of temperature and humidity have been related to body size, and modifications of Bergmann's rule have been made to take into account factors associated with high latitudes and elevations as well as arid habitats (e.g., "latitude effect," Snow 1954; "aridity effect," Hamilton 1961). However, other factors, such as hypoxia, fasting endurance, and life-history traits (resource competition and migration),

are also known to facilitate variation in body size (Calder 1974, 1984; Hopkins and Powell 2001; Millien et al. 2006).

The climate of the Andes changes dramatically from the warm, wet temperate zone of the Colombian Andes to the colder, arid climates characteristic of the Altiplano and Patagonia. Patagonia is cool, dry, and windy, with substantial seasonal and diurnal temperature fluctuations. Birds that inhabit southern Patagonia experience average low temperatures ranging from 3°C (Esquel, Chubut) to 8°C (Rio Gallegos, Santa Cruz). The Andean Altiplano is also semi-arid, with most precipitation falling during the austral summer (December to February; Garreaud et al. 2003), leaving the rest of the year cool, dry, and windy. The average low temperatures at Cusco, Peru (3,248 m), and La Paz (4,012 m) have been reported as 5°C and 1°C, respectively (Canty and Associates 2005). Separated by an average of 115 km from the puna zone of the Andes, the lowlands of the Peruvian coast consist of scattered river valleys and associated wetlands that are also classified as semi-arid (Pearson and Plenge 1974).

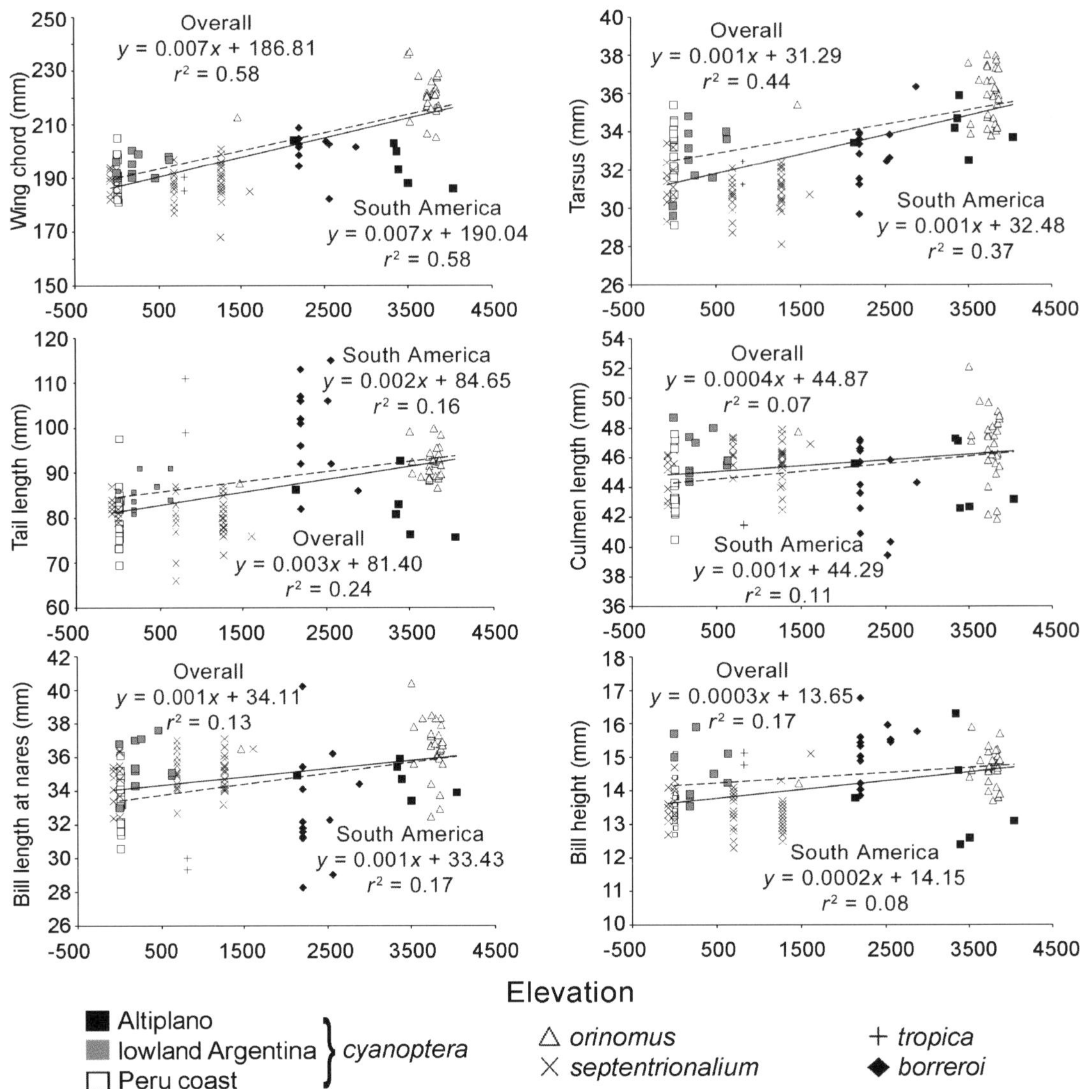

FIG. 7. Relationships between elevation and body-size measurements for male Cinnamon Teal. Dashed regression lines are for South American individuals only.

However, in contrast to the climates of the Altiplano and Patagonia, the Peruvian coast is, on average, 10°C warmer, with temperatures ranging from 15 to 18°C (Canty and Associates 2005). We found that individuals in the warmer, wetter climates of North America (*A. c. septentrionalium*), the Colombian lowlands (*A. c. tropica*), and the Peruvian coast (*A. c. cyanoptera*) had smaller body sizes than those in the central high Andes (*A. c. orinomus* and *A. c. cyanoptera*) and Patagonia (*A. c. cyanoptera*).

Individuals in high-altitude populations of Cinnamon Teal are significantly larger than their closest lowland relatives. The largest subspecies, *A. c. orinomus*, is found exclusively in the central high Andes, with no records of dispersal to adjacent lowland habitats. Individuals collected at mid-elevations (~2,500 m; *A. c. borreroi* and *A. c. cyanoptera* in northwest Argentina) tended to have intermediate body size (Figs. 7–10). High-altitude habitats exert selection pressures that arise from multiple factors (Monge and León-Velarde 1991).

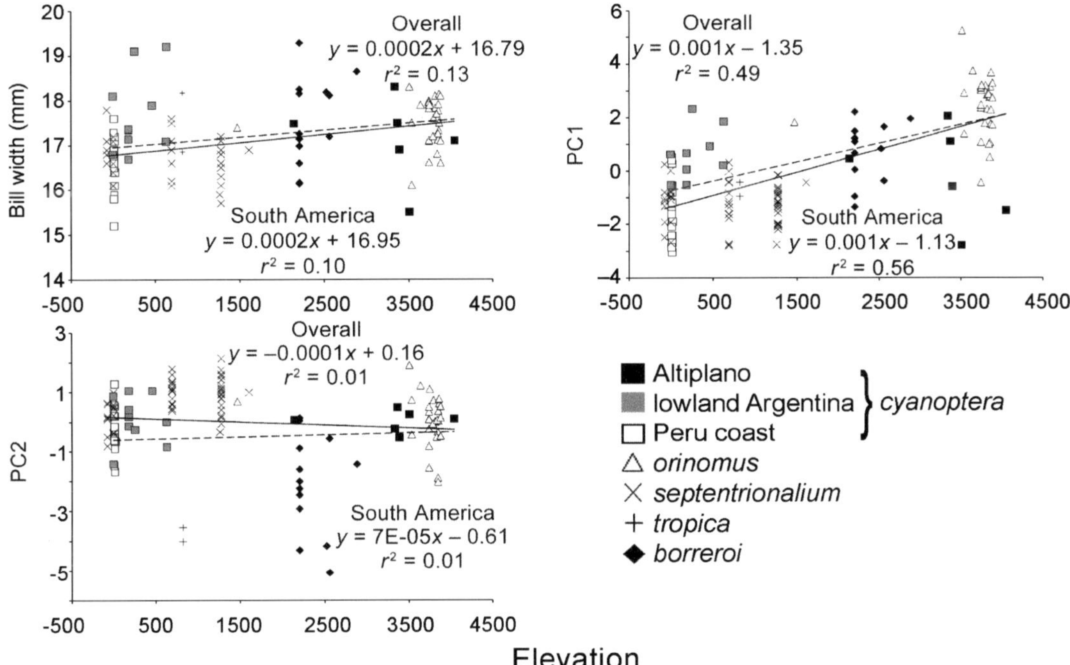

Fig. 8. Relationships between elevation and body-size measurements for male Cinnamon Teal. Dashed regression lines are for South American individuals only.

Besides having a cold, arid climate, these habitats have low air density and the partial pressure of oxygen at 4,000 m is ~60% that at sea level, which may also explain, in part, why high Andean resident populations have larger body size than individuals in populations at lower elevations in the Andes with similar climatic factors (Colombia and Patagonia). Hemoglobin oxygen affinity and body size, for example, have been found to be correlated, such that larger animals tend to have higher affinity (Schmidt-Nielsen and Larimer 1958, Hopkins and Powell 2001). By contrast, smaller-bodied animals tend to have higher metabolic requirements for oxygen, which may favor a higher venous oxygen tension (Schmidt-Nielsen and Larimer 1958, Hopkins and Powell 2001). Other waterfowl species that inhabit similar elevational gradients in the Andes also show a strong correlation between body size and elevation (Blake 1977, Bulgarella et al. 2007). Each highland population also possesses amino acid polymorphisms in the major hemoglobin genes that are likely adaptive (McCracken et al. 2009a, b). Thus, there is an overall trend among South American waterfowl. Larger individuals are found at higher elevations,

whereas the adjacent lowlands are inhabited by smaller conspecifics that also differ in other important traits.

Additionally, there is a general trend for sedentary species to comply more often with Bergmann's rule than migratory species, possibly because nonmigratory species are more affected than migratory species by climatic and other factors such as food availability, in that resident populations are exposed to the same local selection pressures throughout all seasons (Meiri and Dayan 2003). Cinnamon Teal comprise both sedentary and migratory subspecies, with the migratory small-bodied *A. c. septentrionalium* showing few significant correlations with either latitude or elevation. There was a correlation with bill shape (PC2) and elevation among males, which was attributable to a decrease of <0.6 mm in bill width or bill height between Utah (1,275 m) and either Oregon or California (<700 m). However, male breeding-site philopatry is typically very low in dabbling ducks (Anderson et al. 1992). Conversely, South American subspecies, with the exception of the southernmost populations in Argentina, may be predominantly nonmigratory

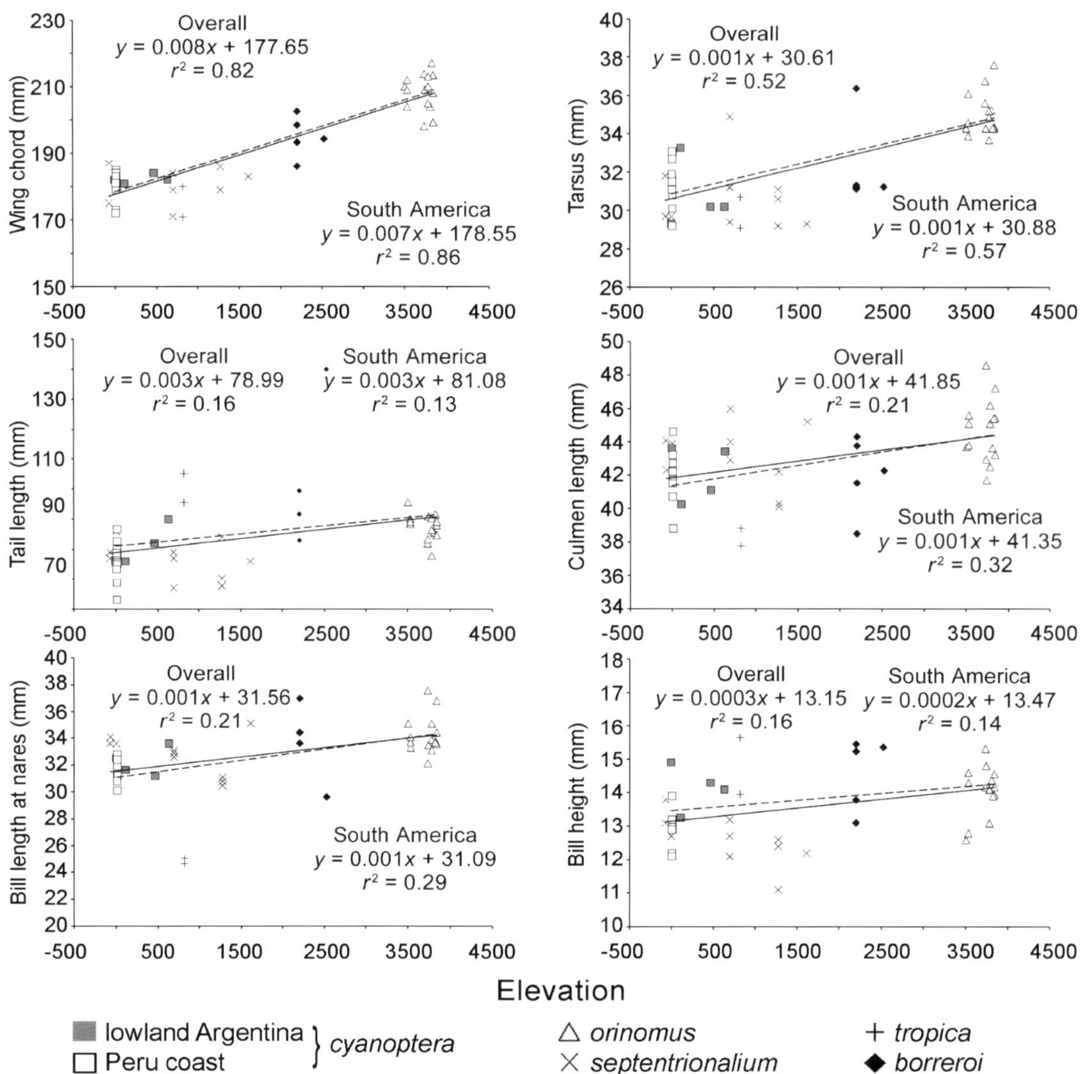

Fig. 9. Relationships between elevation and body-size measurements for female Cinnamon Teal. Dashed regression lines are for South American individuals only.

and show significant correlations between morphological and geographic variables, especially *A. c. cyanoptera*, which occupies a wide range of habitats from coastal Peru to southern Patagonia.

Little information is available on the movements of individual teal between the lowlands and highlands of South America. The lowland subspecies, *A. c. cyanoptera*, occurs in the highlands in small numbers, but the extent of its distribution in the Andes is unknown. We sampled individual *A. c. cyanoptera* in the highlands at only the northern and southern edges of the Altiplano.

Six individuals of this subspecies were collected at 2,141–3,369 m in northwestern Argentina (KGM 442, KGM 1110, KGM 1142) and at 3,393–4,039 m in Peru (REW 118, REW 122, REW 164). There are no records of *A. c. cyanoptera* breeding in the high Andes, and only one individual we collected was in breeding condition (KGM 1142), judging by gonad size (left testis: 30 × 10 mm), even though all individuals were in complete breeding plumage. Two individuals (REW 118, REW 122) from Jauja, Peru (3,506 m), were part of a large group that contained both highland and lowland subspecies.

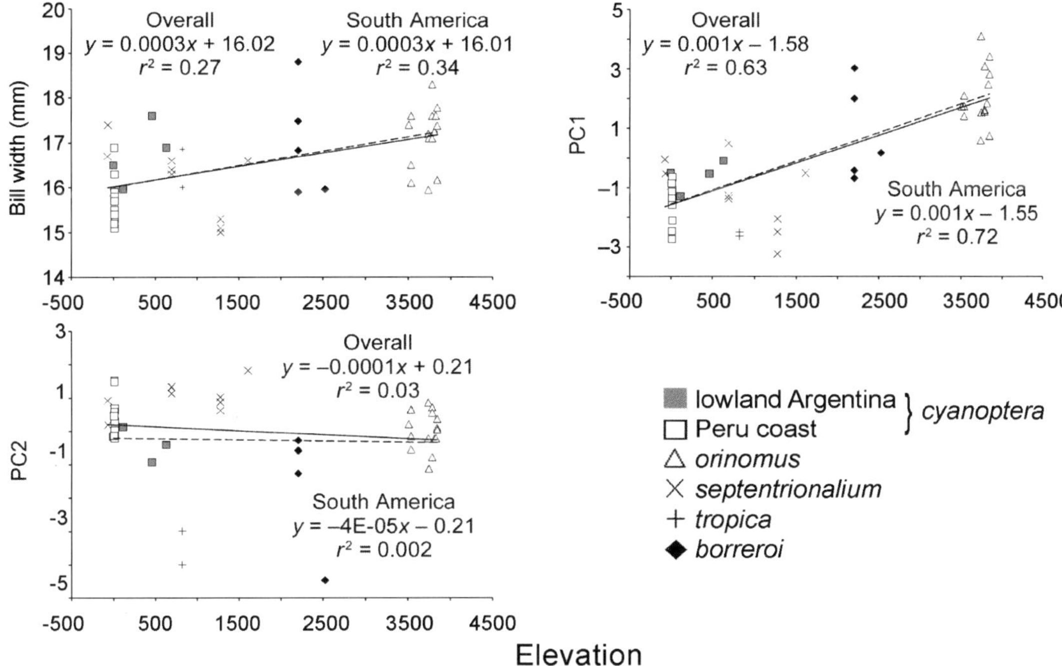

FIG. 10. Relationships between elevation and body-size measurements for female Cinnamon Teal. Dashed regression lines are for South American individuals only.

All other individuals were either solitary or accompanied by one or two other individuals, and no other Cinnamon Teal were found in the surrounding areas. This suggests that these individuals may have been migrants or, more likely, vagrants to these areas rather than permanent residents, as each individual was assigned to the nearest lowland population. In addition, there are no records of *A. c. orinomus* descending to coastal habitats. One *A. c. orinomus* (KGM 441) was collected at 1,468 m in Salta, Argentina, which, to our knowledge, is the lowest elevation reported for this subspecies. Pearson and Plenge (1974) recorded occasional sightings of other Andean waterfowl species (e.g., *A. puna* and *A. flavirostris*) on the coast of Peru, which they attributed to decreased food availability at high elevations during the dry season or competition with seasonal migrants from the south. Water temperature of high Andean lakes (>4,000 m) shows little seasonal variation within the Andean tropical regions, and only the shallow ponds and lakes will freeze or dry up (R. E. Wilson pers. obs.). Cinnamon Teal populations thus face a variety of environmental factors, and phenotypic diversity appears to have arisen from spatial and temporal heterogeneity in selection pressures resulting in adaptations to the local environment.

Subspecies classification.—Morphological (plumage and body size) distinctiveness of individuals in adjacent geographic areas of North America and South America led to the naming of five Cinnamon Teal subspecies (Snyder and Lumsden 1951). However, this classification had not previously been tested. Our analyses (MANOVA and ANOVA) differentiated all subspecies for males, and female *A. c. orinomus* differed from all other subspecies. Discriminant analysis showed high accuracy of subspecies prediction for males of all subspecies and female *A. c. orinomus* and *A. c. cyanoptera*. However, diagnosability of individuals to subspecific groups using the 75% rule (Amadon 1949) showed that few characters reliably distinguished subspecies, excluding *A. c. orinomus*. Low diagnosability among subspecies for females may be attributable, in part, to low sample sizes. The most reliable characters that enabled diagnosis between *A. c. orinomus* and the other subspecies were wing chord, tarsus, and PC1 (overall body-size variable). Low diagnosability of *A. c. cyanoptera* with respect to North American and Colombian subspecies could be attributable to

Table 7. Classification of predicted subspecies of Cinnamon Teal (*Anas cyanoptera*) based on body-size measurements and discriminant analysis with (inside parentheses) and without (outside parentheses) cross-validation. The percentages of individuals that were assigned to their initial subspecific classification are in bold text.

	Predicted			
Initially classified as:	*cyanoptera*	*orinomus*	*septentrionalium*	*borreroi*
Male (*n* = 127)				
cyanoptera	**76.5 (76.5)**	0.0 (0.0)	14.7 (14.7)	8.8 (8.8)
orinomus	0.0 (0.0)	**100 (100)**	0.0 (0.0)	0.0 (0.0)
septentrionalium	8.0 (10.0)	0.0 (0.0)	**92.0 (90.0)**	0.0 (0.0)
borreroi	7.7 (30.8)	0.0 (0.0)	0.0 (0.0)	**92.3 (69.2)**
Total correct:	89.8 (86.6)			
Female (*n* = 44)				
cyanoptera	**85.7 (71.4)**	0.0 (0.0)	0.0 (28.6)	0.0 (0.0)
orinomus	0.0 (0.0)	**100 (100)**	0.0 (0.0)	0.0 (0.0)
septentrionalium	30.0 (50.0)	0.0 (0.0)	**70.0 (50.0)**	0.0 (0.0)
borreroi	0.0 (20.0)	0.0 (40.0)	0.0 (0.0)	**100 (40.0)**
Total correct:	88.6 (72.7)			

within-subspecies variation, given that there were significant mean differences between populations of *A. c. cyanoptera*. When analyzed at the population level, the Argentine and Peruvian coastal populations were diagnosable from *A. c. borreroi* and *A. c. septentrionalium* using bill length measurements or PC1. Upon examination of measurements originally used to define these subspecies, the results are not surprising, because there is considerable overlap in body-size measurements, which indicates that measurements alone may not be sufficient to distinguish subspecies. Other characters, such as plumage coloration and patterns, have been proposed to differentiate subspecies (Snyder and Lumsden 1951). Although the coloration of males within and among subspecies is variable, plumage divergence in color patches that appear identical to the human eye has been reported between *A. c. septentrionalium* and South American subspecies (*A. c. orinomus* and *A. c. cyanoptera*; Wilson et

TABLE 8. Classification of predicted area of origin of individual Cinnamon Teal (*Anas cyanoptera*) based on body-size measurements and discriminant analysis with (inside parentheses) and without (outside parentheses) cross-validation. The percentages of individuals that were assigned to their collection locality are in bold text.

	Predicted				
Initially classified as:	Central high Andes	Argentina	Peruvian Coast	North America	Colombia highlands
Male (*n* = 127)					
Central high Andes	**83.3 (83.3)**	8.3 (8.3)	8.3 (8.3)	0.0 (0.0)	0.0 (0.0)
Argentina	0.0 (0.0)	**80.0 (40.0)**	20.0 (30.0)	0.0 (20.0)	0.0 (10.0)
Peruvian Coast	5.6 (5.6)	0.0 (0.0)	**77.8 (77.8)**	11.1 (11.1)	5.6 (5.6)
North America	0.0 (0.0)	6.0 (8.0)	4.0 (6.0)	**90.0 (86.0)**	0.0 (0.0)
Colombian highlands	7.7 (7.7)	23.1 (23.1)	0.0 (15.4)	0.0 (0.0)	**69.2 (53.8)**
Total correct:	83.5 (77.2)				
Female (*n* = 44)					
Central high Andes	**100 (100)**	0.0 (0.0)	0.0 (0.0)	0.0 (0.0)	0.0 (0.0)
Argentina	0.0 (0.0)	**75.0 (75.0)**	25.0 (25.0)	0.0 (0.0)	0.0 (0.0)
Peruvian Coast	0.0 (0.0)	20.0 (20.0)	**60.0 (60.0)**	20.0 (20.0)	0.0 (0.0)
North America	0.0 (0.0)	10.0 (10.0)	20.0 (40.0)	**70.0 (50.0)**	0.0 (0.0)
Colombian highlands	0.0 (20.0)	0.0 (20.0)	0.0 (0.0)	0.0 (0.0)	**100 (40.0)**
Total correct:	81.8 (70.5)				

TABLE 9. Pairwise diagnosability index values (D_{ij}/D_{ji}) for males of subspecies of Cinnamon Teal (*Anas cyanoptera*). D_{ij} values greater than zero indicate that population *i* is diagnosable from population *j* and are in bold.

	Wing chord	Tarsus	Tail	Nare	Culmen	Bill height	Bill width	PC1
orinomus								
and *cyanoptera*	**37.91/14.22**	**5.57/0.46**	**21.27/5.30**	**5.41**/−0.45	**5.35**/−2.04	**2.50**/−0.02	**2.27**/−0.15	**0.37/4.79**
and *borreroi*	**34.13/7.73**	**6.23/0.58**	−16.81/**9.06**	**11.02/1.82**	**8.30**/−1.03	−1.27/**1.31**	−2.25/**0.65**	−0.50/**3.68**
and *septentrionalium*	**40.98/17.92**	**6.34/2.15**	**18.65/5.85**	**3.25**/−1.64	**3.02**/−3.37	**2.25/0.33**	**1.48**/−0.24	**2.60/5.89**
and *cyanoptera*								
and *borreroi*	−7.06/**17.32**	**3.33**/−2.52	−5.47/**24.28**	**8.55**/−0.66	**6.30**/−2.33	−0.33/**2.85**	−1.47/**1.99**	**3.31**/−1.32
and *septentrionalium*	**13.92**/−7.18	**3.43**/−0.96	**7.30**/−9.40	−0.77/**4.13**	−1.01/**4.67**	**1.31**/−1.21	**0.69**/−1.57	−0.20/**3.53**
and *septentrionalium*								
and *borreroi*	−3.40/**20.39**	−1.64/**3.30**	−4.94/**21.66**	**9.74/1.50**	**7.63/0.01**	**0.02/2.60**	−1.55/**1.19**	**4.40/0.91**

TABLE 10. Pairwise diagnosability index values ($D_{ij}/(D_{ji})$) for females of subspecies of Cinnamon Teal (*Anas cyanoptera*). D_{ij} values greater than zero indicate that population *i* is diagnosable from population *j* and are in bold.

	Wing chord	Tarsus	Tail	Nare	Culmen	Bill height	Bill width	PCI
orinomus								
and *cyanoptera*	**34.35/16.45**	**6.67/1.80**	**24.91/3.07**	**3.84**/−0.64	**5.37**/−1.14	**2.31**/−0.62	**2.63**/−0.13	**0.64/0.27**
and *borreroi*	**32.65/3.85**	**10.63/1.58**	−67.43/**9.29**	**9.47**/−1.25	**9.91**/−0.54	−3.02/**1.69**	**4.18**/−0.74	−4.18/**2.62**
and *septentrionalium*	**38.24/17.13**	**8.46/2.65**	**26.19/4.21**	**4.82**/−1.10	**5.74**/−1.91	**3.01/0.05**	**2.75**/−0.22	**0.69/4.96**
cyanoptera								
and *borreroi*	−5.80/**20.13**	−6.70/**2.96**	−55.88/**24.89**	−7.38/**2.38**	−7.48/**2.35**	−2.22/**2.54**	−2.99/**2.07**	**1.27**/−4.51
and *septentrionalium*	−11.40/**6.83**	**4.53**/−1.89	**14.64**/−12.21	−2.73/**2.22**	−3.32/**3.71**	**2.22**/−0.84	−1.58/**1.55**	−1.14/**2.77**
septentrionalium								
and *borreroi*	−5.11/**24.01**	−5.84/**4.75**	−54.72/**26.17**	−7.84/**3.36**	**8.24**/−2.73	−1.54/**3.24**	−3.09/**2.19**	**7.28**/−0.01

al. 2008). The Colombian subspecies (*A. c. borreroi* and *A. c. tropica*) are typically darker in coloration, with spotting occurring at higher frequency (100% in *A. c. tropica*) than in the other three subspecies, but spotting also can be variable, with substantial overlap among other subspecies (Snyder and Lumsden 1951). The tone of the cinnamon color in males ranges from dark (Colombian subspecies) to pale (*A. c. orinomus*). Females are more difficult to differentiate with plumage, but in general, as with males, Colombian subspecies are darker in color. Thus, we suggest that the current subspecies classification is valid on the basis of body-size measurements (present study) and plumage coloration and as described by Snyder and Lumsden (1951) and Wilson et al. (2008).

Acknowledgments

We thank R. Acero, Y. Arzamendia, D. Blanco, G. Cao, R. Cardón, S. Chavarra, C. Chehébar, R. Clarke, M. Christie, A. Contreras, V. H. Eztellan, M. Funes, S. Goldfeder, I. Gomez, A. Gonzalez, M. Herrera, G. Jarrell, L. Janke, D. Johnson, K. Johnson, C. Kopuchian, P. J. McCracken, R. Miatello, K. Naoki, M. Nores, M. Peck, C. Quiroga, D. Ramadori, K. Ramirez, A. Rojas, P. Tubaro, A. del Valle, M. Vidaurre, and the following agencies in Argentina (Dirección de Fauna Silvestre—Secretaría de Ambiente y Desarrollo Sustentable de la República Argentina, Museo Argentino de Ciencias Naturales, Dirección de Fauna Santa Cruz, Ministerio de la Producción Chubut, Secretaría de Estado de Produccíon Río Negro, Centro de Ecología Aplicada y Dirección Provincial Recursos Faunisticos y Areas Naturales Protegidas Neuquén, Dirección de Ordenamiento Ambiental-Área Técnica de Fauna Cordoba, Secretaría de Medioambiente y Desarrollo Sustentable Salta, Dirección Provincial de Recursos Naturales y Medioambiente Jujuy), Bolivia (Colección Boliviana de Fauna), Canada (Royal Ontario Museum), Peru (Instituto Nacional de Recursos Naturales del Perú, Museo de Historia Natural de la Universidad de San Marcos), and the United States (Ambassador Duck Club, Brown's Park National Wildlife Refuge [NWR], Edwards Air Force Base, Malheur NWR, Smithsonian Institution National Museum of Natural History, Sony Bono Salton Sea NWR, and University of Alaska Fairbanks Department of Mathematics and Statistics). Expedition and laboratory costs were funded by the Institute of Arctic Biology at the University of Alaska Fairbanks, Alaska EPSCoR (NSF EPS-0092040 and EPS-0346770), grants from the Delta Waterfowl Foundation, Frank M. Chapman Fund at the American Museum of Natural History, David Burnett Memorial Award to R.E.W., and National Science Foundation grant DEB-0444748 to K.G.M. C. Lewis-Ames, D. Banks, A. Collins, M. Eaton, F. James, S. Sonsthagen, M. Sorenson, K. Winker, and two anonymous reviewers provided helpful comments on drafts of the manuscript.

Appendix 1. Specimens of *Anas cyanoptera* examined, with collection locality. KGM, JT, and REW specimens are catalogued at University of Alaska Museum, Fairbanks.

A. c. borreroi
COLOMBIA: Dept. Putumayo, Sibundoy
ROM 79230, ROM 79231, ROM 79232, ROM 79233,ROM 79234, ROM 91946, ROM 91947, ROM 91948, ROM 91949, ROM 91950, ROM 91954, SM437473, SM437474
COLOMBIA: Dept. Cundinamarca, La Hererra
ROM 91943, ROM 91953
COLOMBIA: Dept. Cundinamarca, Laguna Fuquene
SM 437475
COLOMBIA: Dept. Cundinamarca, Sabana de Bogota
ROM 91944, SM437472

A. c. tropica
COLOMBIA: Dpto. Valle del Cauca, Vijes
ROM 91957, ROM 91958, ROM 91959, ROM 91960

A. c. septentrionalium
USA: Utah, Weber Co., 41° 14' 59.7" N, 112° 07' 55.8" W, 1,275 m
REW 075
USA: Utah, Salt Lake Co., 40° 50' 50.7" N, 112° 01' 50.9" W, 1,275 m
REW 077, REW 078, REW 079
USA: Oregon, Columbia Co., 45° 45' 18.1" N, 122° 50' 51.4" W, 1 m

---

(*continued*)

APPENDIX 1. (Continued)

REW 797, REW 398, REW 399, REW 400, REW 401, REW 402, REW 403, REW 404, REW 406
USA: California, Imperial Co., 33° 11′ 24.0″N, 115° 35′ 18.5″W, −68 m
REW 411, REW 412, REW 414, REW 416, REW 418, REW 419, REW 421
USA: California, Imperial Co., 33° 11′ 39.0″N, 115° 34′ 46.2″W, −73 m
REW 415, REW 420
USA: California, Kerns Co., 34° 47′ 43.5″N, 118° 07′ 11.3″W, 693 m
REW 422, REW 423, REW 424, REW 425, REW 426, REW 427, REW 428, REW 429, REW 430, REW 431, REW 432,
REW 433, REW 434, REW 435, REW 436, REW 437
USA: Utah, Salt Lake Co., 40° 50′ 45.1″N, 112° 01′ 41.7″W, 1,275 m
REW 438, REW 439, REW 440, REW 441, REW 442, REW 443, REW 444, REW 445, REW 446, REW 447, REW 448,
REW 449, REW 450, REW 451, REW 452, REW 453, REW 454, REW 455, REW 456
USA: Colorado, Moffat Co., 40° 59′ 10.7″N, 108° 59′ 10.5″W, 1,609 m
REW 457, REW 458
USA: Oregon, Harney Co., 48° 43′ 53.7″N, 118° 50′ 25.3″W, 1,260 m
REW 464

A. c. cyanoptera
ARGENTINA: Neuquen, Rio Collon Cura, R.N. 40, 40° 12′ 45″S, 70° 38′ 58″W, 625 m[a]
KGM 268
ARGENTINA: Cordoba, Laguna La Felipa, 33° 04′ 17″S, 63° 31′ 33″W, 184 m[a]
KGM 310, KGM 313, KGM 311, KGM 312
ARGENTINA: Cordoba, S. Canals, 33° 36′ 23″S, 62° 53′ 16″W, 112 m[a]
KGM 322
ARGENTINA: Jujuy, S. Purmamarca, 23° 49′ 13″S, 65° 28′ 34″W, 2,141 m
KGM 442
PERU: Dpto. Lima, S. Huacho, 11° 10′ 12.9″S, 77° 35′ 31.4″W, 15 m
REW 081, REW 082
PERU: Dpto. Junin, Jauja, Laguna de Paca, 11° 44′ 14.5″S, 75° 29′ 32.7″W, 3,506 m
REW 118, REW 122
PERU: Dpto. Ancash, Laguna Conococha, 10° 07′ 10.8″S, 77° 17′ 00.7″W, 4,039 m
REW 164
PERU: Dpto. Lambayeque, ca. Puerto Eten, 06° 54′ 51.9′S, 79° 52′ 22.4″W, 13 m
REW 193, REW 194, REW 195, REW 196
PERU: Dpto. Lambayeque, Playa Monsefu, 06° 54′ 03.7″S, 79° 53′ 42.4″W, 12 m
REW 198, REW 199
PERU: Dpto. La Libertad, Magdalena de Cao, 07° 51′ 54.3″S, 79° 20′ 51.2″W, 23 m
REW 200
PERU: Dpto. Ancash, Chimbote, 09° 07′ 26.0″S, 78° 33′ 11.3″W, 15 m
REW 203, REW 204, REW 205
PERU: Dpto. Ancash, Puerto Huarmey, 10° 05′ 52.0″S, 78° 09′ 10.3″W, 14 m
REW 206
PERU: Dpto. Lima, Albufera de Medio Mundo, 10° 55′ 25.9″S, 77° 40′ 10.8″W, 14 m
REW 207
PERU: Dpto. Ica, Pisco, 13° 41′ 46.8″S, 76° 13′ 07.3″W, 7 m
REW 235
PERU: Dpto. Ica, Pisco, 13° 40′ 47.2″S, 76° 12′ 56.6″W, 9 m
REW 236
PERU: Dpto. Tacna, Ite, 17° 52′ 47.2″S, 71° 01′ 05.9″W, 10 m
REW 298, REW 299, REW 300, REW 301, REW 302, REW 303, REW 304
PERU: Dpto. Arequipa, Punta de Bombon-Islay, 17° 11′ 31.9″S, 71° 46′ 19.4″W, 8 m
REW 305, REW 306
PERU: Dpto. Lima, 2 km N. La Laguna, 12° 33′ 13.0″S, 76° 42′ 42.1″W, 9 m
REW 315, REW 316, REW 317

(*continued*)

APPENDIX 1. (Continued)

ARGENTINA: Chubut, Laguna Terraplen, 42°59′50.7″S, 71°30′55.1″W, 630 m
KGM 712, KGM 713
ARGENTINA: Santa Cruz, Estancia La Angostura, 48°38′33.9″S, 70°38′37.3″W, 460 m
KGM 766, KGM 767
ARGENTINA: Santa Cruz, ca. Punta Loyola, 51°37′35.7″S, 69°00′59.4″W, −3 m
KGM 797, KGM 798
ARGENTINA: Santa Cruz, ca. Punta Loyola, 51°36′54.9″S, 68°59′26.6″W, 0 m
KGM 799
ARGENTINA: Chubut, S. Lago Colhue Huapi, 45°38′49.6″S, 68°56′45.1″W, 256 m
KGM 808
ARGENTINA: Catamarca, Antofogasta de la Sierra, Laguna La Alumbrera, 26°06′46.4″S 67°25′26.7″W, 3,338 m
KGM 1110
ARGENTINA: Catamarca, Embalse Las Cortaderas, 27°33′21.2″S, 68°08′41.9″, 3,369 m
KGM 1142

A. c. orinomus
ARGENTINA: Salta, NE La Caldera, 24°33′01″S, 65°22′15″W, 1,468 m
KGM 441
BOLIVIA: Dpto. La Paz, Lago Titicaca, 16°11′45″S, 68°37′28″W, 3,808 m
KGM 485, KGM 486, KGM 487
BOLIVIA: Dpto. La Paz, Lago Titicaca, 16°20′13″S, 68°41′20″W, 3,854 m
KGM 499
BOLIVIA: Dpto. Oruro, Lago Uru Uru, 18°02′03″S, 67°08′46″W, 3,735 m
KGM 527, KGM 528, KGM 529, KGM 530, KGM 531, KGM 532, KGM 533, KGM 534, KGM 535
BOLIVIA: Dpto. La Paz, Lago Titicaca, 16°25′28″S, 68°51′43″W, 3,850 m
KGM 557
BOLIVIA: Dpto. La Paz, Lago Titicaca, Cohani, 16°21′03″S, 68°37′40″W, 3,839 m
KGM 559, KGM 560
BOLIVIA: Dpto. La Paz, Lago Titicaca, Cohani, 16°21′02″S, 68°37′48″W, 3,840 m
KGM 561, KGM 562
BOLIVIA: Dpto. La Paz, Lago Titicaca, Cohani, 16°21′07″S, 68°38′06″W, 3,845 m
KGM 563, KGM 564, KGM 565, KGM 566
PERU: Dpto. Junin, Jauja, Laguna de Paca, 11°44′14.5″S, 75°29′32.7″W, 3,506 m
REW 125, REW 126
PERU: Dpto. Cusco, Laguna Chacan, 13°26′02.6″S, 72°07′49.6″W, 3,533 m
REW 238, REW 239, REW 240, REW 241, REW 242
PERU: Dpto. Cusco, ca. Chinchero, 13°25′49.3″S, 72°03′41.7″W, 3,789 m
REW 248
PERU: Dpto. Cusco, Urubamba Valley, 13°25′22.9″S, 72°02′38.2″W, 3,743 m
REW 253, REW 254
PERU: Dpto. Cusco, ca. Laguna Pomacanchi, 14°06′51.9″S, 71°27′56.6″W, 3,781 m
REW 255, REW 256, REW 257, REW 258, REW 259
PERU: Dpto. Puno, Lago Titicaca, Jaru Jaru, 15°59′05.6″S, 69°36′24.3″W, 3,824 m
REW 268, REW 269
PERU: Dpto. Puno, Lago Titicaca, ca. Puno, 15°52′01.2″S, 69°56′21.3″W, 3,830 m
REW 271
PERU: Dpto. Puno, Lago Umayo, Sillvstani, 15°42′45.8″S, 70°09′00.0″W, 3,853 m
REW 272
PERU: Dpto. Puno, Deustva, 15°33′50.0″S, 70°14′33.1″W, 3,871 m
REW 284, REW 285, REW 286

---

[a] These elevation values are interpolated from the U.S. Geological Survey's GTOPO30 digital elevation model (available at eros.usgs.gov/); all other elevations were measured with a GPS receiver.

APPENDIX 2. Dry body-size measurements (mm) for three subspecies of Cinnamon Teal (*A. cyanoptera cyanoptera*).

	A. c. orinomus[a]			*A. c. cyanoptera*[a]			*A. c. septentrionalium*[a]		
	Mean	SE	Range	Mean	SE	Range	Mean	SE	Range
Male									
Wing chord	215.4	1.05	200–229	186.8	1.39	176–201	185.5	1.0	163–199
Tarsus bone	34.75	0.30	32.2–37.3	32.31	0.21	30.1–34.1	30.32	0.14	27.7–32.1
Tail	96.07	1.08	82.0–108.0	86.52	1.22	75–102	78.75	0.53	66.0–85.0
Bill length–nare	36.61	0.32	32.4–39.6	34.32	0.37	30.9–37.6	35.05	0.16	32.0–36.8
Culmen	47.19	0.44	42.0–52.4	44.36	0.49	40.1–49.3	44.70	0.20	41.7–47.1
Bill height	14.36	0.23	12.5–17.6	13.34	0.14	11.5–15.2	12.58	0.10	11.0–14.4
Bill width	16.49	0.13	14.9–17.4	16.05	0.18	14.1–17.7	15.58	0.15	12.6–17.4
Female									
Wing chord	202.4	1.7	193–217	177.9	2.0	167–191	178.5	1.8	169–186
Tarsus bone	33.59	0.54	30.6–37.2	31.29	0.45	29.3–33.5	30.29	0.43	28.8–32.8
Tail	91.87	1.48	84.0–102.0	83.94	2.85	71.3–102.0	77.10	1.86	69.0–88.0
Bill length–nare	34.16	0.44	32.2–37.9	31.83	0.34	29.9–33.1	32.83	0.49	30.5–35.4
Culmen	44.57	0.55	41.7–49.8	41.41	0.62	38.6–43.8	42.37	0.54	39.7–44.9
Bill height	13.82	0.24	12.5–15.5	12.80	0.29	11.1–14.1	11.83	0.28	10.2–13.1
Bill width	15.70	0.28	13.6–17.2	15.04	0.16	14.2–16.2	15.07	0.39	13.1–17.3

[a] Sample sizes: *A. c. orinomus*, 29 males and 14 females; *A. c. cyanoptera*, 27 males and 10 females; *A. c. septentrionalium*, 47 males and 10 females.

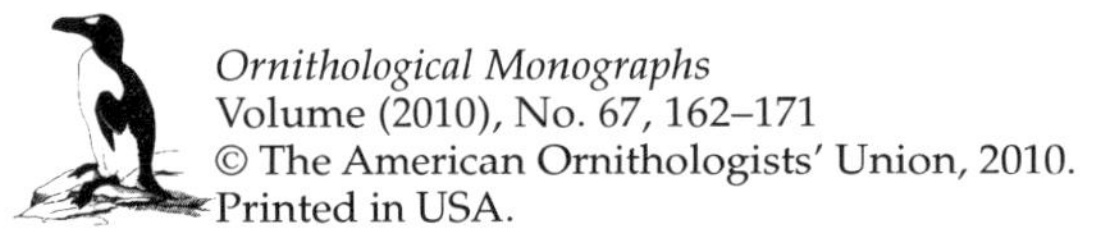

Ornithological Monographs
Volume (2010), No. 67, 162–171
© The American Ornithologists' Union, 2010.
Printed in USA.

CHAPTER 13

ALASKA SONG SPARROWS (*MELOSPIZA MELODIA*) DEMONSTRATE THAT GENETIC MARKER AND METHOD OF ANALYSIS MATTER IN SUBSPECIES ASSESSMENTS

CHRISTIN L. PRUETT[1,2,3] AND KEVIN WINKER[2]

[1]*Florida Institute of Technology, Department of Biological Sciences, 150 W. University Boulevard, Melbourne, Florida 32901, USA; and*
[2]*University of Alaska Museum, 907 Yukon Drive, Fairbanks, Alaska 99775, USA*

ABSTRACT.—We examined genetic and morphological characteristics of the Song Sparrows (*Melospiza melodia*) of northwestern North America, which have a relatively large number of phenotypically described subspecies ($n = 6$ in this region). Mitochondrial DNA (mtDNA) sequences showed little information about these subspecies, with no reciprocal monophyly evident. However, differences in body mass and microsatellite allele frequencies supported continued recognition of subspecific units for taxonomy and conservation. Song Sparrow subspecies in this region are probably representative of many recently diverged populations that have not been isolated long enough for complete lineage sorting using mtDNA markers, yet which have evolved differences that are likely to be genetically based. We emphasize the importance of using multiple lines of evidence, genetic and morphological, in assessing subspecific status, lest we overlook important biological diversity that has accrued below the level of full species.

Key words: body mass, *Melospiza melodia*, microsatellite, mitochondrial DNA, population genetics, Song Sparrow.

Las Poblaciones de Alaska de *Melospiza melodia* Demuestran que los Marcadores Genéticos y los Métodos de Análisis Afectan la Evaluación de las Subespecies

RESUMEN.—Examinamos características genéticas y morfológicas en poblaciones del noroeste de Norteamérica de *Melospiza melodia*, una especie que tiene un número relativamente grande de subespecies descritas fenotípicamente ($n = 6$ en esta región). Las secuencias de ADN mitocondrial (ADNmt) mostraron poca información sobre estas subespecies, que no presentan monofilia recíproca evidente. Sin embargo, las diferencias en tamaño corporal y las frecuencias alélicas de microsatélites continúan apoyando el reconocimiento de unidades subespecíficas para taxonomía y conservación. Las subespecies de M. melodia en esta región probablemente representan varias poblaciones que se diferenciaron recientemente y que no han estado aisladas por suficiente tiempo para alcanzar la separación completa de linajes del ADNmt, pero que se han diferenciado evolutivamente en rasgos que probablemente tienen una base genética. Enfatizamos la importancia de usar líneas de evidencia múltiples, genéticas y morfológicas, al evaluar el estatus de las subespecies. De lo contrario, pasaremos por alto una diversidad biológica importante que se ha acumulado por debajo del nivel de las especies.

[3]E-mail: cpruett@fit.edu

Ornithological Monographs, Number 67, pages 162–171. ISBN: 978-0-943610-86-3. © 2010 by The American Ornithologists' Union.

THE USEFULNESS OF subspecies as units in conservation and taxonomy has been a subject of ongoing debate (Zink 2004, Remsen 2005, Phillimore and Owens 2006). Some researchers have stressed the continuing use of subspecies as valid taxonomic units (Patten and Unitt 2002, Phillimore and Owens 2006), whereas others have advocated the complete elimination of subspecies, elevating groups that are reciprocally monophyletic (assessed using mitochondrial DNA [mtDNA] sequences) to full species (Zink 2004). By using a single-criterion approach (even if using cladistic criteria on morphological characters), many readily identifiable subspecies that are based on phenotypic (morphological) characters but that do not show complete mitochondrial (or phenotypic) monophyly would be ignored in conservation decisions (Zink and Dittmann 1993a, Zink et al. 2000). The single-locus genetic approach advocated by Zink (2004), although often useful when assessing historical relationships (Avise 2000, Zink and Barrowclough 2008), can be problematic when examining the uniqueness of populations that have recently diverged, have large effective population sizes, or have rapidly evolved differences adapted to varying environments (Greenberg et al. 1998, Moritz 2002, Bulgin et al. 2003).

We present a case study of Song Sparrows (*Melospiza melodia*) found in northwestern North America, a group of populations that is relatively rich in phenotypically based subspecies (Patten and Pruett 2009) and has a recent colonization history over much of its distribution (Zink and Dittmann 1993a, Fry and Zink 1998). We examined eight breeding populations of Song Sparrows, representing six subspecies, using body-mass measurements, mtDNA cytochrome-*b* sequences, and eight microsatellite loci to determine the validity of a single-locus mtDNA approach for subspecies assessment. Regarding genetic differentiation between and among subspecies that are based on phenotype, we consider how the use of different genetic markers and different methods used to analyze these data might give different answers.

METHODS

Phenotype.—Song Sparrow subspecies ($n = 6$) in northwestern North America are described and recognizable on the basis of plumage and mensural characters (Patten and Pruett 2009); the most pronounced attribute of populations in this region is the increasing body size of individuals from east to west in Alaska (Gabrielson and Lincoln 1951). Although intraspecific plumage variation may be adaptive (Zink and Remsen 1986, Mumme et al. 2006), minor variations among populations raise legitimate questions about just how different some described subspecies really are in an evolutionary, adaptive context. Dramatic changes in body size among populations arguably provide better evidence of localized adaptive variation within a species, and it is here that we focused our measure of phenotype in this species (although some body-size differences could be the result of developmental plasticity; West-Eberhard 2003). We measured the body mass of 268 male Song Sparrows collected during the breeding season (Appendix). We grouped birds into subspecies (Patten and Pruett 2009) on the basis of collection locality and plumage, and the mean and standard error of body mass were calculated for each group. We used one-tailed *t*-tests to determine whether there were differences in body mass between neighboring subspecies (Fig. 1). We recognize that body mass alone is not a sufficient character to separate these subspecies using the established threshold criterion of the 75% rule (Amadon 1949, Patten and Unitt 2002) but consider that differences among populations in such a fundamental attribute may have important biological significance.

Population genetics.—We extracted whole genomic DNA from the tissues of 205 Song Sparrows (Glenn 1997) from eight breeding populations that correspond to six named subspecies (Fig. 1 and Table 1). We collected birds during the breeding season. We amplified most of the mtDNA cytochrome-*b* gene (1,137 bp) and cycle-sequenced amplifications using four primer pairs per individual for a subset of the extracted tissues (Table 1). We used the following primers: L14851 (Kornegay et al. 1993), H16064 (Harshman 1996), L15350 (Klicka and Zink 1997), and H15424 (Hackett 1996). We sequenced amplified products in both directions using an ABI 373A or 3100 automated sequencer (Applied Biosystems, Foster City, California). We deposited all sequences in GenBank (Table 1). We amplified and genotyped eight microsatellite loci for all individuals (Table 1) as presented in Pruett and Winker (2005). We used this data set for microsatellite analyses.

We used maximum-likelihood (PAUP*, version 4.0b10; Swofford 2003) and Bayesian analyses (MRBAYES, version 3.1.2; Huelsenbeck and Ronquist 2001) to construct phylogenetic trees. We determined the most appropriate model and

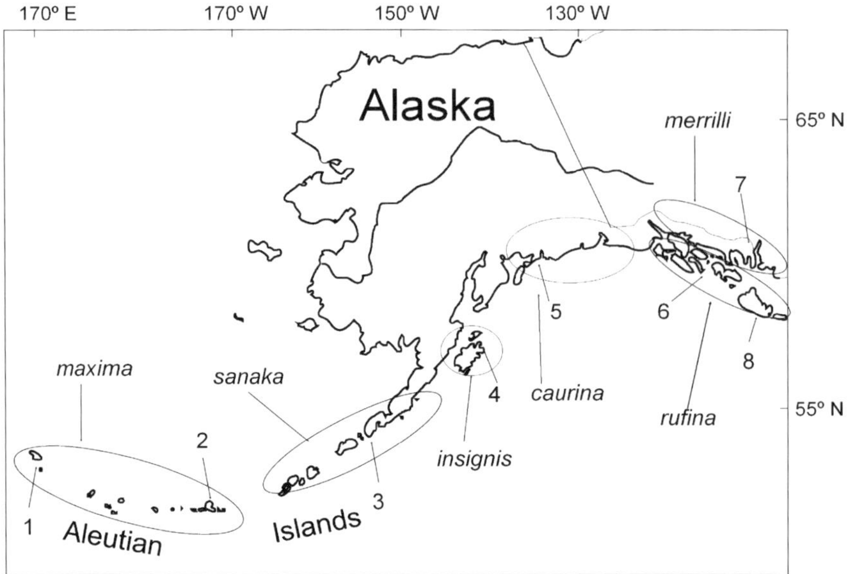

FIG. 1. Map of Song Sparrow (*Melospiza melodia*) subspecies and collection locations used in genetic data analyses. Numbers correspond to (1) Attu Island, (2) Adak Island, (3) Alaska Peninsula, (4) Kodiak Island, (5) Copper River Delta, (6) Alexander Archipelago, (7) Hyder, and (8) Queen Charlotte Islands, British Columbia.

parameter estimates for analyses using Akaike's information criterion in MODELTEST, version 3.06 (Posada and Crandall 1998) and PAUP*. We used the Hasegawa, Kishino, and Yano model (Hasegawa et al. 1985) with the shape of the gamma distribution accounting for substitution rate heterogeneity (HKY+G). We evaluated bootstrap support for the likelihood tree by resampling the data matrix 100 times (Felsenstein 1985). For the Bayesian analysis, we used four independent runs with random starting trees to ensure that the Markov chain converged at optimal likelihood values. We sampled trees every 10,000 generations, and we ran each analysis for 8 million generations. We discarded trees sampled before the Markov chain reached a plateau as burn-in

TABLE 1. Sampling location, number of individuals sequenced, number of individuals genotyped, and GenBank accession numbers for Song Sparrows used in this study. Museum voucher numbers are provided on GenBank.

Location	Sequenced (*n*)	Genotyped (*n*)	GenBank accession numbers
Attu Island, Aleutian Islands, Alaska	10	30 [a]	AY156386–395
Adak Island, Aleutian Islands, Alaska	10	30	AY156396–405
Alaska Peninsula, Alaska	14	21 [b]	AY156406–411, 156162–165, 450608–611
Kodiak Island, Alaska	4	22	AY156166–169
Cordova, Copper River Delta, Alaska	10	30	AY156412–421
Alexander Archipelago, Alaska	5	30 [c]	AY156174–178
Hyder, Alaska	5	18	AY156161, 422–425
Queen Charlotte Islands, British Columbia	4	24	AY156170–173
Massachusetts	2	0	AY156179–180

[a] Includes individuals from Attu Island (27) and Shemya Island (3).
[b] Includes individuals from King Cove (10), Shumagin Islands (9), Unalaska Island (2), and Amak Island (3).
[c] Includes individuals from Prince of Wales Island (17), Gravina Island (8), Revillagigedo Island (2), Heceta Island (2), and Warren Island (1).

and used the remaining trees to approximate the posterior probability of the phylogeny (Huelsenbeck and Ronquist 2001). We imported trees into PAUP* and constructed a majority-rule consensus tree. We determined the posterior probabilities of clades as the percentage of occurrence of each clade among all sampled trees (Huelsenbeck and Ronquist 2001). We used Lincoln's Sparrow (*Melospiza lincolnii*; GenBank AY156181) as an outgroup in both phylogenetic analyses. We developed haplotype networks using NETWORK, version 4.5.10 (Fluxus Technology, Clare, United Kingdom; Bandelt et al. 1999) to compare with phylogenetic trees.

We performed tests for linkage disequilibrium and Hardy-Weinberg equilibrium using GDA (Lewis and Zaykin 2001). We used MICRO-CHECKER (van Oosterhout et al. 2004) to test for the presence of null alleles, stuttering, and large-allele drop-out. We estimated genetic distances based on the proportion of shared alleles between individuals (Bowcock et al. 1994) using MICROSAT (Minch et al. 1995). We constructed a neighbor-joining tree using individuals as operational taxonomic units (OTUs) by using the NEIGHBOR subroutine in PHYLIP, version 3.5 (Felsenstein 1993), and TREEVIEW, version 1.5 (Page 1996).

RESULTS

Body mass.—We found a pronounced east-to-west increase in individual body mass among subspecies from southeast Alaska and British Columbia to the western extreme of the species' distribution in the Near Islands of the Aleutian archipelago (Fig. 2). Going from west to east (Fig. 1), male *M. m. maxima* weighed, on average, 0.94–22.0 g ($\bar{x}$ [± SE] = 45.74 ± 0.38 g) more than any other subspecies (Fig. 2) but did not weigh significantly more than birds from their neighboring subspecies, *M. m. sanaka* (P = 0.06; $\bar{x}$ = 44.8 ± 0.43 g). *Melospiza m. sanaka* weighed, on average, 5.98 g more than the next-largest and neighboring subspecies, *M. m. insignis* (P < 0.001; $\bar{x}$ = 38.82 ± 0.60 g). Male *M. m. insignis* weighed, on average, 10.38 g more than male *M. m. caurina* (P < 0.001, $\bar{x}$ = 28.44 ± 0.45 g), and male *M. m. caurina* weighed, on average, 1.15 g more than male *M. m. rufina* (P = 0.02, $\bar{x}$ = 27.29 ± 0.32 g). Finally, male *rufina* weighed, on average, 3.55 g more than male *merrilli* (P < 0.001; $\bar{x}$ = 23.74 ± 0.35 g).

MtDNA analyses.—We found that maximum-likelihood bootstrap and Bayesian trees had similar topologies; in Figure 3, we present the maximum-likelihood phylogram, with individuals with identical haplotypes found on the same branch. We

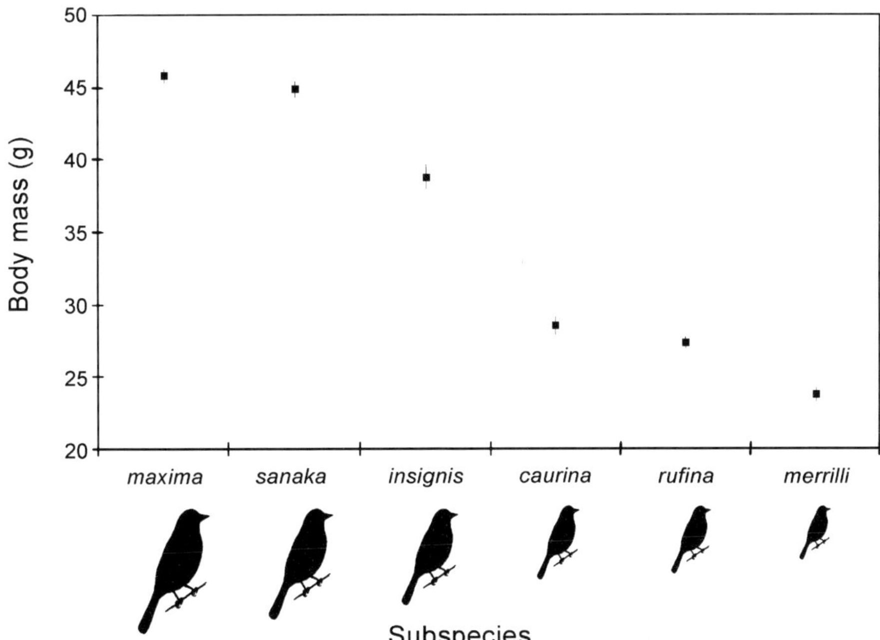

FIG. 2. Mean and standard error of body mass of each Song Sparrow subspecies.

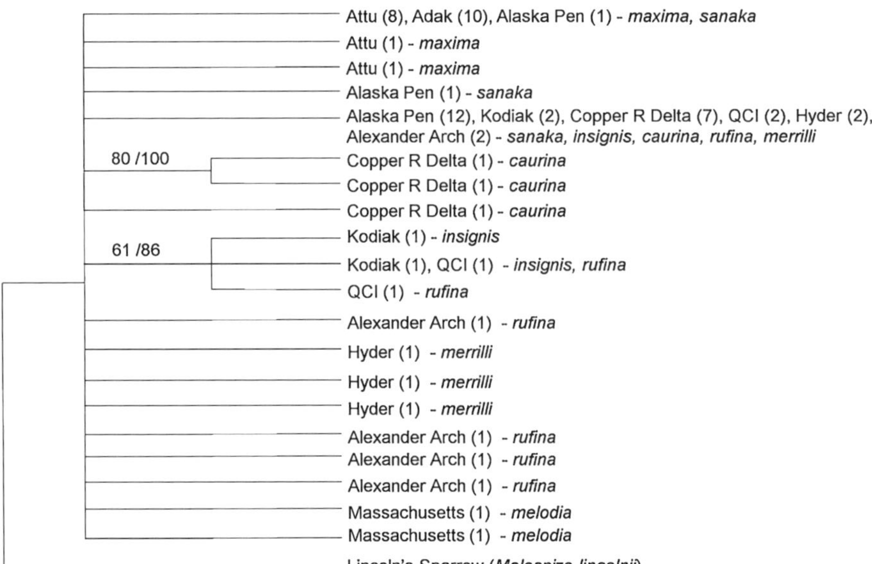

Fig. 3. Maximum-likelihood tree of cytochrome-*b* haplotypes of Song Sparrows. Bootstrap and posterior probability of nodes are presented above branches. Nodes with <50% bootstrap support were collapsed.

determined that bootstrap replicates and Bayesian posterior probabilities were similar for the branches that were moderately to strongly supported (>50% bootstrap and >90% posterior probability; Fig. 3) and that none of the subspecies or locations were reciprocally monophyletic (Fig. 3). We examined the mtDNA haplotype network and found some structure among cytochrome-*b* sequences corresponding to at least one subspecies, *M. m.maxima*, but found that monophyly was not present (Fig. 4). We found that one individual *M. m. sanaka* had the most common haplotype for *maxima* and that the

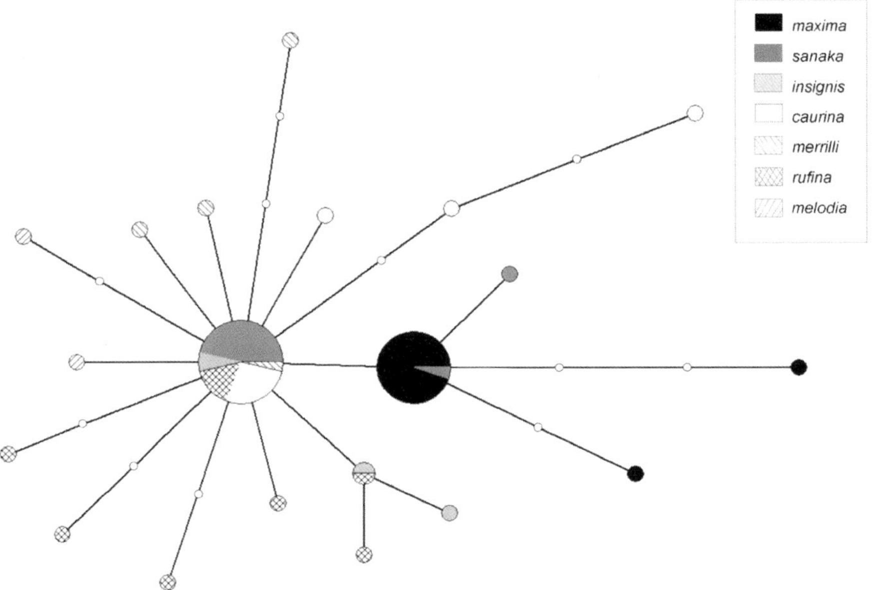

FIG. 4. Network of relationships among cytochrome-*b* haplotypes of Alaska Song Sparrows. Small white circles are missing haplotypes.

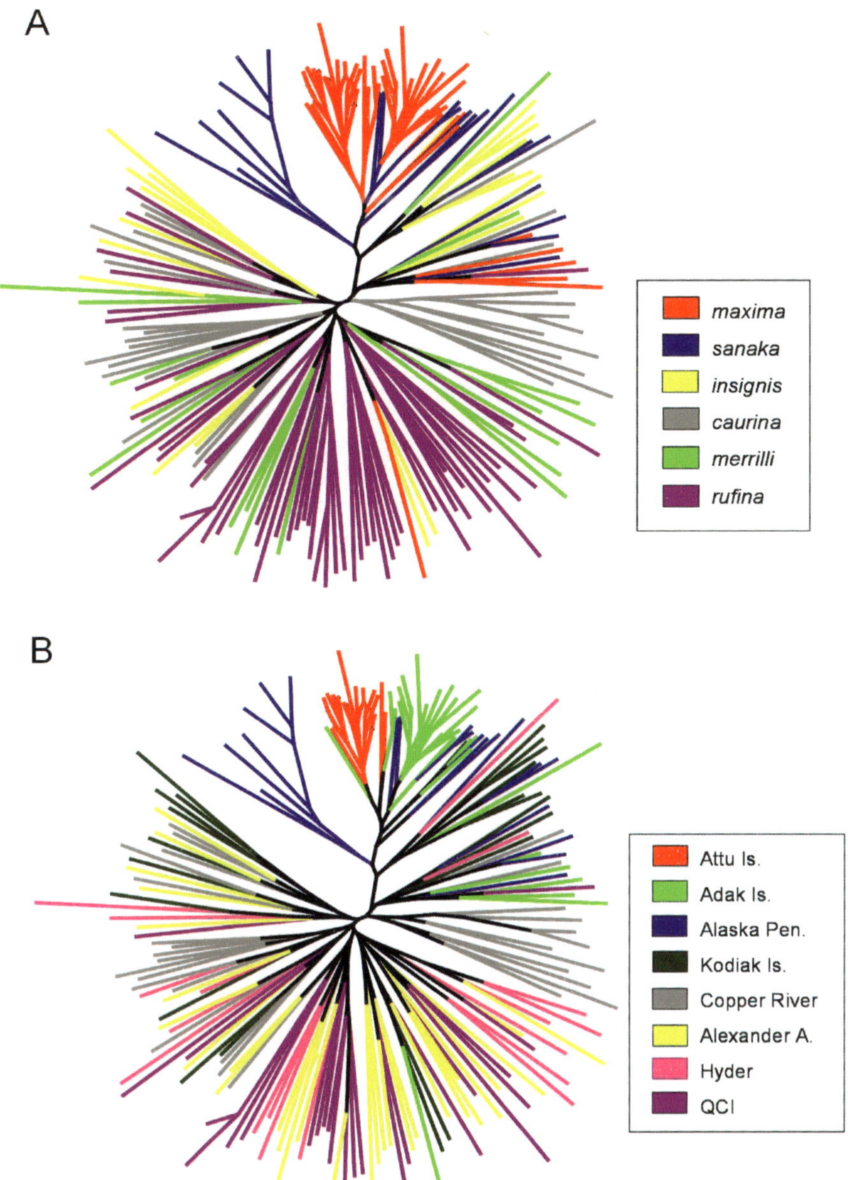

FIG. 5. Neighbor-joining trees based on the proportion of shared alleles among individual Song Sparrows genotyped at eight microsatellite loci, showing the relationships among (A) subspecies and (B) sampling locations.

most frequent haplotype in the entire data set was in individuals of all of the Alaska subspecies except *maxima* (Fig. 4).

Microsatellite analyses.—We determined that all loci were in linkage equilibrium, but two (Mme1 from Attu Island and Mme2 from Kodiak Island) were deficient in heterozygotes after adjustments for multiple comparisons (Pruett and Winker 2005). However, we found no evidence of genotyping artifacts such as null alleles, stuttering, or large-allele drop-out.

We constructed a genetic distance tree based on the proportion of shared alleles among individuals and found clustering of locations and subspecies in several cases. Individuals from Attu Island within the subspecies *maxima* clustered together (Fig. 5). However, one grouping of the subspecies as a whole was not clearly defined (Fig. 5).

Individuals in the subspecies *caurina* were primarily in two clusters, and populations from the most western subspecies, *sanaka* and *maxima*, clustered together. Our findings match results based on genetic clustering analyses (STRUCTURE; Falush et al. 2003), wherein Pruett and Winker (2005) found nine genetic groups among northwestern Song Sparrow populations. These clusters represented each of the eight subspecies plus the population on Attu Island. We used microsatellite distance trees with individuals as OTUs to provide another way to visualize the relationships among individuals and found that there is gene flow among units but that population structure exists at several levels, even within one subspecies (e.g., *maxima*). It is noteworthy that these distance trees seem to provide substantially more information about populations and individuals than the mtDNA phylogenies.

Discussion

On the basis of single-locus mtDNA criteria for diagnosing taxonomic units, Song Sparrows would be identified as a single species with no subspecies in northwestern North America, corroborating Zink and Dittmann (1993a) and Fry and Zink (1998). Yet body-mass data alone show that among-population differences are profound, representing an effective doubling of body mass from *merrilli* to *maxima* (Fig. 2). Although mass data do not enable separation of all of these subspecies using the 75% rule (Amadon 1949, Patten and Unitt 2002) and this character appears to form a step cline (Fig. 2), diagnosability using plumage characters achieves this threshold for each (not shown).

Despite subspecies-level phenotypic variation (e.g., Fig. 2), we found, on the basis of mtDNA cytochrome-*b* sequence data, a lack of reciprocal monophyly and remarkably little structure for the subspecies or populations in this region. For example, birds from the eastern portion of the United States (Massachusetts; subspecies *M. m. melodia*) are not separable from those found in the Aleutian Islands by mtDNA phylogenetic tree-building methods and are separated only by a single mutation from the most common Alaskan haplotype (Fig. 4). It is important to note that these birds are separated by >3,000 km, differ by 20–25 g in body mass, and have distinctive differences in plumage coloration (Patten and Pruett 2009). These facts alone should eliminate the possibility

that there is ongoing gene flow between these locations; thus, incomplete lineage sorting and the retention of ancestral haplotypes would appear to be the best explanation for the sharing of these haplotypes across such geographic space.

Although these findings support previous conclusions about the phylogeography of Song Sparrows (Hare and Shields 1992, Zink and Dittmann 1993a, Fry and Zink 1998), with a likely recent range expansion throughout much of North America, they reveal very little about the current subspecific status of Song Sparrows in Alaska. If we used haplotype frequency differences to diagnose subspecies, only the subspecies *maxima* differed from the remaining five subspecies (Fig. 4). In addition, researchers who used the mtDNA control region (an area thought to have the highest mutation rate within the mitochondrial genome; Baker and Marshall 1997) and mtDNA restriction fragment-length polymorphisms (RFLPs) did not find monophyly or frequency differences between Alaskan birds and those in the rest of North America (Hare and Shields 1992, Zink and Dittmann 1993a, Fry and Zink 1998). Thus, the mutation rate of mtDNA appears to be too slow to provide a useful measure of current isolation (if we consider only complete or near-complete lineage sorting). In this case, this locus provides little or no information about adaptive variation among these populations, despite small current population sizes and evidence of historical population reductions (Pruett and Winker 2005).

Studies that have used rapidly mutating genetic markers (i.e., microsatellites; Goldstein and Schlötterer 1999) have found accurate assessments of recent isolation and current gene flow (Cornuet et al. 1999, Johnson et al. 2003, Berry et al. 2004, Paetkau et al. 2004, Underwood et al. 2007) that might provide a means of inferring population history (Sun et al. 2009). Although the fixation of neutral nuclear alleles through drift is not likely in a short period (Zink and Barrowclough 2008), higher mutation rates in microsatellites lead to greater allelic variation, enabling shifts in allele frequencies to be readily observable and, thus, informative in understanding whether groups are diagnosably different. Genetic analyses based on microsatellites show very limited or nonexistent gene flow among many of these populations and subspecies (Pruett et al. 2008a, b), supporting the idea that differences in plumage color, body size, and mass are found in isolated populations that are on independent evolutionary paths. In addition,

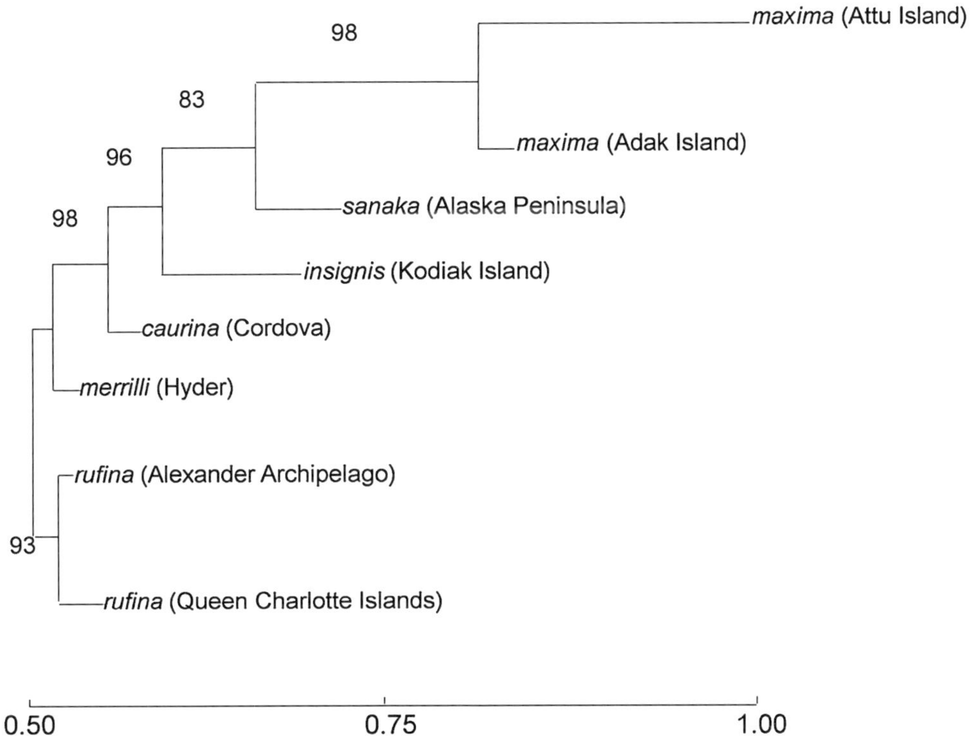

FIG. 6. Neighbor-joining tree based on population allele frequencies, based on tree presented by Pruett and Winker (2005).

a genetic distance tree based on population-level differences in allele frequencies shows grouping of populations that are the same subspecies, with these nodes having high bootstrap support (Fig. 6). These relationships are also revealed using clustering analyses (Pruett and Winker 2005, Pruett et al. 2008a) such as STRUCTURE (Falush et al. 2003) and, to some extent, in the individual-based microsatellite genotype distance trees (Fig. 5). These findings support the traditional subspecies assessed using phenotypic differences that are readily observable in this group (e.g., Fig. 2; Patten and Pruett 2009). These results indicate that many Song Sparrow subspecies likely represent units on different evolutionary trajectories. Yet, in part because gene flow is ongoing (albeit at very low levels in some cases; Pruett 2002; Pruett and Winker 2005; Pruett et al. 2008a, b), these units are not full species, but rather quintessential subspecies (Fig. 5).

The northwestern Song Sparrow subspecies we examined are morphologically and genetically distinctive, just not diagnosable (e.g., reciprocally monophyletic) using mtDNA sequences.

Should wildlife managers and conservation biologists ignore such clear and striking differences because of the recent timing of isolation? If so, then we likely put too much emphasis (1) on the timing of population-level divergences and (2) on the stochastic aspects of genetic drift in relation to effective population size occurring during divergence, rather than taking into account the effects of current isolation and ongoing adaptive evolution (Moritz 2002). For these reasons and others (such as sampling error, neutral or near-neutral aspects of contemporary genetic data, and the importance of gene flow), we conclude that mtDNA data cannot be considered definitive (whether one is using monophyly or distance thresholds) for biodiversity assessment (Winker 2009).

The problem of using a single genetic measure to identify units, as shown clearly here with Song Sparrows, is likely to occur in other morphologically diverse taxa found in temperate and high-latitude areas. Many novel environments have opened for colonization within the past 10,000 years (Pielou 1991), and it is not surprising that birds moved into these areas and evolved unique adaptations to new environments. These populations, often identified as subspecies, represent a large amount of the biodiversity found in nontropical areas and should be appropriately managed and conserved. For example, Song Sparrows found on islands contain a large proportion of the unique alleles found in this group, and these island populations are also differentiated morphologically and genetically (Wilson et al. 2009). They also exhibit differences in behavior, in that some populations are at least partially migratory and others are sedentary (Pruett et al. 2008b). Thus, several northwestern Song Sparrow populations (e.g, Attu Island birds) should be managed as unique evolutionary units (e.g., Pruett et al. 2008a).

Phylogeography and its workhorse mtDNA sequence data do not provide a complete view of evolutionary genetics, nor does this approach provide the complete toolkit required to understand the distribution of genetic variation in time and space. Microsatellites are also imperfect, but they can provide important indicators of evolutionary divergence on shorter time-scales. And because adaptive evolution can occur fairly rapidly in some populations (e.g., Clegg et al. 2008),

reliance on a single-locus approach or a single analytic methodology risks overlooking or inadvertently minimizing important population-level differentiation. We advocate using multiple genetic markers in addition to phenotypic characters to determine taxonomic and management units. (For a case in which a Song Sparrow subspecies was found to be invalid using these criteria, see Pruett et al. 2004; for a case in which morphology alone was insufficient to describe unique populations, see Pruett et al. 2008a). Without examining additional genetic markers in groups identified by phenotypic characters, we would overlook much of the important biodiversity found in Song Sparrows.

ACKNOWLEDGMENTS

This project was supported by the University of Alaska Museum, the National Geographic Society, the U.S. Department of Agriculture (USDA-ARS), the National Science Foundation (DEB-9981915), the Center for Global Change and Arctic System Research, and an anonymous donor. We thank J. Johnson, F. James, and an anonymous reviewer for comments on the manuscript and C. Topp for Song Sparrow silhouettes.

APPENDIX. Song Sparrows (*Melospiza melodia*) in the collections of the University of Alaska Museum, Fairbanks (UAM), used for body-mass measurements in this study.

Subspecies	Collection location	UAM catalogue numbers
M. m. maxima	Alaska, Aleutian Is., Adak Is.	8461, 10040, 10041, 10165, 10170, 10179, 10185, 10186, 10188, 10942, 10946, 10948, 11175–11178, 11229, 11267, 11268, 11280, 11511, 11557–11562, 11827, 12143, 13059, 13186, 13289, 13290, 14611, 15298, 15305, 15307, 15312, 15315, 15319, 15321, 15326, 15339–15341, 15344, 15353, 15366, 15367, 15369
	Alaska, Aleutian Is., Amlia Is.	11289–11297, 13160
	Alaska, Aleutian Is., Attu Is.	7223–7225, 7228, 7650, 7651, 8091, 8092, 8097, 8099, 8100, 8102, 8107, 8127, 8128, 8308, 8416, 8418, 8462, 8463, 8609, 8611, 8612, 8774, 8775, 9302–9305, 10072, 10073, 11173, 11174, 11224–11228, 11242, 11270, 11277, 11556, 11790, 11828, 11846–11849, 12094, 12141, 13056, 13140, 14165, 15115–15117, 15299, 15301, 15302, 15304, 15313, 15320, 15325, 15330, 15331, 15342, 15343, 15346, 15352, 15354, 15356, 15358, 15363, 15378, 19345, 19346, 19351, 20518,
	Alaska, Aleutian Is., Buldir Is.	8781
	Alaska, Aleutian Is., Igitkin Is.	15323
	Alaska, Aleutian Is., Kanaga Is.	8779
	Alaska, Aleutian Is., Kasatochi Is.	13173
	Alaska, Aleutian Is., Semisopochnoi Is.	9602
	Alaska, Aleutian Is., Shemya Is.	9418, 10944, 11171, 11243, 11244, 11363, 11825, 13128, 19048
	Alaska, Aleutian Is., Tanaga Is.	15318
M. m. sanaka	Alaska, Alaska Peninsula, King Cove	10091, 11362, 11365, 11366, 11381, 11389, 11823
	Alaska, Aleutian Is., Bogoslof Is.	11236–11238
	Alaska, Aleutian Is., Ugamak Is.	15334, 15335, 15384, 18514
	Alaska, Aleutian Is., Samalga Is.	15374, 15377
	Alaska, Shumagin Is., Popof Is.	10171, 10187, 11276, 11379, 11390, 12142
M. m. insignis	Alaska, Kodiak Archipelago, Kodiak Is.	7522, 8776, 8777, 8807, 11871, 12139, 14001–14003, 14009, 14010
M. m. caurina	Alaska, Copper River Delta	8922, 10652, 11046, 11101, 11103, 11104, 11141, 11180, 11182, 11210, 11223, 11231, 11234, 11272, 11361, 11382–11384
M. m. rufina	Alaska, Alexander Archipelago, Gravina Is.	13438, 13886, 13887, 13941, 13944, 15426, 15433, 17575
	Alaska, Alexander Archipelago, Heceta Is.	13241, 13288
	Alaska, Alexander Archipelago, Prince of Wales Is.	11712, 11824, 12455, 13463, 13912, 13913, 13915, 13916, 13936–13938, 13943, 13945, 13952, 13953, 14658, 14714, 14942–14944, 15086, 15087, 15193
	Alaska, Alexander Archipelago, Revillagigedo Is.	11516, 11517
	British Columbia, Queen Charlotte Is., Graham Is.	11179, 11544, 11546–11550, 11552, 11553, 13079, 13080, 13908–13910, 13919, 13920
M. m. merrilli	Alaska, Hyder	7341–7343, 7346, 8115, 8379, 8447, 8449, 8607, 10159, 13921, 15338, 18103

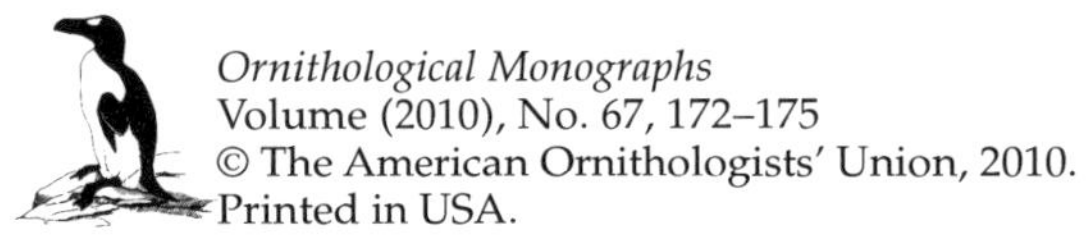

Ornithological Monographs
Volume (2010), No. 67, 172–175
© The American Ornithologists' Union, 2010.
Printed in USA.

CHAPTER 14

AVIAN SUBSPECIES: SUMMARY AND PROSPECTUS

SUSAN M. HAIG[1,3] AND KEVIN WINKER[2,4]

[1]*U.S. Geological Survey Forest and Rangeland Ecosystem Science Center,
3200 SW Jefferson Way, Corvallis, Oregon 97331, USA; and*
[2]*University of Alaska Museum 907 Yukon Drive, Fairbanks, Alaska 99775, USA*

ABSTRACT.—The 14 papers in this monograph represent the first broad-based evaluation of avian subspecies in decades and one of few, if any, multifaceted treatments of subspecies for any taxon. As such, there are multiple points of agreement and disagreement. Most authors consider the concept of subspecies a valid taxonomic category for units below the species level. All authors point to the need to reexamine taxa with modern methods to confirm their identity as subspecies. All authors also agree that the best approach to recognizing a subspecies is to include multiple characters (e.g., an mtDNA study alone will not suffice). However, issues regarding the reconciliation of data sets in which we expect evolutionary rates to differ, how various methods are implemented and compared, and the statistical analyses used have not been resolved. We conclude by calling for renewed interest in examining avian subspecies that have not had modern approaches applied to their classification. Each species evaluated will add to an improved understanding of avian diversity and its generation and will be a significant contribution to conservation.

Key words: biodiversity, conservation, debates, taxonomy, techniques.

Subespecies de Aves: Síntesis y Perspectivas

RESUMEN.—Los 14 artículos que conforman esta monografía representan la primera evaluación general de las subespecies de aves en décadas y uno de los pocos, si no el único, tratamiento multifacético de las subespecies para cualquier taxón. Como tal, existen muchos puntos de acuerdo y desacuerdo. La mayoría de los autores consideran el concepto de subespecie como una categoría taxonómica válida para unidades bajo el nivel de especie. Todos los autores señalan la necesidad de reexaminar los taxones con métodos modernos para confirmar su identidad como subespecies. Todos los autores también concuerdan en que la mejor aproximación para reconocer una subespecie es incluir múltiples caracteres (e.g., un estudio con sólo ADNmt no sería suficiente). Sin embargo, no se han resuelto asuntos relacionados con la unificación de conjuntos de datos para los cuales esperamos que las tasas evolutivas sean diferentes, con la forma en que diferentes métodos son implementados y comparados, y con los diferentes análisis estadísticos que son usados. Concluimos haciendo un llamado a renovar el interés en estudiar las subespecies de aves para las cuales aún no se han aplicado aproximaciones modernas de clasificación. Cada especie evaluada adicionará conocimiento importante para el entendimiento de la diversidad de las aves y los procesos que la generaron, y será una contribución significativa para la conservación.

[3]E-mail: susan_haig@usgs.gov
[4]E-mail: kevin.winker@alaska.edu

Ornithological Monographs, Number 67, pages 172–175. ISBN: 978-0-943610-86-3. © 2010 by The American Ornithologists' Union. All rights reserved. Please direct all requests for permission to photocopy or reproduce article content through the University of California Press's Rights and Permissions website, http://www.ucpressjournals.com/reprintInfo.asp. DOI: 10.1525/om.2010.67.1.172.

Dᴜʀɪɴɢ ᴛʜᴇ ᴘᴀsᴛ two decades, the annual number of publications in organismal biology that include the topic of subspecies has approximately doubled (Fig. 1), a trend likely to continue given our increasing knowledge of biodiversity, technological advances, and efforts to successfully manage and conserve species at risk. Thus, it seems clear that the concept of subspecies and the biological variation that it encompasses will retain importance for a long time to come.

Our motivation for addressing subspecies was to provide a counter-perspective to ongoing criticisms of the concept by proponents of the phylogenetic species concept. We felt that subspecies needed to be constructively addressed within the framework of the biological species concept, which, despite debate, remains the dominant species paradigm from both legal and research perspectives. Further, in organizing the original American Ornithologists' Union symposium on this subject in 2008, we found an overwhelmingly enthusiastic response for constructively addressing subspecies and less interest in debating species concepts and whether subspecies should be done away with entirely. Insofar as we are discussing biological variation, putting aside labels, we trust that most will have found something of interest herein.

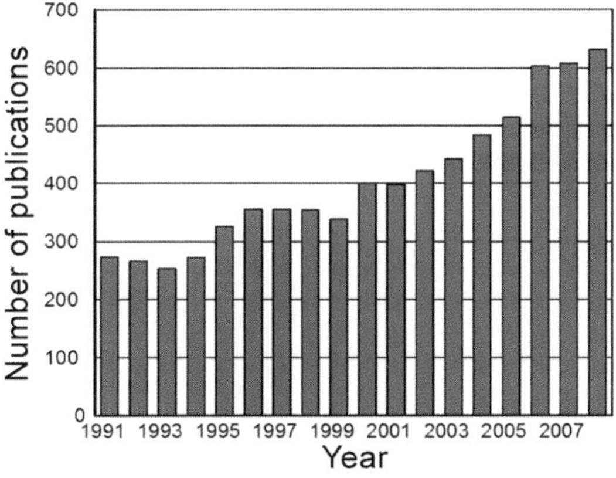

Fɪɢ. 1. The annual number of publications including subspecies as a topic found on the Web of Science (1991–2008; http://thomsonreuters.com/products_services/science/science_products/a-z/web_of_science) in the disciplines zoology, plant sciences, ecology, evolutionary biology, genetics and heredity, veterinary sciences, entomology, ornithology, biodiversity, conservation, marine and freshwater biology, and agronomy. This same pattern also occurred when only ornithology was considered (not shown).

Among the 13 preceding chapters, readers will find that authors generally agree that subspecies are a useful, albeit difficult, taxonomic category. Although subspecies are problematic almost by definition, a confounding issue that these authors acknowledged is that many, if not most, avian subspecies need a modern reconsideration of their classification. So many subspecies were described before the advent of modern statistics, sampling, molecular methods, and the quantification of phenotypic traits that they need to be revisited to determine whether the patterns described earlier actually hold true. Thus, the criticism that many avian subspecies do not represent significant geographic variation needs to be tempered with the realization that we have a great deal of updating to carry out before accepting such claims as valid. A modern treatment will most likely result in many avian subspecies being determined invalid and either lumped or perhaps downgraded to categories such as grades or distinct population segments. Furthermore, we will probably find that many avian subspecies are actually full biological species. Use of more representative data sets and modern methodology to determine species limits among allopatric populations will accelerate recognition of these taxa.

We do not advocate discarding historical work just because it was not done to current standards. Consider, for example, that Darwin (1859) provided a solid foundation for evolutionary biology without using modern genetics or statistical tests, and that Linnaeus's simple original description of *Corvus corax* in 1758 remains valid. We also do not wish to imply that the descriptive science of biodiversity can only go forward using sophisticated analyses; any volume of *Zootaxa* or the *Proceedings of the Biological Society of Washington* demonstrates that classic descriptive taxonomy retains an important place in modern biology. Nevertheless, such descriptions represent the beginning of the process of biodiversity science, and it is just that, a process—the erection and testing of hypotheses using a series of approaches and more data (and more specimens) until the true situation has been robustly inferred. The history of taxonomy and systematics shows that this process is usually neither short nor easy for most lineages. Most named subspecies are stalled somewhere along this lengthy process, and some even remain to be named. So we are not likely to arrive quickly at a stable subspecies-level taxonomy even if legions of taxonomists take up the charge. Specimen shortages alone preclude this, even in the comparatively well-studied class Aves (e.g., Stoeckle and Winker 2009). Nor are there any shortcuts evident among the tools of statistics or genetics, even though statistical and genetic analyses are integral parts of subspecific research, as examples in this volume show.

Continued improvements in the development and implementation of new and robust statistical analyses are needed to evaluate subspecies. For example, the 75% rule used by many today is simply a guideline, not a formal statistical test, and it does not adequately address the issue of clines, which requires new statistical approaches. In this volume, James, Patten, Phillimore, and others provide new approaches (or ideas) to address the need for improved statistics, but each will need follow-up. We do not expect a silver bullet to appear; the challenge remains in evaluating both the biological and statistical significance of results in the light of guidelines for categorizing populations. More work is needed on the latter, as well.

For example, new approaches are needed to determine the lower limits of valid subspecies. Authors in this volume agree that multiple characters are important to consider in subspecific diagnosis, yet reconciling the differences we expect to find among our measurements of different characters can make overall interpretations challenging. For example, how do we reconcile the discord between mtDNA and phenotypic evolutionary rates discussed by Oyler-McCance, Pérez-Emán et al., Pruett and Winker, and others? And how do we reconcile plumage and morphological differentiation when both have a genetic component that current (putatively) neutral molecular tools are almost certainly not sampling? The difficulty stems from comparing results in population or subspecific divisions found using one approach with results from another approach when the factors being compared change on totally different time scales (because of the different rates of evolutionary phenomena such as selection and genetic drift). Thus, any strict subspecific diagnosis will have to have a subjective element, as does diagnosis of most taxa.

Although several of the chapters show why multiple data sets are important in assessing taxonomic designations, use of molecular methods alone to improve our understanding of variation among populations is increasing. Reconciling such approaches, often done without phenotypic data, with a taxonomy based on phenotype will continue to be challenging. Scientists dealing with this issue for marine mammals have proposed the concept of demographically independent populations (B. Taylor, National Oceanic and Atmospheric Administration, pers. comm.). Demographically independent populations correspond to ecological time and are defined as a unit in which internal population dynamics are far more important for maintaining unit integrity than external dynamics. Although there is no broadly accepted amount of dispersal to define a demographically independent population, marine mammalogists have decided that dispersal on the order of 1% or so per year is important for demographic differentiation in marine mammals. At this level of interchange, there is no expectation for development of a recognizable phylogeographic signal, but instead frequency differences in haplotypes or a small number of private microsatellite alleles would suffice as evidence for defining demographically independent populations. They consider subspecies to be in the gray area between demographically independent populations and species.

One growing challenge to updating subspecies descriptions is to consider the situation that occurs when what was historically a smooth cline of variation among populations has been anthropogenically broken up into allopatric segments that now possess all the attributes of diagnosable subspecies (because of the loss of intermediate populations). Do we modify taxonomy accordingly? How might this affect management, if at all? This pattern promises to become more prominent as the effects of habitat fragmentation and climate change become more pronounced throughout the world.

Even if all agree that we need to revisit subspecific classifications, how will this be undertaken? Professional societies responsible for maintaining lists of biodiversity need to catch up to (and keep up with) the management and conservation needs of agencies, countries, and societies. Because this work is often done on a volunteer basis and is rarely considered cutting-edge science at universities and museums, the priorities of taxonomists and of professional biodiversity managers often differ—this is one of the reasons why this gap has developed. Bridging it again will require some creativity from both sides. Recognition of the problem, as illustrated here, is a promising first step.

We treat subspecies as discrete taxonomic categories, although we recognize that the real situation is too complex to be fully captured in this simple way. Subspecies address the geographic component of

variation and differentiation, and although definitions and diagnoses may have to vary among cases,
they will be scientific and repeatable if the criteria
in each case are made explicit. As scientists, professional societies, agencies, universities, and museums renew their commitment to this topic and
readdress subspecies using modern approaches
and make revisions accordingly, we are certain that
the outcome will be renewed acceptance of the concept of taxonomic units below the species level. We
acknowledge that this acceptance will be gradual,
and it will proceed largely on a species-by-species
basis, as the case studies in this volume illustrate.

ACKNOWLEDGMENTS

We thank the authors for their patience and willing-
ness to update manuscripts following each of the many
reviews they dealt with. We are grateful to F. James
for working so tirelessly on the introduction and pro-
viding detailed reviews of each chapter. We thank B.
Johnson for preparing the Literature Cited and D. Pratt
for providing his beautiful cover art of the fantails.
We also thank the many reviewers, including P. Arcese,
R. Banks, P. Beerli, B. Bowen, S. Chambers, C. Cicero,
M. Fitzpatrick, A. Hecht, G. Hill, F. James, J. Johnson,
D. MacDonald, D. Mindell, M. Patten, A. T. Peterson,
C. Phillips, J. V. Remsen, R. Ricklefs, C. Schuler, P. Tubaro,
P. Unitt, R. Waples, R. Wilson, and R. Zink. We are grateful to M. Morrison for his skill and patience in seeing
this monograph through its many phases, M. Penrose
for his skills in the monograph's production, R. Earles
for copyediting, and C. D. Cadena, C. Cornelius, and
L. R. Malizia for Spanish abstract translation. Finally,
we are grateful to the U.S. Geological Survey Forest
and Rangeland Ecosystem Science Center for providing
funding for the color cover and plates. The findings and
conclusions in this monograph are those of the authors
and do not necessarily represent the views of the U.S.
Geological Survey.

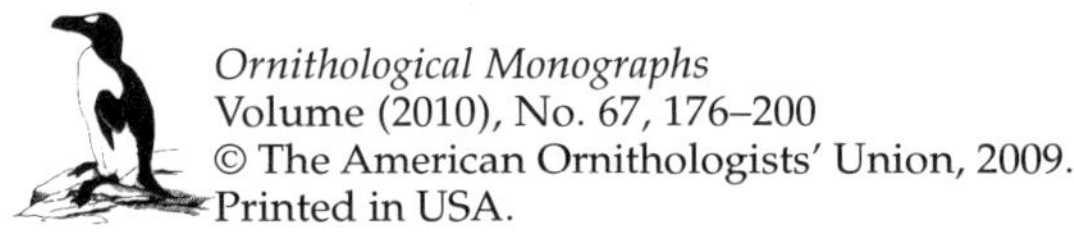*Ornithological Monographs*
Volume (2010), No. 67, 176–200
© The American Ornithologists' Union, 2009.
Printed in USA.

LITERATURE CITED

AGAPOW, P.-M., O. R. P. BININDA-EMONDS, K. A. CRANDALL, J. L GITTLEMAN, G. M. MACE, J. C. MARSHALL, AND A. PURVIS. 2004. The impact of species concept on biodiversity studies. Quarterly Review of Biology 79:161–179.

ALDRICH, J. W., AND F. C. JAMES. 1991. Ecogeographic variation in the American Robin (*Turdus migratorius*). Auk 108:230–249.

ALLEN, J. A. 1871. On the mammals and winter birds of east Florida, with an examination of certain assumed specific characters in birds, and a sketch of the bird-faunæ of eastern North America. Bulletin of the Museum of Comparative Zoology 2:161–450.

AMADON, D. 1949. The seventy-five per cent rule for subspecies. Condor 51:250–258.

AMADON, D. 1950. The Hawaiian honeycreepers (Aves, Drepaniidae). Bulletin of the American Museum of Natural History 95:155–262.

AMADON, D. 1966. The superspecies concept. Systematic Zoology 15:245–249.

AMADON, D., AND L. L. SHORT. 1976. Treatment of subspecies approaching species status. Systematic Zoology 25:161–167.

AMADON, D., AND L. L. SHORT. 1992. Taxonomy of lower categories—Suggested guidelines. Bulletin of the British Ornithologists' Club 112A:11–38.

AMAR, A., F. AMIDON, B. ARROYO, J. A. ESSELSTYN, AND A. P. MARSHALL. 2008. Population trends of the forest bird community on the Pacific island of Rota, Mariana Islands. Condor 110:421–427.

AMERICAN ORNITHOLOGISTS' UNION. 1886. The Code of Nomenclature and Check-list of North American Birds Adopted by the American Ornithologists' Union. American Ornithologists' Union, New York.

AMERICAN ORNITHOLOGISTS' UNION. 1895. Check-list of North American Birds, 2nd ed. American Ornithologists' Union, New York.

AMERICAN ORNITHOLOGISTS' UNION. 1910. Check-list of North American Birds, 3rd ed. American Ornithologists' Union, New York.

AMERICAN ORNITHOLOGISTS' UNION. 1931. Check-list of North American Birds, 4th ed. American Ornithologists' Union, New York.

AMERICAN ORNITHOLOGISTS' UNION. 1955. Thirtieth supplement to the American Ornithologists' Union *Check-list of North American Birds*. Auk 72:292–295.

AMERICAN ORNITHOLOGISTS' UNION. 1957. Check-list of North American Birds, 5th ed. American Ornithologists' Union, Baltimore, Maryland.

AMERICAN ORNITHOLOGISTS' UNION. 1986. Check-list of North American Birds, 6th ed. American Ornithologists' Union, Washington, D.C.

AMERICAN ORNITHOLOGISTS' UNION. 1998. Check-list of North American Birds. 7th ed. American Ornithologists' Union, Washington, D.C.

AMERSON, A. B., JR., W. A. WHISTLER, AND T. D. SCHWANER. 1982. Wildlife and wildlife habitat of American Samoa. II. Accounts of flora and fauna. U.S. Department of Interior, Fish and Wildlife Service, Washington, D.C.

ANDERSON, B. W. 2007. Breeding birds, wintering birds, and subspecies of White-cheeked Geese. Bulletin of the Revegetation and Wildlife Management Center, vol. 2.

ANDERSON, D. R., K. P. BURNHAM, AND W. L. THOMPSON. 2000. Null hypothesis testing: Problems, prevalence, and an alternative. Journal of Wildlife Management 64:912–923.

ANDERSON, M. G., J. M. RHYMER, AND F. C. ROHWER. 1992. Philopatry, dispersal, and the genetic structure of waterfowl populations. Pages 365–395 *in* Ecology and Management of Breeding Waterfowl (B. D. J. Batt, A. D. Afton, M. G. Anderson, C. D. Ankney, D. H. Johnson, J. A. Kadlec, and G. L. Krapu, Eds.). University of Minnesota Press, Minneapolis.

ASHTON, K. G. 2002. Patterns of within-species body size variation of birds: Strong evidence for Bergmann's rule. Global Ecology and Biogeography 11:505–523.

ATWOOD, J. L. 1988. Speciation and geographic variation in Black-tailed Gnatcatchers. Ornithological Monographs, no. 42.

AVISE, J. C. 1989. Gene trees and organismal histories: A phylogenetic approach to population biology. Evolution 43:1192–1208.

AVISE, J. C. 2000. Phylogeography: The History and Formation of Species. Harvard University Press, Cambridge, Massachusetts.

AVISE, J. C. 2004. Molecular Markers, Natural History, and Evolution, 2nd ed. Sinauer Associates, Sunderland, Massachusetts.

AVISE, J. C. 2006. Evolutionary Pathways in Nature: A Phylogenetic Approach. Cambridge University Press, Cambridge, United Kingdom.

AVISE, J. C., J. ARNOLD, R. M. BALL, JR., E. BERMINGHAM, T. LAMB, J. E. NEIGEL, C. A. REEB, AND N. C. SAUNDERS. 1987. Intraspecific phylogeography: The mitochondrial DNA bridge between population genetics and systematics. Annual Review of Ecology and Systematics 18:489–522.

AVISE, J. C., AND R. M. BALL, JR. 1990. Principles of genealogical concordance in species concepts and biological taxonomy. Oxford Surveys in Evolutionary Biology 7:45–67.

AVISE, J. C., AND K. WOLLENBERG. 1997. Phylogenetics and the origin of species. Proceedings of the National Academy of Sciences USA 94:7748–7755.

BACON, J. P., AND N. J. STRAUSFELD. 1986. The dipteran "giant fibre" pathway: Neurons and signals. Journal of Comparative Physiology A 158:529–548.

BADYAEV, A. V. 2005. Maternal inheritance and rapid evolution of sexual size dimorphism: Passive effects or active strategies? American Naturalist 166:S17–30.

BADYAEV, A. V., AND G. E. HILL. 2000. The evolution of sexual dimorphism in the House Finch. I. Population divergence in morphological covariance structure. Evolution 54:1784–1794.

BAILEY, R. C., AND J. BYRNES. 1990. A new, old method for assessing measurement error in both univariate and multivariate morphometric studies. Systematic Zoology 39:124–130.

BAIRD, S. F. 1858. Catalogue of North American Birds. Smithsonian Institution, Washington, D.C.

BAKER, A. J., AND H. D. MARSHALL.1997. Mitochondrial control region sequences as tools for understanding evolution. Pages 51–83 in Avian Molecular Evolution and Systematics (D. Mindell, Ed.). Academic Press, San Diego, California.

BAKER, R. H. 1951. The avifauna of Micronesia, its origin, evolution, and distribution. University of Kansas Publications, Museum of Natural History, no. 3.

BALDWIN, S. P., H. C. OBERHOLSER, AND L. G. WORLEY. 1931. Measurements of birds. Scientific Publications of the Cleveland Museum of Natural History, no. 2.

BALL, R. M., JR., AND J. C. AVISE. 1992. Mitochondrial DNA phylogeographic differentiation among avian populations and the evolutionary significance of subspecies. Auk 109:626–636.

BALLARD, J. W. O., AND M. C. WHITLOCK. 2004. The incomplete natural history of mitochondria. Molecular Ecology 13:729–744.

BANDELT, H.-J., P. FORSTER, AND A. RÖHL. 1999. Median-joining networks for inferring intraspecific phylogenies. Molecular Biology and Evolution 16:37–48.

BANKS, R. C., AND R. C. LAYBOURNE. 1977. Plumage sequence and taxonomy of Laysan and Nihoa finches. Condor 79:343–348.

BARRACLOUGH, T. G., AND S. NEE. 2001. Phylogenetics and speciation. Trends in Ecology and Evolution 16:391–399.

BARROWCLOUGH, G. F. 1980. Genetic and phenotypic differentiation in a wood warbler (genus *Dendroica*) hybrid zone. Auk 97:655–668.

BARROWCLOUGH, G. F. 1982. Geographic variation, predictiveness, and subspecies. Auk 99:601–603.

BARROWCLOUGH, G. F., J. G. GROTH, L. A. MERTZ, AND R. J. GUTIÉRREZ. 2004. Phylogeographic structure, gene flow and species status in Blue Grouse (*Dendragapus obscurus*). Molecular Ecology 13:1911–1922.

BARROWCLOUGH, G. F., AND R. M. ZINK. 2009. Funds enough, and time: mtDNA, nuDNA and the discovery of divergence. Molecular Ecology 18:2934–2936.

BEATTIE, M. 1994. Endangered and threatened wildlife and plants; 1-year finding for a petition to list the Pacific Coast population of the Cactus Wren under the Endangered Species Act. 59 Federal Register 45659, FR Doc. 94-21785.

BEHLE, W. H. 1973. Clinal variation in the White-throated Swifts from Utah and the Rocky Mountain region. Auk 90:299–306.

BELLIURE, J., G. SORCI, A. P. MØLLER, AND J. CLOBERT. 2000. Dispersal distances predict subspecies richness in birds. Journal of Evolutionary Biology 13:480–487.

BELLROSE, F. C. 1980. Ducks, Geese and Swans of North America. Stackpole Books, Harrisburg, Pennsylvania.

BENEDICT, N. G., S. J. OYLER-McCANCE, S. E. TAYLOR, C. E. BRAUN, AND T. W. QUINN. 2003. Evaluation of the eastern (*Centrocercus urophasianus urophasianus*) and western (*Centrocercus urophasianus phaios*) subspecies of Sage-Grouse using mitochondrial control-region sequence data. Conservation Genetics 4:301–310.

BENKMAN, C. W. 1989. On the evolution and ecology of island populations of crossbills. Evolution 43:1324–1330.

BENKMAN, C. W. 1993a. Adaptation to single resources and the evolution of crossbill (*Loxia*) diversity. Ecological Monographs 63:305–325.

BENKMAN, C. W. 1993b. The evolution, ecology, and decline of the Red Crossbill of Newfoundland. American Birds 47:225–229.

BENKMAN, C. W. 1994. Comments on the ecology and status of the Hispaniolan Crossbill (*Loxia Ieucoptera* [sic] *megaplaga*), with recommendations for its conservation. Caribbean Journal of Science 30:250–254.

BENKMAN, C. W. 2007. Red Crossbill types in Colorado: Their ecology, evolution and distribution. Colorado Birds 41:153–163.

BENKMAN, C. W., J. W. SMITH, P. C. KEENAN, T. L. PARCHMAN, AND L. SANTISTEBAN. 2009. A new species of the Red Crossbill (Fringillidae: *Loxia*) from Idaho. Condor 111:169–176.

BENSCH, S., T. ANDERSSON, AND S. ÅKESSON. 1999. Morphological and molecular variation across a migratory divide in Willow Warblers, *Phylloscopus trochilus*. Evolution 53:1925–1935.

BENTON, M. J. 2000. Stems, nodes, crown clades, and rank-free lists: Is Linnaeus dead? Biological Reviews 75:633–648.

BENTON, M. J. 2007. The PhyloCode: Beating a dead horse? Acta Palaeontologica Polonica 52:651–655.

BERGMANN, C. 1847. Ueber die Verhältnisse der Wärmeökonomie der Thiere zu ihrer Grösse. Gottinger Studien 3:595–708.

BERRY, O., M. D. TOCHER, AND S. D. SARRE. 2004. Can assignment tests measure dispersal? Molecular Ecology 13:551–561.

BERTHOLD, P., A. J. HELBIG, G. MOHR, AND U. QUERNER. 1992. Rapid microevolution of migratory behaviour in a wild bird species. Nature 360:668–670.

BICKFORD, D., D. J. LOHMAN, N. S. SODHI, P. K. L. NG, R. MEIER, K. WINKER, K. K. INGRAM, AND I. DAS. 2007. Cryptic species as a window on diversity and conservation. Trends in Ecology and Evolution 22: 148–155.

BIERREGAARD, R. O., JR. 1988. Morphological data from understory birds in *terra firme* forest in the central Amazonian Basin. Revista Brasileira de Biologia 48:169–178.

BINFORD, L. C. 1965. Two new subspecies of birds from Oaxaca, Mexico. Occasional Papers of the Museum of Zoology, Louisiana State University 30:1–6.

BLACKWELDER, R. E. 1954. The open season on taxonomists. Systematic Zoology 3:177–181.

BLAKE, E. R. 1977. Manual of Neotropical Birds, vol. 1. University of Chicago Press, Chicago, Illinois.

BOCK, W. J. 1966. An approach to the functional analysis of bill shape. Auk 83:10–51.

BOKMA, F. 2003. Testing for differences in rates of speciation between higher taxa. Evolution 57:2469–2474.

BOLES, W. E. 2006. Family Rhipiduridae (fantails). *In* Handbook of the Birds of the World, vol. 11: Old World Flycatchers to Old World Warblers (J. del Hoyo, A. Elliott, and D. A. Christie, Eds.). Lynx Edicions, Barcelona, Spain.

BORGMEIER, T. 1957. Basic questions of systematics. Systematic Zoology 6:53–69.

BORTOLOTTI, G. R. 2006. Natural selection and coloration: Protection, concealment, advertisement, or deception? Pages 3–35 *in* Bird Coloration, vol. 2: Function and Evolution (G. E. Hill and K. J. McGraw, Eds.). Harvard University Press, Cambridge, Massachusetts.

BOWCOCK, A. M., A. RUIZ-LINARES, J. TOMFOHRDE, E. MINCH, J. R. KIDD, AND L. L. CAVALLI-SFORZA. 1994. High resolution of human evolutionary trees with polymorphic microsatellites. Nature 368:455–457.

BRENNAN, L. A. 1999. Northern Bobwhite (*Colinus virginianus*). *In* The Birds of North America, no. 397 (A. Poole and F. Gill, Eds.). Academy of Natural Sciences, Philadelphia, and American Ornithologists' Union, Washington, D.C.

BRENT, R. 1973. Algorithms for Minimization without Derivatives. Prentice-Hall, Englewood Cliffs, New Jersey.

BROOKE, M. DE L., S. H. M. BUTCHART, S. T. GARNETT, G. M. CROWLEY, N. B. MANTILLA-BENIERS, AND A. J. STATTERSFIELD. 2008. Rates of movement of threatened bird species between IUCN Red List categories and toward extinction. Conservation Biology 22:417–427.

BROOKFIELD, J. 2002. Review of *Genes, Categories, and Species* by Jody Hey. Genetic Research 79:107–108.

BROWNING, M. R. 1990. Taxa of North American birds described from 1957 to 1987. Proceedings of the Biological Society of Washington 103:432–451.

BROWNING, M. R. 1994. A taxonomic review of *Dendroica petechia* (Yellow Warbler) (Aves: Parulinae). Proceedings of the Biological Society of Washington 107:27–51.

BRUMFIELD, R. T. 2005. Mitochondrial variation in Bolivian populations of the Variable Antshrike (*Thamnophilus caerulescens*). Auk 122:414–432.

BUCHANAN, K. L., M. R. EVANS, A. R. GOLDSMITH, D. M. BRYANT, AND L. V. ROWE. 2001. Testosterone influences basal metabolic rate in male House Sparrows: A new cost of dominance signaling? Proceedings of the Royal Society of London, Series B 268:1337–1344.

BUERKLE, C. A. 2000. Morphological variation among migratory and non-migratory populations of Prairie Warblers. Wilson Bulletin 112:99–107.

BULGARELLA, M., R. E. WILSON, C. KOPUCHIAN, T. H. VALQUI, AND K. G. McCRACKEN. 2007. Elevational variation in body size of Crested Ducks (*Lophonetta specularioides*) from the central high Andes, Mendoza, and Patagonia. Ornitología Neotropical 18:587–602.

BULGIN, N. L., H. L. GIBBS, P. VICKERY, AND A. J. BAKER. 2003. Ancestral polymorphisms in genetic markers obscure detection of evolutionarily distinct populations in the endangered Florida Grasshopper Sparrow (*Ammodramus savannarum floridanus*). Molecular Ecology 12:831–844.

BULLOCK, T. H. 1985. Comparative neuroethology of startle, rapid escape, and giant fiber-mediated responses. Pages 2–12 *in* Neural Mechanisms of Startle Behavior (R. C. Eaton, Ed.). Plenum Press, London.

BURBRINK, F. T., R. LAWSON, AND J. B. SLOWINSKI. 2000. Mitochondrial DNA phylogeography of the polytypic North American rat snake (*Elaphe obsoleta*): A critique of the subspecies concept. Evolution 54:2107–2118.

BURN, H. 2006. [Color plates 24 MONARCHIDAE (*Arses, Myiagra*) and 25 MONARCHIDAE (*Myiagra, Lamprolia*); pp. 314–315, 318–331]. *In* Handbook of Birds of the World, vol. 11: Old World Flycatchers to Old World Warblers (J. del Hoyo, A. Elliott, and D. A. Christie, Eds.). Lynx Edicions, Barcelona, Spain.

BURNEY, C. W., AND R. T. BRUMFIELD. 2009. Ecology predicts levels of genetic differentiation in Neotropical birds. American Naturalist 174:358–368.

BURNEY, D. A., H. F. JAMES, L. PIGOTT BURNEY, S. L. OLSON, W. KIKUCHI, W. L. WAGNER, M. BURNEY, D. McCLOSKEY, D. KIKUCHI, F. V. GRADY, AND OTHERS. 2001. Fossil evidence for a diverse biota from Kauai and its transformation since human arrival. Ecological Monographs 71:615–641.

BURTT, E. H., JR. 1981. The adaptiveness of animal colors. BioScience 31:723–729.

BURTT, E. H., JR., AND J. M. ICHIDA. 2004. Gloger's rule, feather-degrading bacteria, and color variation among Song Sparrows. Condor 106:681–686.

BUSH, M. B., D. R. PIPERNO, P. A. COLINVAUX, L. KRISSEK, P. E. DE OLIVEIRA, L. A. KRISSEK, M. C. MILLER, AND W. E. ROWE. 1992. A 14300-yr paleoecological profile of a lowland tropical lake in Panama. Ecological Monographs 62:251–275.

CABE, P. R., AND D. N. ALSTAD. 1994. Interpreting population differentiation in terms of drift and selection. Evolutionary Ecology 8:489–492.

CABOT, J., AND C. URDIALES. 2005. The subspecific status of Sardinian Warblers *Sylvia melanocephala* in the Canary Islands with the description of a new subspecies from Western Sahara. Bulletin of the British Ornithologists' Club 125:230–240.

CADENA, C. D., J. KLICKA, AND R. E. RICKLEFS. 2007. Evolutionary differentiation in the Neotropical montane region: Molecular phylogenetics and phylogeography of *Buarremon* brush-finches (Aves, Emberizidae). Molecular Phylogenetics and Evolution 44:993–1016.

CALDER, W. A., III. 1974. Consequences of body size for avian energetics. Pages 86–151 *in* Avian Energetics (R. A. Paynter, Jr., Ed.). Nuttall Ornithological Club, Cambridge, Massachusetts.

CALDER, W. A., III. 1984. Size, Function and Life History. Harvard University Press, Cambridge, Massachusetts.

CANTY AND ASSOCIATES. 2005. Weatherbase. [Online.] Available at www.weatherbase.com.

CARLING, M. D., AND R. T. BRUMFIELD. 2008. Integrating phylogenetic and population genetic analyses of multiple loci to test species divergence hypotheses in *Passerina* buntings. Genetics 178:363–377.

CARLQUIST, S. 1974. Island Biology. Columbia University Press, New York.

CARROLL, S. B. 2008. Evo-devo and an expanding evolutionary synthesis: A genetic theory of morphological evolution. Cell 134:25–36.

CARSTENSEN, D. W., AND J. M. OLESEN. 2009. Wallacea and its nectarivorous birds: Nestedness and modules. Journal of Biogeography 36:1540–1550.

CASSIN, J. 1856. Illustrations of the Birds of California, Texas, Oregon, British and Russian America. J.B. Lippincott, Philadelphia.

CHIKARA, O. 2002. Little-known and neglected distinctive (sub)species of southern Japan. Oriental Bird Club Bulletin 35:26–31.

CHUI, C. K. S., AND S. M. DOUCET. 2009. A test of ecological and sexual selection hypotheses for geographical variation in coloration and morphology of Golden-crowned Kinglets (*Regulus satrapa*). Journal of Biogeography 36:1945–1957.

CIBOIS, A., J.-C. THIBAULT, AND E. PASQUET. 2004. Biogeography of eastern Polynesian monarchs (*Pomarea*): An endemic genus close to extinction. Condor 106:837–851.

CIBOIS, A., J.-C. THIBAULT, AND E. PASQUET. 2007. Uniform phenotype conceals double colonization by reed-warblers of a remote Pacific archipelago. Journal of Biogeography 34:1150–1166.

CICERO, C. 1996. Sibling species of titmice in the *Parus inornatus* complex (Aves: Paridae). University of California Publications in Zoology, no. 128.

CICERO, C., AND N. K. JOHNSON. 2006. Diagnosability of subspecies: Lessons from Sage Sparrows (*Amphispiza belli*) for analysis of geographic variation in birds. Auk 123:266–274.

CICERO, C., AND N. K. JOHNSON. 2007. Narrow contact of desert Sage Sparrows (*Amphispiza belli nevadensis* and *A. b. canescens*) in Owens Valley, eastern California: Evidence from mitochondrial DNA, morphology, and GIS-based niche models. Pages 78–95 *in* Festschrift for Ned K. Johnson: Geographic Variation and Evolution in Birds (C. Cicero and J. V. Remsen, Jr., Eds.). Ornithological Monographs, no. 63.

CLARE, E. L., B. K. LIM, M. D. ENGSTROM, J. L. EGER, AND P. D. N. HEBERT. 2007. DNA barcoding of Neotropical bats: Species identification and discovery within Guyana. Molecular Ecology Notes 7:184–190.

CLAYTON, N. S. 1990. Assortative mating in Zebra Finch subspecies, *Taeniopygia guttata guttata* and *T. g. castanotis*. Philosophical Transactions of the Royal Society of London, Series B: Biological Sciences 330:351–370.

CLEGG, S. M., F. D. FRENTIU, J. KIKKAWA, G. TAVECCHIA, AND I. P. F. OWENS. 2008. 4000 years of phenotypic change in an island bird: Heterogeneity of selection over three microevolutionary timescales. Evolution 62:2393–2410.

CLEMENT, M., D. POSADA, AND K. A. CRANDALL. 2000. TCS: A computer program to estimate gene genealogies. Molecular Ecology 9:1657–1659.

CLEMENTS, J. F. 2000. Birds of the World: A Checklist, 5th ed. Ibis Publishing Company, Vista, California.

CLEMENTS, J. F. 2007. The Clements Checklist of Birds of the World, 6th ed. Cornell Lab of Ornithology, Ithaca, New York.

COHEN, J. 1988. Statistical Power Analysis for the Behavioral Sciences, 2nd ed. Erlbaum, Hillsdale, New Jersey.

COKINOS, C. 2000. Hope Is the Thing with Feathers. J.P. Tarcher/Putnam, New York.

COLLAR, N. J. 1996. Species concepts and conservation: A response to Hazevoet. Bird Conservation International 6:197–200.

COLLAR, N. J. 1997. Taxonomy and conservation: Chicken and egg. Bulletin of the British Ornithologists' Club 117:122–136.

COLLAR, N. J. 2005. How many bird species are there in Asia? Oriental Bird Club Bulletin 38:20–30.

COLLAR, N. J. 2006a. A partial revision of the Asian Babblers (Timaliidae). Forktail 22:85–112.

COLLAR, N. J. 2006b. A taxonomic reappraisal of the Black-browed Barbet *Megalaima oorti*. Forktail 22:170–173.

COLLAR, N. J. 2007a. Philippine bird taxonomy and conservation: A commentary on Peterson (2006). Bird Conservation International 17:103–113.

COLLAR, N. J. 2007b. Taxonomic notes on some insular *Loriculus* hanging-parrots. Bulletin of the British Ornithologists' Club 127:97–107.

COLLAR, N. J. 2008. Subjectivity and space in evaluating species limits: A response to Peterson and Moyle. Forktail 24:112–113.

COLLAR, N. J., M. J. CROSBY, AND A. J. STATTERSFIELD. 1994. Birds to Watch 2: The World List of Threatened Birds. BirdLife International, Cambridge, United Kingdom.

COLLINS-SCHRAMM, H. E., B. CHIMA, T. MORII, K. WAH, Y. FIGUEROA, L. A. CRISWELL, R. L. HANSON, W. C. KNOWLER, G. SILVA, J. W. BELMONT, AND M. F. SELDIN. 2004. Mexican American ancestry-informative markers: Examination of population structure and marker characteristics in European Americans, Mexican Americans, Amerindians and Asians. Human Genetics 114:263–271.

COMMITTEE ON THE STATUS OF ENDANGERED WILDLIFE IN CANADA (COSEWIC). 2005. Guidelines for recognizing Designatable Units below the species level (Appendix F5 in the COSEWIC O&P Manual). Committee on the Status of Endangered Wildlife in Canada, Ottawa. [Online.] Available at www. dfo-mpo.gc.ca/csas/Csas/Schedule-Horraire/ Details/2005/11_Nov/COSEWIC_DU_guidelines_ EN.pdf.

COMMITTEE ON THE STATUS OF ENDANGERED WILDLIFE IN CANADA (COSEWIC). 2008. Guidelines for recognizing Designatable Units below the species level (Appendix F5 in the COSEWIC O&P Manual). Committee on the Status of Endangered Wildlife in Canada, Ottawa.

CONNELLY, J. W., J. W. GRATSON, AND K. P. REESE. 1998. Sharp-tailed Grouse (*Tympanuchus phasianellus*). *In* The Birds of North America, no. 354 (A. Poole and F. Gill, Eds.). Academy of Natural Sciences, Philadelphia, and American Ornithologists' Union, Washington, D.C.

COOPER, A., J. RHYMER, H. F. JAMES, S. L. OLSON, C. E. MCINTOSH, M. D. SORENSON, AND R. C. FLEISCHER. 1996. Ancient DNA and island endemics. Nature 381:484.

CORNUET, J. M., S. PIRY, G. LUIKART, A. ESTOUP, AND M. SOLIGNAC. 1999. New methods employing multilocus genotypes to select or exclude populations as origins of individuals. Genetics 153:1989–2000.

CORY, C. B., AND C. E. HELLMAYR. 1925. Catalogue of the Birds of the Americas and the Adjacent Islands, vol. 13, part 4. Field Museum of Natural History, Chicago.

COUES, E. 1866. List of the birds of Fort Whipple, Arizona: With which are incorporated all other species ascertained to inhabit the Territory; with brief critical and field notes, descriptions of new species, etc. Proceedings of the Academy of Natural Sciences of Philadelphia 18:38–100.

COUES, E. 1871. Progress of American ornithology. American Naturalist 5:364–373.

COUES, E. 1872. Key to North American Birds. Estes and Lauriat, Boston.

COUES, E. 1884. On the application of trinomial nomenclature to zoology. Zoologist 8:241–247.

COYNE, J. A., AND H. A. ORR. 2004. Speciation. Sinauer Associates, Sunderland, Massachusetts.

CRACRAFT, J. 1983. Species concepts and speciation analysis. Pages 159–187 *in* Current Ornithology, vol. 1 (R. F. Johnston, Ed.). Plenum Press, New York.

CRACRAFT, J. 1985. Historical biogeography and patterns of differentiation within the South American avifauna: Areas of endemism. Pages 49–84 *in* Neotropical Ornithology (P. A. Buckley, M. S. Foster, E. S. Morton, R. S. Ridgely, and F. G. Buckley, Eds.). Ornithological Monographs, no. 36.

CRACRAFT, J. 1992. The species of the birds-of-paradise (Paradisaeidae): Applying the phylogenetic species concept to a complex pattern of diversification. Cladistics 8:1–43.

CRACRAFT, J. 1997. Species concepts in systematics and conservation biology—An ornithological viewpoint. Pages 325–339 *in* Species: The Units of Biodiversity (M. F. Claridge, H. A. Dawah, and M. R. Wilson, Eds.). Chapman & Hall, New York.

CRACRAFT, J. 2000. Species concepts in theoretical and applied biology: A systematic debate with consequences. Pages 3–14 *in* Species Concepts and Phylogenetic Theory: A Debate (Q. D. Wheeler and R. Meier, Eds.). Columbia University Press, New York.

CRAIG, R. J., AND E. TAISACAN. 1994. Notes on the ecology and population decline of the Rota Bridled White-eye. Wilson Bulletin 106:165–169.

CRAMP, S., K. E. L. SIMMONS, AND C. M. PERRINS, Eds. 1977–1994. Handbook of the Birds of Europe, the Middle East and Africa, vols. 1–9. Oxford University Press, Oxford, United Kingdom.

CRANDALL, K. A., O. R. P. BININDA-EMONDS, G. M. MACE, AND R. K. WAYNE. 2000. Considering evolutionary processes in conservation biology. Trends in Ecology and Evolution 15:290–295.

CUNNINGHAM, C. W. 1997. Is congruence between data partitions a reliable predictor of phylogenetic accuracy? Empirically testing an iterative procedure for choosing among phylogenetic methods. Systematic Biology 46:464–478.

CURSON, J., D. QUINN, AND D. BEADLE. 1994. Warblers of the Americas: An Identification Guide. Houghton Mifflin, Boston.

CUTRIGHT, P. R., AND M. J. BRODHEAD. 1981. Elliott Coues: Naturalist and Frontier Historian. University of Illinois Press, Urbana.

DALE, J. 2006. Intraspecific variation in coloration. Pages 36–86 *in* Bird Coloration, vol. 2: Function and Evolution (G. E. Hill and K. J. McGraw, Eds.). Harvard University Press, Cambridge, Massachusetts.

DANIELS, S. R., N. J. L. HEIDEMAN, M. G. J. HENDRICKS, M. E. MOKONE, AND K. A. CRANDALL. 2005. Unraveling evolutionary lineages in the limbless fossorial skink genus *Acontias* (Sauria: Scincidae): Are subspecies equivalent systematic units? Molecular Phylogenetics and Evolution 34:645–654.

DARWIN, C. 1859. On the Origin of Species by Means of Natural Selection, or the Preservation of Favoured Races in the Struggle for Life. John Murray, London.

DARWIN, C. 1895. On the Origin of Species by Means of Natural Selection, or the Preservation of Favoured Races in the Struggle for Life, 6th ed. John Murray, London.

DAVIS, J. I., AND K. C. NIXON. 1992. Populations, genetic variation, and the delimitation of phylogenetic species. Systematic Biology 41:421–435.

DAYRAT, B., P. D. CANTINO, J. A. CLARKE, AND K. DE QUEIROZ. 2008. Species names in the PhyloCode: The approach adopted by the International Society for Phylogenetic Nomenclature. Systematic Biology 57:507–514.

DEGNAN, J. H., AND N. A. ROSENBERG. 2009. Gene tree discordance, phylogenetic inference and the multispecies coalescent. Trends in Ecology and Evolution 24:332–340.

DELACOUR, J. 1956. The Waterfowl of the World, vol. 2. Country Life Limited, London.

DEL HOYO, J., A. ELLIOTT, J. SARGATAL, AND D. A. CHRISTIE, Eds. 1992–2008. Handbook of the Birds of the World, vols. 1–13. Lynx Edicions, Barcelona, Spain.

D'ELIA, J., M. ZWARTJES, AND S. MCCARTHY. 2008. Considering legal viability and societal values when deciding what to conserve under the U.S. Endangered Species Act. Conservation Biology 22:1072–1074.

DE QUEIROZ, K. 1998. The general lineage concept of species, species criteria, and the process of speciation: A conceptual unification and terminological recommendations. Pages 57–75 in Endless Forms: Species and Speciation (D. J. Howard and S. H. Berlocher, Eds.). Oxford University Press, Oxford, United Kingdom.

DE QUEIROZ, K. 2005a. Different species problems and their resolution. BioEssays 27:1263–1269.

DE QUEIROZ, K. 2005b. Ernst Mayr and the modern concept of species. Proceedings of the National Academy of Sciences USA 102:6600–6607.

DE QUEIROZ, K., AND M. J. DONOGHUE. 1988. Phylogenetic systematics and the species problem. Cladistics 4:317–338.

DE QUEIROZ, K., AND M. J. DONOGHUE. 1990. Phylogenetic systematics revisited. Cladistics 6:83–90.

DE QUEIROZ, K., AND J. GAUTHIER. 1992. Phylogenetic taxonomy. Annual Review of Ecology and Systematics 23:449–480.

DESALLE, R., M. G. EGAN, AND M. SIDDALL. 2005. The unholy trinity: Taxonomy, species delimitation and DNA barcoding. Philosophical Transactions of the Royal Society of London, Series B 360:1905–1916.

D'HORTA, F. M., J. M. CARDOSO DA SILVA, AND C. C. RIBAS. 2008. Species limits and hybridization zones in *Icterus cayanensis-chrysocephalus* group (Aves: Icteridae). Biological Journal of the Linnean Society 95:583–597.

DIAMOND, J. M., M. E. GILPIN, AND E. MAYR. 1976. Species-distance relation for birds of the Solomon Archipelago, and the paradox of the great speciators.

Proceedings of the National Academy of Sciences USA 73:2160–2164.

DICKERMAN, R. W., AND J. GUSTAFSON. 1996. The Prince of Wales Spruce Grouse: A new subspecies from southeastern Alaska. Western Birds 27:41–47.

DICKINSON, E. C., Ed. 2003. The Howard and Moore Complete Checklist of the Birds of the World, 3rd ed. Princeton University Press, Princeton, New Jersey.

DI GIACOMO, A. G. 1995. Two new species for Argentine avifauna. Hornero 14:77–78.

DIJKSTRA, L. H., AND D. J. JELLYMAN. 1999. Is the subspecies classification of freshwater eels *Anguilla australis australis* Richardson and *A. a. schmidtii* Phillips still valid? Marine and Freshwater Research 50:261–263.

DILLON, S., AND J. FJELDSÅ. 2005. The implications of different species concepts for describing biodiversity patterns and assessing conservation needs for African birds. Ecography 28:682–692.

DIMCHEFF, D. E., S. V. DROVETSKI, AND D. P. MINDELL. 2002. Phylogeny of Tetraoninae and other galliform birds using mitochondrial 12S and ND2 genes. Molecular Phylogenetics and Evolution 24:203–215.

DINGLE, C., W. HALFWERK, AND H. SLABBEKOORN. 2008. Habitat-dependent song divergence at subspecies level in the Grey-breasted Wood-wren. Journal of Evolutionary Biology 21:1079–1089.

DIZON, A. E., C. LOCKYER, W. F. PERRIN, D. P. DEMASTER, AND J. SISSON. 1992. Rethinking the stock concept: A phylogeographic approach. Conservation Biology 6:24–36.

DONOGHUE, M. J. 1985. A critique of the biological species concept and recommendations for a phylogenetic alternative. Bryologist 88:172–181.

DROVETSKI, S. V. 2002. Molecular phylogeny of grouse: Individual and combined performance of W-linked, autosomal, and mitochondrial loci. Systematic Biology 51:930–945.

DROVETSKI, S. V., R. M. ZINK, AND N. A. MODE. 2009. Patchy distributions belie morphological and genetic homogeneity in rosy-finches. Molecular Phylogenetics and Evolution 50:437–445.

DROVETSKI, S. V., R. M. ZINK, S. ROHWER, I. V. FADEEV, E. V. NESTEROV, I. KARAGODIN, E. A. KOBLIK, AND Y. A. RED'KIN. 2004. Complex biogeographic history of a Holarctic passerine. Proceedings of the Royal Society of London, Series B 271:545–551.

DRUMMOND, A. J., S. Y. W. HO, M. J. PHILLIPS, AND A. RAMBAUT. 2006. Relaxed phylogenetics and dating with confidence. PLoS Biology 4:e88.

DRUMMOND, A. J., AND A. RAMBAUT. 2007. BEAST: Bayesian evolutionary analysis by sampling trees. BMC Evolutionary Biology 7:214.

DRUMMOND, A. J., A. RAMBAUT, B. SHAPIRO, AND O. G. PYBUS. 2005. Bayesian coalescent inference of past population dynamics from molecular sequences. Molecular Biology and Evolution 22:1185–1192.

Dubois, A. 1871. Conspectus systematicus et geographicus Avium Europæarum. C. Muquardt, H. Merzbach, Brussels.

Dumbacher, J. P., and R. C. Fleischer 2001. Phylogenetic evidence for color pattern convergence in toxic pitohuis: Müllerian mimicry in birds? Proceedings of the Royal Society of London, Series B 268:1971–1976.

Dunning, J. B. 2008. CRC Handbook of Avian Body Masses. CRC Press, Boca Raton, Florida.

Eckert, A. J., and B. C. Carstens. 2008. Does gene flow destroy phylogenetic signal? The performance of three methods for estimating species phylogenies in the presence of gene flow. Molecular Phylogenetics and Evolution 49:832–842.

Edelaar, P. 2008. Assortative mating also indicates that Common Crossbill *Loxia curvirostra* vocal types are species. Journal of Avian Biology 39:9–12.

Edwards, S. V. 2009. Is a new and general theory of molecular systematics emerging? Evolution 63:1–19.

Edwards, S. V., and P. Beerli. 2000. Perspective: Gene divergence, population divergence, and the variance in coalescence time in phylogeographic studies. Evolution 54:1839–1854.

Edwards, S. V., and S. Bensch. 2009. Looking forwards or looking backwards in avian phylogeography? A comment on Zink and Barrowclough 2008. Molecular Ecology 18:2930–2933.

Edwards, S. V., S. B. Kingan, J. D. Calkins, C. N. Balakrishnan, W. B. Jennings, W. J. Swanson, and M. D. Sorensen. 2005. Speciation in birds: Genes, geography, and sexual selection. Proceedings of the National Academy of Sciences USA 102:6550–6557.

Edwards, S. V., L. Liu, and D. K. Pearl. 2007. High-resolution species trees without concatenation. Proceedings of the National Academy of Sciences USA 104:5936–5941.

Eggert, L. S. 1996. A phylogenetic approach to management of coastal California Cactus Wrens (*Campylorhynchus brunneicapillus*). M.S. thesis, San Diego State University, San Diego, California.

Elliott, D. G. 1914. In memoriam: Philip Lutley Sclater. Auk 31:1–13.

Ellsworth, D. L., R. L. Honeycutt, N. J. Silvy, K. D. Rittenhouse, and M. H. Smith. 1994. Mitochondrial-DNA and nuclear-gene differentiation in North American prairie grouse (Genus *Tympanuchus*). Auk 111:661–671.

Endler, J. A. 1977. Geographic variation, speciation and clines. Princeton University Press, Princeton, New Jersey.

Endler, J. A., and P. W. Mielke, Jr. 2005. Comparing entire colour patterns as birds see them. Biological Journal of the Linnean Society 86:405–431.

Engelmoer, M., and C. S. Roselaar. 1998. Geographical Variation in Waders. Kluwer, Dordrecht, The Netherlands.

Evarts, S. 2005. Cinnamon Teal *Anas cyanoptera*. Pages 549–553 *in* Ducks, Geese, and Swans (J. Kear, Ed.). Oxford University Press, Oxford, United Kingdom.

Fallon, S. M. 2007. Genetic data and the listing of species under the U.S. Endangered Species Act. Conservation Biology 21:1186–1195.

Falush, D., M. Stephens, and J. K. Pritchard. 2003. Inference of population structure using multilocus genotype data: Linked loci and correlated allele frequencies. Genetics 164:1567–1587.

Fancy, S. G., and T. J. Snetsinger. 2001. What caused the population decline of the Bridled White-eye on Rota, Mariana Islands? Pages 274–280 *in* Evolution, Ecology, Conservation, and Management of Hawaiian Birds: A Vanishing Avifauna (J. M. Scott, S. Conant, and C. Van Riper III, Eds.). Studies in Avian Biology, no. 22.

Farris, J. S., M. Kallersio, A. G. Kluge, and C. Bult. 1995. Constructing a significance test for incongruence. Systematic Biology 44:570–572.

Felsenstein, J. 1985. Confidence limits on phylogenies: An approach using the bootstrap. Evolution 39:783–791.

Felsenstein, J. 1993. PHYLIP (phylogeny inference package), version 3.5. [Online.] Available at evolution. gs.washington.edu/phylip.html.

Filardi, C. E., and R. G. Moyle. 2005. Single origin of a pan-Pacific bird group and upstream colonization of Australasia. Nature 438:216–219.

Filardi, C. E., and C. E. Smith. 2005. Molecular phylogenetics of monarch flycatchers (genus *Monarcha*) with emphasis on Solomon Island endemics. Molecular Phylogenetics and Evolution 37:776–788.

Fisher, A. K. 1898. Rank of the Sage Sparrow. Auk 15:190.

Fisher, R. A. 1930. The Genetical Theory of Natural Selection. Clarendon Press, Oxford, United Kingdom.

Fitzpatrick, J. W. 1980. Foraging behavior of Neotropical tyrant flycatchers. Condor 82:43–57.

Fitzpatrick, J. W. 1985. Form, foraging behavior, and adaptive radiation in the Tyrannidae. Pages 447–470 *in* Neotropical Ornithology (P. A. Buckley, M. S. Foster, E. S. Morton, R. S. Ridgely, and F. G. Buckley, Eds.). Ornithological Monographs, no. 36.

Fitzpatrick, J. W. 2004. Ochre-bellied Flycatcher, *Mionectes oleagineus*. Page 309 *in* Handbook of Birds of the World, vol. 9: Cotingas to Pipits and Wagtails (J. del Hoyo, A. Elliott, and D. A. Christie, Eds.). Lynx Edicions, Barcelona, Spain.

Fjeldså, J., and N. Krabbe. 1990. Birds of the High Andes. Apollo Books, Svendborg.

Fleischer, R. C., G. Fuller, and D. B. Ledig. 1995. Genetic structure of endangered Clapper Rail (*Rallus longirostris*) populations in southern California. Conservation Biology 9:1234–1243.

Fleischer, R. C., H. F. James, and S. L. Olson. 2008. Convergent evolution of Hawaiian and Australo-Pacific Honeyeaters from distant songbird ancestors. Current Biology 18:1927–1931.

FLEISCHER, R. C., J. J. KIRCHMAN, J. P. DUMBACHER, L. BEVIER, C. DOVE, N. C. ROTZEL, S. V. EDWARDS, M. LAMMERTINK, K. J. MIGLIA, AND W. S. MOORE. 2006. Mid-Pleistocene divergence of Cuban and North American Ivory-billed Woodpeckers. Biology Letters 2:466–469.

FLEISCHER, R. C., AND C. E. MCINTOSH. 2001. Molecular systematics and biogeography of the Hawaiian avifauna. Pages 51–60 *in* Evolution, Ecology, Conservation, and Management of Hawaiian Birds: A Vanishing Avifauna (J. M. Scott, S. Conant, and C. Van Riper III, Eds.). Studies in Avian Biology, no. 22.

FLEISCHER, R. C., C. E. MCINTOSH, AND C. L. TARR. 1998. Evolution on a volcanic conveyor belt: Using phylogeographic reconstructions and K-Ar-based ages of the Hawaiian Islands to estimate molecular evolutionary rates. Molecular Ecology 7:533–545.

FLEISCHER, R. C., B. SLIKAS, J. BEADELL, C. ATKINS, C. E. MCINTOSH, AND S. CONANT. 2007. Genetic variability and taxonomic status of the Nihoa and Laysan millerbirds. Condor 109:954–962.

FLEISCHER, R. C., C. L. TARR, H. F. JAMES, B. SLIKAS, AND C. E. MCINTOSH. 2001. Phylogenetic placement of the Poouli, *Melamprosops phaeosoma*, based on mitochondrial DNA sequence and osteological characters. Pages 98–104 *in* Evolution, Ecology, Conservation, and Management of Hawaiian Birds: A Vanishing Avifauna (J. M. Scott, S. Conant, and C. Van Riper III, Eds.). Studies in Avian Biology, no. 22.

FRANCIS, R. I. C. C., AND R. H. MATTLIN. 1986. A possible pitfall in the morphometric application of discriminant function analysis: Measurement bias. Marine Biology 93:311–313.

FRASER, D. J., AND L. BERNATCHEZ. 2001. Adaptive evolutionary conservation: Towards a unified concept for defining conservation units. Molecular Ecology 10:2741–2752.

FREELAND, J. R., AND P. T. BOAG. 1999. Phylogenetics of Darwin's finches: Paraphyly in the tree-finches and two divergent lineages in the Warbler Finch. Auk 116:577–588.

FREEMAN, S., AND J. C. HERRON. 2007. Evolutionary Analysis, 4th ed. Pearson Prentice Hall, Upper Saddle River, New Jersey.

FRY, A. J., AND R. M. ZINK. 1998. Geographic analysis of nucleotide diversity and Song Sparrow (Aves: Emberizidae) population history. Molecular Ecology 7:1303–1313.

FU, Y. X. 1997. Statistical tests of neutrality of mutations against population growth, hitchhiking and background selection. Genetics 147:915–925.

FUNK, D. J., AND K. OMLAND. 2003. Species-level paraphyly and polyphyly: Frequency, causes, and consequences, with insights from animal mitochondrial DNA. Annual Review of Ecology, Evolution, and Systematics 34:397–423.

FUNK, W. C., E. D. FORSMAN, T. D. MULLINS, AND S. M. HAIG. 2008. Introgression and dispersal among Spotted Owl (*Strix occidentalis*) subspecies. Ecological Applications 1:161–171.

FUNK, W. C., T. D. MULLINS, AND S. M. HAIG. 2007. Conservation genetics of Snowy Plovers (*Charadrius alexandrinus*) in the Western Hemisphere: Population genetic structure and delineation of subspecies. Conservation Genetics 8:1287–1309.

FUTUYMA, D. J. 1979. Evolutionary Biology. Sinauer Associates, Sunderland, Massachusetts.

FUTUYMA, D. J. 1998. Evolutionary Biology, 3rd ed. Sinauer Associates, Sunderland, Massachusetts.

FUTUYMA, D. J. 2005. Evolution. Sinauer Associates, Sunderland, Massachusetts.

GABBIANI, F., H. G. KRAPP, AND G. LAURENT. 1999. Computation of object approach by a wide-field, motion-sensitive neuron. Journal of Neuroscience 19:1122–1141.

GABRIELSON, I. N., AND F. C. LINCOLN. 1951. The races of Song Sparrows in Alaska. Condor 53:250–255.

GALATOWITSCH, M. L., AND R. L. MUMME. 2004. Escape behavior of Neotropical homopterans in response to a flush-pursuit predator. Biotropica 36:586–595.

GAMAUF, A., J.-O. GJERSHAUG, N. RØV, K. KVALØY, AND E. HARING. 2005. Species or subspecies? The dilemma of taxonomic ranking of some south-east Asian hawk-eagles (genus *Spizaetus*). Bird Conservation International 15:99–117.

GAMMONLEY, J. H. 1996. Cinnamon Teal (*Anas cyanoptera*). *In* The Birds of North America, no. 217 (A. Poole and F. Gill, Eds.). Academy of Natural Sciences, Philadelphia, and American Ornithologists' Union, Washington, D.C.

GARCÍA-MORENO, J., N. CORTÉS, G. M. GARCÍA-DERAS, AND B. E. HERNÁNDEZ-BAÑOS. 2006. Local origin and diversification among *Lampornis* hummingbirds: A Mesoamerican taxon. Molecular Phylogenetics and Evolution 38:488–498.

GARCÍA-MORENO, J., A. G. NAVARRO-SIGÜENZA, A. T. PETERSON, AND L. A. SÁNCHEZ-GONZÁLEZ. 2004. Genetic variation coincides with geographic structure in the Common Bush-tanager (*Chlorospingus ophthalmicus*) complex from Mexico. Molecular Phylogenetics and Evolution 33:186–196.

GARNETT, S. T., AND L. CHRISTIDIS. 2007. Implications of changing species definitions for conservation purposes. Bird Conservation International 17:187–195.

GARREAUD, R., M. VUILLE, AND A. C. CLEMENT. 2003. The climate of the Altiplano: Observed current conditions and mechanisms of past changes. Palaeogeography, Palaeoclimatology, Palaeoecology 194:5–22.

GATH, I., AND A. B. GEVA. 1989. Unsupervised optimal fuzzy clustering. IEEE Transactions on Pattern Analysis and Machine Intelligence 11:773–781.

GENNER, M. J., AND G. F. TURNER. 2005. The mbuna cichlids of Lake Malawi: A model for rapid speciation and adaptive radiation. Fish and Fisheries 6:1–34.

GENTRY, A. H. 2001. Patrones de diversidad y composición florística en los bosques de las montañas neotropicales. Pages 85–123 *in* Bosques Nublados del Neotrópico (M. Kappelle and A. D. Brown, Eds.). Instituto Nacional de Biodiversidad, InBio, Santo Domingo de Heredia, Costa Rica.

GILL, F. B. 1982. Might there be a resurrection of the subspecies? Auk 99:598–599.

GILL, F. B. 2007. Ornithology, 3rd ed. W.H. Freeman, New York.

GILL, F. B., AND M. WRIGHT. 2006. Birds of the World: Recommended English Names. Princeton University Press, Princeton, New Jersey.

GIPPOLITI, S., AND G. AMORI. 2007. The problem of subspecies and biased taxonomy in conservation lists: The case of mammals. Folia Zoologica 56:113–117.

GLENN, T. 1997. Genetic bottlenecks in long-lived vertebrates: Mitochondrial and microsatellite DNA variation in American alligators and whooping cranes. Ph.D. dissertation, University of Maryland, College Park.

GOETZMANN, W. H. 1986. New Lands, New Men: America and the Second Great Age of Discovery. Viking, New York.

GOLDSTEIN, D. B., AND C. SCHLÖTTERER. 1999. Microsatellites: Evolution and Applications. Oxford University Press, Oxford, United Kingdom.

GOLDSTEIN, P. Z., R. DESALLE, G. AMATO, AND A. P. VOGLER. 2000. Conservation genetics at the species boundary. Conservation Biology 14:120–131.

GONZALEZ, G., G. SORCI, A. P. MØLLER, P. NINNI, C. HAUSSY, AND F. DE LOPE. 1999. Immunocompetence and condition-dependent sexual advertisement in male House Sparrows (*Passer domesticus*). Journal of Animal Ecology 73:1225–1234.

GONZÁLEZ-FORERO, M. 2009. Removing ambiguity from the biological species concept. Journal of Theoretical Biology 256:76–80.

GONZÁLEZ-JOSÉ, R., I. ESCAPA, W. A. NEVES, R. CÚNEO, AND H. M. PUCCIARELLI. 2008. Cladistic analysis of continuous modularized traits provides phylogenetic signals in *Homo* evolution. Nature 453:775–778.

GOULD, S. J., AND R. C. LEWONTIN. 1979. The spandrels of San Marco and the Panglossian paradigm: A critique of the adaptationist programme. Proceedings of the Royal Society of London, Series B 205:581–598.

GRANGE, W. 1940. A comparison of the displays and vocal performances of the Greater Prairie-Chicken, Lesser Prairie-Chicken, Sharp-tailed Grouse and Sooty Grouse. Passenger Pigeon 2:127–133.

GRANT, P. R., AND B. R. GRANT. 2006. Species before speciation is complete. Annals of the Missouri Botanical Garden 93:94–102.

GRANT, P. R., AND B. R. GRANT. 2008. How and Why Species Multiply: The Radiation of Darwin's Finches. Princeton University Press, Princeton, New Jersey.

GREENBERG, R., P. J. CORDERO, S. DROEGE, AND R. C. FLEISCHER. 1998. Morphological adaptation with no mitochondrial DNA differentiation in the coastal plain Swamp Sparrow. Auk 115:706–712.

GREENLAW, J. S. 1996. Eastern Towhee (*Pipilo erythrophthalmus*). *In* The Birds of North America, no. 262 (A. Poole and F. Gill, Eds.). Academy of Natural Sciences, Philadelphia, and American Ornithologists' Union, Washington, D.C.

GREENLAW, J. S., AND G. E. WOOLFENDEN. 2007. Wintering distributions and migration of Saltmarsh and Nelson's sharp-tailed sparrows. Wilson Journal of Ornithology 119:361–377.

GREGORIUS, H. R. 1980. The probability of losing an allele when diploid genotypes are sampled. Biometrics 36:643–652.

GREGORY, P. A. 2006. Species accounts for *Arses, Myiagra, Lamprolia,* and *Machaerirhynchus.* Pages 315–325 *in* Handbook of Birds of the World, vol. 11: Old World Flycatchers to Old World Warblers (J. del Hoyo, A. Elliott, and D. A. Christie, Eds.). Lynx Edicions, Barcelona, Spain.

GRINNELL, J. 1898a. Birds of the Pacific Slope of Los Angeles County. G.A. Swerdfiger, Pasadena, California.

GRINNELL, J. 1898b. Rank of the Sage Sparrow. Auk 15:58–59.

GRINNELL, J. 1905. The California Sage Sparrow. Condor 7:18–19.

GRINNELL, J., AND A. H. MILLER. 1944. The distribution of the birds of California. Pacific Coast Avifauna, no. 27.

GROVES, C. P. 1986. Systematics of the great apes. Page 187–217 *in* Comparative Primate Biology, I: Systematics, Evolution, and Anatomy (D. R. Swindler and J. Erwin, Eds.). Alan R. Liss, New York.

GUIMERÀ, R., AND L. A. N. AMARAL. 2005. Functional cartography of complex metabolic networks. Nature 433:895–900.

GUTIÉRREZ, R. J., G. F. BARROWCLOUGH, AND J. G. GROTH. 2000. A classification of the grouse (Aves: Tetraoninae) based on mitochondrial DNA sequences. Wildlife Biology 6:205–211.

GUYOT, I., AND J.-C. THIBAULT. 1987. Les oiseaux terrestres des Iles Wallis-et-Futuna (Pacifique sud-ouest). L'Oiseau et R. F. O. 57:226–250.

HACKETT, S. J. 1996. Molecular phylogenetics and biogeography of tanagers in the genus *Ramphocelus* (Aves). Molecular Phylogenetics and Evolution 5:368–382.

HAFFER, J. 1985. Avian zoogeography of the Neotropical lowlands. Pages 113–146 *in* Neotropical Ornithology (P. A. Buckley, M. S. Foster, E. S. Morton, R. S. Ridgely, and F. G. Buckley, Eds.). Ornithological Monographs, no. 36.

HAFFER, J. 2001. Ornithological research traditions in central Europe during the 19th and 20th centuries. Journal für Ornithologie 142:27–93.

HAFFER, J. 2007. Ornithology, Evolution, and Philosophy: The Life and Science of Ernst Mayr 1904–2005. Springer, Berlin.

HAFFER, J., AND J. W. FITZPATRICK. 1985. Geographic variation in some Amazonian forest birds. Pages 147–168 *in* Neotropical Ornithology (P. A. Buckley, M. S. Foster, E. S. Morton, R. S. Ridgely, and F. G. Buckley, Eds.). Ornithological Monographs, no. 36.

HAIG, S. M., E. A. BEEVER, S. M. CHAMBERS, H. M. DRAHEIM, B. D. DUGGER, S. DUNHAM, E. ELLIOTT-SMITH, J. B. FONTAINE, D. C. KESLER, B. J. KNAUS, AND OTHERS. 2006. Taxonomic considerations in listing subspecies under the U.S. Endangered Species Act. Conservation Biology 20:1584–1594.

HAIG, S. M., AND J. D'ELIA. This volume. Avian subspecies and the U.S. endangered species act. Pages 170–173 *in* Avian Subspecies (K. Winker and S. M. Haig, Eds.). Ornithological Monographs, no. 67.

HAIG, S. M., T. D. MULLINS, AND E. D. FORSMAN. 2004. Subspecific relationships and genetic structure in the Spotted Owl. Conservation Genetics 5:683–705.

HALL, E. R., AND K. R. KELSON. 1959. The Mammals of North America, vol. 1. Roland Press, New York.

HAMILTON, T. H. 1961. The adaptive significances of intraspecific trends of variation in wing length and body size among bird species. Evolution 15:180–195.

HANSON, H. C. 2006. The White-cheeked Geese: Taxonomy, Ecophysiographic Relationships, Biogeography, and Evolutionary Considerations, vol. 1: Eastern Taxa. Avvar Books, Blythe, California.

HARDING, R. M. 1996. New phylogenies: An introductory look at the coalescent. Pages 15–22 *in* New Uses for New Phylogenies (P. H. Harvey, A. J. L. Brown, J. M. Smith, and S. Nee, Eds.). Oxford University Press, Oxford, United Kingdom.

HARE, M. P., AND G. F. SHIELDS. 1992. Mitochondrial-DNA variation in the polytypic Alaskan Song Sparrow. Auk 109:126–132.

HARMON, L. J., J. T. WEIR, C. BROCK, R. E. GLOR, AND W. CHALLENGER. 2008. GEIGER: Investigating evolutionary radiations. Bioinformatics 24:129–131.

HARRIS, J. D., AND E. FROUFE. 2004. Taxonomic inflation: Species concept or historical geopolitical bias? Trends in Ecology and Evolution 20:6–7.

HARRISON, R. G. 1993. Hybrids and hybrid zones: Historical perspective. Pages 3–12 *in* Hybrid Zones and the Evolutionary Process (R. G. Harrison, Ed.). Oxford University Press, New York.

HARSHMAN, J. 1996. Phylogeny, evolutionary rates, and ducks. Ph.D. dissertation, University of Chicago, Chicago, Illinois.

HARVEY, P. H., R. M. MAY, AND S. NEE. 1994. Phylogenies without fossils. Evolution 48:523–529.

HASEGAWA, M., H. KISHINO, AND T. YANO. 1985. Dating of the human-ape splitting by a molecular clock of mitochondrial DNA. Journal of Molecular Evolution 22:160–174.

HASKELL, D. G., AND A. ADHIKARI. 2009. Darwin's manufactory hypothesis is confirmed and predicts the extinction risk of extant birds. PLoS One 4:e5460.

HATSOPOULOS, N., F. GABBIANI, AND G. LAURENT. 1995. Elementary computation of object approach by a wide-field visual neuron. Science 270:1000–1003.

HAZEVOET, C. J. 1995. The Birds of the Cape Verde Islands. British Ornithologists' Union, Tring, United Kingdom.

HAZEVOET, C. J. 1996. Conservation and species lists: Taxonomic neglect promotes the extinction of endemic birds, as exemplified by taxa from eastern Atlantic islands. Bird Conservation International 6:181–196.

HEBERT, P. D. N., S. RATNASINGHAM, AND J. R. DEWAARD. 2003. Barcoding animal life: Cytochrome *c* oxidase subunit 1 divergences among closely related species. Proceedings of the Royal Society of London, Series B 270 (Supplement 1):S96–S99.

HEBERT, P. D. N., M. Y. STOECKLE, T. S. ZEMLAK, AND C. M. FRANCIS. 2004. Identification of birds through DNA barcodes. PLoS Biology 2:1657–1663.

HELBIG, A. J., A. G. KNOX, D. T. PARKIN, G. SANGSTER, AND M. COLLINSON. 2002. Guidelines for assigning species rank. Ibis 144:518–525.

HENNIG, W. 1966. Phylogenetic Systematics. University of Illinois Press, Urbana.

HEWITT, G. 2000. The genetic legacy of the Quaternary ice ages. Nature 405:907–913.

HEY, J. 2001. The mind of the species problem. Trends in Ecology and Evolution 16:326–329.

HEY, J., R. S. WAPLES, M. L. ARNOLD, R. K. BUTLIN, AND R.G. HARRISON. 2003. Understanding and confronting species uncertainty in biology and conservation. Trends in Ecology and Evolution 18:597–603.

HILL, G. E. 1993. Geographic variation in the carotenoid plumage pigmentation of male House Finches (*Carpodacus mexicanus*). Biological Journal of the Linnean Society 49:63–86.

HILL, G. E. 2006. Environmental regulation of ornamental coloration. Pages 507–560 *in* Bird Coloration, vol. 1: Mechanisms and Measurements (G. E. Hill and K. J. McGraw, Eds.). Harvard University Press, Cambridge, Massachusetts.

HIRSCHFELD, E., Ed. 2008. Rare Birds Yearbook 2009. MagDig Media, Shrewsbury, United Kingdom.

HOEKSTRA, H., R. J. HIRSCHMANN, R. A. BUNDY, P. A. INSEL, AND J. P. CROSSLAND. 2006. A single amino acid mutation contributes to adaptive beach mouse color pattern. Science 313:101–104.

HOEKSTRA, H. E., J. G. KRENZ, AND M. W. NACHMAN. 2005. Local adaptation in the rock pocket mouse (*Chaetodipus intermedius*): Natural selection and phylogenetic history of populations. Heredity 94:217–228.

HOLDEREGGER, R., AND H. H. WAGNER. 2006. A brief guide to landscape genetics. Landscape Ecology 21:793–796.

HOLMQVIST, M. H., AND M. V. SRINIVASAN. 1991. A visually evoked escape response of the housefly. Journal of Comparative Physiology A 169:451–459.

HOPKINS, S. R., AND F. L. POWELL. 2001. Common themes of adaptation to hypoxia. Advances in Experimental Medicine and Biology 502:153–167.

Hostert, E. E. 1997. Reinforcement: A new perspective on an old controversy. Evolution 51:697–702.

Hudson, R. R., and J. A. Coyne. 2002. Mathematical consequences of the genealogical species concept. Evolution 56:1557–1565.

Huelsenbeck, J. P., and F. Ronquist. 2001. MRBAYES: Bayesian inference of phylogenetic trees. Bioinformatics 17:754–755.

Huler, S. 2004. Defining the Wind: The Beaufort Scale, and How a Nineteenth-Century Admiral Turned Science into Poetry. Crown, New York.

Hull, J. M., W. K. Savage, J. L. Bollmer, R. T. Kimball, P. G. Parker, N. K. Whiteman, and H. B. Ernest. 2008. On the origin of the Galápagos Hawk: An examination of phenotypic differentiation and mitochondrial paraphyly. Biological Journal of the Linnean Society 95:779–789.

Hupp, J. W., and C. E. Braun. 1991. Geographic variation among Sage Grouse in Colorado. Wilson Bulletin 103:255–261.

Huxley, J. S. 1942. Evolution, the Modern Synthesis. Allen & Unwin, London.

Inger, R. F. 1961. Problems in the application of the subspecies concept in vertebrate taxonomy. Pages 262–285 in Vertebrate Speciation (W. F. Blair, Ed.). University of Texas Press, Austin.

International Code of Zoological Nomenclature. 1999. International Code of Zoological Nomenclature. International Trust for Zoological Nomenclature, London.

Irestedt, M., J. Fjeldså, U. S. Johansson, and P. G. P. Ericson. 2002. Systematic relationships and biogeography of the tracheophone suboscines (Aves: Passeriformes). Molecular Phylogenetics and Evolution 23:499–512.

Irwin, D. E. 2009. Incipient ring speciation revealed by a migratory divide. Molecular Ecology 18:2923–2925.

Irwin, D. E., S. Bensch, J. H. Irwin, and T. D. Price. 2005. Speciation by distance in a ring species. Science 307:414–416.

Irwin, D. E., S. Bensch, and T. Price. 2001. Speciation in a ring. Nature 409:333–337.

Isaac, N. J. B., J. Mallet, and G. M. Mace. 2004. Taxonomic inflation: Its influence on macroecology and conservation. Trends in Ecology and Evolution 19:464–469.

Isler, M. L., P. R. Isler, and B. M. Whitney. 1998. Use of vocalizations to establish species limits in antbirds (Passeriformes: Thamnophilidae). Auk 115:577–590.

IUCN Standards and Petitions Working Group. 2008. Guidelines for Using the IUCN Red List Categories and Criteria. Version 7.0. Prepared by the Standards and Petitions Working Group of the IUCN SSC Biodiversity Assessments Sub-Committee in August 2008.

Jabłonski, P. G. 1999. A rare predator exploits prey escape behavior: The role of tail-fanning and plumage contrast in foraging of the Painted Redstart (*Myioborus pictus*). Behavioral Ecology 10:7–14.

Jabłonski, P. G. 2001. Sensory exploitation of prey: Manipulation of the initial direction of prey escapes by a conspicuous 'rare enemy.' Proceedings of the Royal Society of London, Series B 268:1017–1022.

Jabłonski, P. G., K. Laseter, R. L. Mumme, M. Borowiecz, J. P. Cygan, J. Pereira, and E. Sergiej. 2006a. Habitat-specific sensory-exploitative signals in birds: Propensity of dipteran prey to cause evolution of plumage variation in flush-pursuit insectivores. Evolution 60:2633–2642.

Jabłonski, P. G, S. D. Lee, and L. Jerzak. 2006b. Innate plasticity of a predatory behavior: Nonlearned context dependence of avian flush-displays. Behavioral Ecology 17:925–932.

Jabłonski, P. G., and C. McInerney. 2005. Prey escape direction is influenced by the pivoting displays of flush-pursuing birds. Ethology 111:381–396.

Jabłonski, P. G., and N. J. Strausfeld. 2000. Exploitation of an ancient escape circuit by an avian predator: Prey sensitivity to model predator display in the field. Brain, Behavior and Evolution 56:94–106.

Jabłonski, P. G., and N. J. Strausfeld. 2001. Exploitation of an ancient escape circuit by an avian predator: Relationships between taxon-specific prey escape circuits and the sensitivity to visual cues from the predator. Brain, Behavior and Evolution 58:218–240.

James, F. C. 1968. A more precise definition of Bergmann's rule. American Zoologist 8:815–816.

James, F. C. 1970. Geographic size variation in birds and its relationship to climate. Ecology 51:365–390.

James, F. C. 1983. The environmental component of geographic variation in the size and shape of birds: Transplant experiments with blackbirds. Science 221:184–186.

James, F. C. 1991. Complementary descriptive and experimental studies of clinal variation in birds. American Zoologist 31:694–706.

James, H. F. 2004. The osteology and phylogeny of the Hawaiian finch radiation (Fringillidae: Drepanidini), including extinct taxa. Zoological Journal of the Linnean Society 141:207–255.

James, H. F., and S. L. Olson. 1991. Descriptions of thirty-two new species of birds from the Hawaiian Islands: Part II. Passeriformes. Ornithological Monographs, no. 46.

James, H. F., and S. L. Olson. 2003. A giant new species of Nukupu'u (Fringillidae: Drepanidini: *Hemignathus*) from the island of Hawaii. Auk 120:970–981.

James, H. F., and S. L. Olson. 2005. The diversity and biogeography of koa-finches (Drepanidini: *Rhodacanthis*), with descriptions of two new species. Zoological Journal of the Linnean Society 144:527–541.

James, H. F., and S. L. Olson. 2006. A new species of Hawaiian finch (Drepanidini: *Loxioides*) from Makauwahi Cave, Kaua'i. Auk 123:335–344.

JEWETT, S. G., W. P. TAYLOR, W. T. SHAW, AND J. W. ALDRICH. 1953. Birds of Washington State. University of Washington Press, Seattle.

JIGUET, F. 2002. Taxonomy of the Kelp Gull *Larus dominicanus* Lichtenstein inferred from biometrics and wing plumage pattern, including two previously undescribed subspecies. Bulletin of the British Ornithologists' Club 122:50–71.

JOHNSGARD, P. A. 1978. Ducks, Geese, and Swans of the World. University of Nebraska Press, Lincoln.

JOHNSGARD, P. A. 1983. The Grouse of the World. University of Nebraska Press, Lincoln.

JOHNSGARD, P. A. 2002. Grassland Grouse and their Conservation. Smithsonian Institution Press, Washington, D.C.

JOHNSON, J. A. 2008. Recent range expansion and divergence among North American prairie grouse. Journal of Heredity 99:165–173.

JOHNSON, J. A., J. E. TOEPFER, AND P. O. DUNN. 2003. Contrasting patterns of mitochondrial and microsatellite population structure in fragmented populations of Greater Prairie-Chickens. Molecular Ecology 12:3335–3347.

JOHNSON, N. K. 1980. Character variation and evolution of sibling species in the *Empidonax difficilis-flavescens* complex (Aves: Tyrannidae). University California Publications Zoology, no. 112.

JOHNSON, N. K. 1982. Retain subspecies—At least for the time being. Auk 99:605–606.

JOHNSON, N. K., AND J. A. MARTEN. 1992. Macrogeographic patterns of morphometric and genetic variation in the Sage Sparrow complex. Condor 94:1–19.

JOHNSON, N. K., J. V. REMSEN, JR., AND C. CICERO. 1999. Resolution of the debate over species concepts in ornithology: A new comprehensive biologic species concept. Pages 1470–1482 *in* Proceedings of the 22nd International Ornithological Congress, Durban (N. J. Adams and R. H. Slotow, Eds.). BirdLife South Africa, Johannesburg.

JOHNSON, N. K., AND R. M. ZINK. 1985. Genetic evidence for relationships among the Red-eyed, Yellow-green, and Chivi Vireos. Wilson Bulletin 97:421–435.

JOHNSON, R. E. 2002. Black Rosy-Finch (*Leucosticte atrata*). *In* The Birds of North America, no. 678 (A. Poole and F. Gill, Eds.). Birds of North America, Philadelphia.

JOHNSON, R. E., P. HENDRICKS, D. L. PATTIE, AND K. B. HUNTER. 2002. Brown-capped Rosy-Finch (*Leucosticte australis*). *In* The Birds of North America, no. 536 (A. Poole and F. Gill, Eds.). Birds of North America, Philadelphia.

JOHNSTON, R. F., AND R. K. SELANDER. 1964. House Sparrows: Rapid evolution of races in North America. Science 144:548–550.

JONES, R. E. 1964. The specific distinctness of the Greater and Lesser prairie-chickens. Auk 81:65–73.

KAHN, N. W., C. E. BRAUN, J. R. YOUNG, S. WOOD, D. R. MATA, AND T. W. QUINN. 1999. Molecular analysis of genetic variation among large and small-bodied Sage Grouse using mitochondrial control-region sequences. Auk 116:819–824.

KARL, S. A., R. M. ZINK, AND J. R. JEHL, JR. 1987. Allozyme analysis of the California Gull (*Larus californicus*). Auk 104:767–769.

KARR, J. R., AND F. C. JAMES. 1975. Eco-morphological configurations and convergent evolution in species and communities. Pages 258–291 *in* Ecology and Evolution of Communities (M. L. Cody and J. M. Diamond, Eds.). Belknap Press, Cambridge, Massachusetts.

KENDALL, D. G. 1948. On the generalized "birth-and-death" process. Annals of Mathematical Statistics 19:1–15.

KENESEI, T., B. BALASKO, AND J. ABONYI. 2006. A MATLAB toolbox and its web based variant for fuzzy cluster analysis. [Online.] Available at bmf.hu/conferences/huci2006/56_Kenesei.pdf.

KERR, K. C. R., M. Y. STOECKLE, C. J. DOVE, L. A. WEIGT, C. M. FRANCIS, AND P. D. N. HEBERT. 2007. Comprehensive DNA barcode coverage of North American birds. Molecular Ecology Notes 7:535–543.

KEYSERLING, A., AND J. H. BLASIUS. 1840. Die Wirbelthiere Europas. F. Vieweg und Sohn, Braunschweig, Germany.

KIMURA, M. 1983. The Neutral Theory of Molecular Evolution. Cambridge University Press, Cambridge, United Kingdom.

KLICKA, J. T., AND R. M. ZINK. 1997. The importance of recent ice ages in speciation: A failed paradigm. Science 277:1666–1669.

KNOX, A. G. 2007. Order or chaos? Taxonomy and the British list over the last 100 years. British Birds 100:609–623.

KORNEGAY, J. R., T. D. KOCHER, L. A. WILLIAMS, AND A. C. WILSON. 1993. Pathways of lysozyme evolution inferred from the sequences of cytochrome *b* in birds. Journal of Molecular Evolution 37:367–379.

KRÜGER, O., M. D. SORENSON, AND N. B. DAVIES. 2009. Does coevolution promote species richness in parasitic cuckoos? Proceedings of the Royal Society of London, Series B 276:3871–3879.

LANYON, W. E. 1967. Revision and probable evolution of the *Myiarchus* flycatchers of the West Indies. Bulletin of the American Museum of Natural History 136:331–370.

LARSSON, K., AND P. FORSLUND 1992. Genetic and social inheritance of body and egg size in the Barnacle Goose (*Branta leucopsis*). Evolution 46:235–244.

LAWRENCE, G. N. 1864. Catalogue of birds collected at the island of Sombrero, W. I., with observations by A. A. Julien. Annals of the Lyceum of Natural History, New York 8:92–106.

LEAFLOOR, J. O., C. D. ANKNEY, AND D. H. RUSCH. 1998. Environmental effects on body size of Canada Geese. Auk 115:26–33.

LEGGE, J. T., R. ROUSH, R. DeSALLE, A. P. VOGLER, AND B. MAY. 1996. Genetic criteria for establishing evolutionarily significant units in Cryan's Buckmoth. Conservation Biology 10:85–98.

LEINONEN, T., R. B. O'HARA, J. M. CANO, AND J. MERILÄ. 2008. Comparative studies of quantitative trait and neutral marker divergence: A meta-analysis. Journal of Evolutionary Biology 21:1–17.

LEISLER, B., AND H. WINKLER. 1985. Ecomorphology. Pages 155–186 *in* Current Ornithology, vol. 2 (R. F. Johnston, Ed.). Plenum Press, New York.

LEMON, W. C., AND R. H. BARTH, JR. 1992. The effects of feeding rate on reproductive success in the Zebra Finch, *Taeniopygia guttata*. Animal Behaviour 44:851–857.

LENTINO, M., E. BONACCORSO, M. A. GARCÍA, C. PORTAS, E. A. FERNÁNDEZ, AND R. RIVERO. 2003. Longevity records of wild birds in the Henri Pittier National Park, Venezuela. Ornitología Neotropical 14:545–548.

LEWIS, P. O., AND D. ZAYKIN. 2001. Genetic Data Analysis: Computer program for the analysis of allelic data. Version 1.0 (d16c). [Online.] Available at lewis.eeb.uconn.edu/lewishome/software.html.

LI, J. Z., D. M. ABSHER, H. TANG, A. M. SOUTHWICK, A. M. CASTO, S. RAMACHANDRAN, H. M. CANN, G. S. BARSH, M. FELDMAN, L. L. CAVALLI-SFORZA, AND R. M. MYERS. 2008. Worldwide human relationships inferred from genome-wide patterns of variation. Science 319:1100–1104.

LIU, Z., AND R. GEORGE. 2005. Mining weather data using fuzzy cluster analysis. Pages 105–119 *in* Fuzzy Modeling with Spatial Information for Geographic Problems (F. E. Petry, V. B. Robinson, and M. A. Cobb, Eds.). Springer, Berlin.

LOUGHEED, S. C., T. W. ARNOLD, AND R. C. BAILEY. 1991. Measurement error of external and skeletal variables in birds and its effect on principal components. Auk 108:432–436.

LUCCHINI, V., J. HÖGLUND, S. KLAUS, J. SWENSON, AND E. RANDI. 2001. Historical biogeography and a mitochondrial DNA phylogeny of grouse and ptarmigan. Molecular Phylogenetics and Evolution 20:149–162.

MACDOUGALL-SHACKLETON, S. A., R. E. JOHNSON, AND T. P. HAHN. 2000. Gray-crowned Rosy-Finch (*Leucosticte tephrocotis*). *In* The Birds of North America, no. 559 (A. Poole and F. Gill, Eds.). Birds of North America, Philadelphia.

MACE, G. M. 2004. The role of taxonomy in species conservation. Philosophical Transactions of the Royal Society of London, Series B 359:711–719.

MADGE, S., AND H. BURN. 1988. Waterfowl: An Identification Guide to the Ducks, Geese, and Swans of the World. Houghton Mifflin, Boston.

MAGALLÓN, S., AND M. J. SANDERSON. 2001. Absolute diversification rates in angiosperm clades. Evolution 55:1762–1780.

MALHI, R. S., H. M. MORTENSEN, J. A. ESHLEMAN, B. M. KEMP, J. G. LORENZ, F. A. KAESTLE, J. R. JOHNSON, C. GORODEZKY, AND D. G. SMITH. 2003. Native American mtDNA prehistory in the American South-west. American Journal of Physical Anthropology 120:108–124.

MALLET, J. 1995. A species definition for the modern synthesis. Trends in Ecology and Evolution 10:294–299.

MALLET, J. 2007. Subspecies, semispecies, superspecies. Pages 1–5 *in* Encyclopedia of Biodiversity (S. A. Levin, Ed.). Elsevier, Oxford, United Kingdom.

MANEL, S., M. K. SCHWARTZ, G. LUIKART, AND P. TABERLET. 2003. Landscape genetics: Combining landscape ecology and population genetics. Trends in Ecology and Evolution 18:189–197.

MANLY, B. F. J. 2000. Multivariate Statistical Methods: A Primer, 2nd ed. Chapman & Hall/CRC, Boca Raton, Florida.

MANNI, F., AND E. GUÉRARD. 2004. Barrier version 2.2 User's Manual. Population genetics team, Musée de l'Homme, Paris.

MANNI, F., E. GUÉRARD, AND E. HEYER. 2004. Geographic patterns of (genetic, morphologic, linguistic) variation: How barriers can be detected by using Monmonier's algorithm. Human Biology 76:173–190.

MAO, X., A. W. BIGHAM, R. MEI, G. GUTIERREZ, K. M WEISS, T. D. BRUTSAERT, F. LEON-VELARDE, L. G. MOORE, E. VARGAS, P. M. MCKEIGUE, AND OTHERS. 2007. A genomewide admixture mapping panel for Hispanic/Latino populations. American Journal of Human Genetics 80:1171–1178.

MARANTZ, C. A. 1992. Evolutionary implications of vocal and morphological variation in the woodcreeper genus *Dendrocolaptes* (Aves: Dendrocolaptidae). M.S. thesis, Louisiana State University, Baton Rouge.

MARANTZ, C. A. 1997. Geographic variation in plumage patterns in the woodcreeper genus *Dendrocolaptes* (Dendrocolaptidae). Pages 399–429 *in* Studies in Neotropical Ornithology Honoring Ted Parker (J. V. Remsen, Jr., Ed.). Ornithological Monographs, no. 48.

MARANTZ, C. A., A. ALEIXO, L. R. BEVIER, AND M. A. PATTEN. 2003. Family Dendrocolaptidae (Woodcreepers). Pages 358–447 *in* Handbook of the Birds of the World, vol. 8: Broadbills to Tapaculos (J. del Hoyo, A. Elliott, and D. A. Christie, Eds.). Lynx Edicions, Barcelona, Spain.

MARANTZ, C. A., AND M. A. PATTEN. This volume. Quantifying subspecies analysis: A case study of morphometric variation and subspecies in the woodcreeper genus *Dendrocolaptes*. Pages 170–173 *in* Avian Subspecies (K. Winker and S. M. Haig, Eds.). Ornithological Monographs, no. 67.

MARCHANT, S., P. J. HIGGINS, S. J. J. F. DAVIES, J. M. PETER, W. K. STEELE, AND S. J. COWLING, Eds. 1990–2006. Handbook of Australian, New Zealand and Antarctic Birds, vols. 1–7. Oxford University Press, Melbourne, Australia.

MARGOT, J.-L. 2006. What makes a planet? [Online.] Available at www.astro.cornell.edu/~jlm/planet.html.

MARSHALL, J. T., Jr. 1967. Parallel variation in North and Middle American screech-owls. Monographs of the Western Foundation of Vertebrate Zoology, no. 1.

MARTHINSEN, G., L. WENNERBERG, AND J. T. LIFJELD. 2007. Phylogeography and subspecies taxonomy of Dunlins (*Calidris alpina*) in western Palearctic analysed by DNA microsatellites and amplified fragment length polymorphism markers. Biological Journal of the Linnean Society 92:713–726.

MARTIN, J. W., AND B. A. CARLSON. 1998. Sage Sparrow (*Amphispiza belli*). The Birds of North America Online (A. Poole, Ed.). Cornell Lab of Ornithology, Ithaca, New York. [Online.] Available at bna.birds.cornell.edu.bnaproxy.birds.cornell.edu/bna/species/326.

MARTIN, P. R., AND J. J. TEWKSBURY. 2008. Latitudinal variation in subspecific diversification of birds. Evolution 62:2775–2788.

MAURER, B. A. 1994. Geographical Population Analysis: Tools for the Analysis of Biodiversity. Blackwell Science, Oxford, United Kingdom.

MAYR, E. 1933. Birds collected during the Whitney South Sea Expedition. XXIV. Notes on Polynesian Flycatchers and a revision of the genus *Clytorhynchus* Elliot. American Museum Novitates, no. 628.

MAYR, E. 1940. Speciation phenomena in birds. American Naturalist 74:249–278.

MAYR, E. 1942a. Birds collected during the Whitney South Sea Expedition. XLVIII. Notes on the Polynesian species of *Aplonis*. American Museum Novitates, no. 1166.

MAYR, E. 1942b. Systematics and the Origin of Species from the Viewpoint of a Zoologist. Columbia University Press, New York.

MAYR, E. 1945. Birds of the Southwest Pacific: A Field Guide to the Birds of the Area between Samoa, New Caledonia, and Micronesia. Macmillan, New York.

MAYR, E. 1951. Speciation in birds: Progress report on the years 1938–1950. Pages 91–131 *in* Proceedings of the Xth International Ornithological Congress (S. Hörstadius, Ed.). Almquist & Wiksell, Uppsala, Sweden.

MAYR, E. 1956. Geographical character gradients and climatic adaptation. Evolution 10:105–108.

MAYR, E. 1963. Animal Species and Evolution. Belknap Press of Harvard University Press, Cambridge, Massachusetts.

MAYR, E. 1969. Principles of Systematic Zoology. McGraw-Hill, New York.

MAYR, E. 1982a. Of what use are subspecies? Auk 99:593–595.

MAYR, E. 1982b. The Growth of Biological Thought: Diversity, Evolution, and Inheritance. Belknap Press of Harvard University Press, Cambridge, Massachusetts.

MAYR, E. 1983. How to carry out the adaptationist program? American Naturalist 121:324–334.

MAYR, E. 1993. Fifty years of research on species and speciation. Proceedings of the California Academy of Sciences 48:131–140.

MAYR, E. 1996. What is a species, and what is not? Philosophy of Science 63:262–277.

MAYR, E. 2000a. A critique from the biological species concept perspective: What is a Species, and What is Not? Pages 93–100 *in* Species Concepts and Phylogenetic Theory: A Debate (Q. D. Wheeler and R. Meier, Eds.). Columbia University Press, New York.

MAYR, E. 2000b. The biological species concept. Pages 17–29 *in* Species Concepts and Phylogenetic Theory: A Debate (Q. D. Wheeler and R. Meier, Eds.). Columbia University Press, New York.

MAYR, E., and P. D. Ashlock. 1991. Principles of Systematic Zoology, 2nd ed. McGraw-Hill, New York.

MAYR, E., AND J. DIAMOND. 2001. The Birds of Northern Melanesia: Speciation, Ecology, and Biogeography. Oxford University Press, Oxford, United Kingdom.

MAYR, E., E. G. LINSLEY, AND R. L. USINGER. 1953. Methods and Principles of Systematic Zoology. McGraw-Hill, New York.

MAYR, E., AND M. MOYNIHAN. 1946. Birds collected during the Whitney South Sea Expedition. 56. Evolution in the *Rhipidura rufifrons* group. American Museum Novitates, no. 1321.

MAYR, E., AND L. L. SHORT. 1970. Species taxa of North American birds: A contribution to comparative systematics. Publications of the Nuttall Ornithological Club, no. 9.

McCRACKEN, K. G., C. P. BARGER, M. BULGARELLA, K. P. JOHNSON, M. K. KUHNER, A. V. MOORE, J. L. PETERS, J. TRUCCO, T. H. VALQUI, K. WINKER, AND R. E. WILSON. 2009a. Signatures of high-altitude adaptation in the major hemoglobin of five species of Andean dabbling ducks. American Naturalist 174:631–650.

McCRACKEN, K. G., C. P. BARGER, M. BULGARELLA, K. P. JOHNSON, S. A. SONSTHAGEN, J. TRUCCO, T. H. VALQUI, R. E. WILSON, K. WINKER, AND M. D. SORENSON. 2009b. Parallel evolution in the major hemoglobin genes of eight Andean waterfowl. Molecular Ecology 18:3992–4005.

McGRAW, K. J. 2006a. Mechanics of carotenoid-based coloration. Pages 177–242 *in* Bird Coloration, vol. l: Mechanisms and Measurements (G. E. Hill and K. J. McGraw, Eds.). Harvard University Press, Cambridge, Massachusetts.

McGRAW, K. J. 2006b. Mechanics of melanin-based coloration. Pp. 243–294 *in* Bird Coloration, vol. 1: Mechanisms and Measurements (G. E. Hill and K. J. McGraw, Eds.). Harvard University Press, Cambridge, Massachusetts.

McGRAW, K. J., E. A. MACKILLOP, J. DALE, AND M. E. HAUBER. 2002. Different colors reveal different information: How nutritional stress affects the expression of melanin- and structurally based ornamental plumage. Journal of Experimental Biology 205:3747–3755.

McKay, B. D. 2008. Phenotypic variation is clinal in the Yellow-throated Warbler. Condor 110:569–574.

McKitrick, M. C., and R. M. Zink. 1988. Species concepts in ornithology. Condor 90:1–14.

McNab, B. K. 1971. On the ecological significance of Bergmann's rule. Ecology 52:845–854.

McPeek, M. A. 2008. The ecological dynamics of clade diversification and community assembly. American Naturalist 172:E270–E284.

Meiri, S., and T. Dayan. 2003. On the validity of Bergmann's rule. Journal of Biogeography 30:331–351.

Mennill, D. J. 2001. Song characteristics and singing behavior of the Mangrove Warbler (*Dendroica petechia bryanti*). Journal of Field Ornithology 72:327–337.

Merilä, J., and J. D. Fry. 1998. Genetic variation and causes of genotype–environment interaction in the body size of Blue Tit (*Parus caeruleus*). Genetics 148:1233–1244.

Meyer, A. 1993. Phylogenetic relationships and evolutionary processes in East African cichlid fishes. Trends in Ecology and Evolution 8:279–284.

Meyer de Schauensee, R. M. 1970. A Guide to the Birds of South America. Livingston, Wynnewood, Pennsylvania.

Milá, B., T. B. Smith, and R. K. Wayne. 2007. Speciation and rapid phenotypic differentiation in the Yellow-rumped Warbler *Dendroica coronata* complex. Molecular Ecology 16:159–173.

Miller, A. H. 1955. Concepts and problems of avian systematics in relation to the evolutionary processes. Pages 1–22 *in* Recent Studies in Avian Biology (A. Wolfson, Ed.). University of Illinois Press, Urbana.

Miller, M. J., E. Bermingham, J. Klicka, P. Escalante, F. S. Raposo do Amaral, J. T. Weir, and K. Winker. 2008. Out of Amazonia again and again: Episodic crossing of the Andes promotes diversification in a lowland forest flycatcher. Proceedings of the Royal Society of London, Series B 275:1133–1142.

Miller, M. J., E. Bermingham, and R. E. Ricklefs. 2007. Historical biogeography of the New World solitaires (*Myadestes* spp.). Auk 124:868–885.

Millien, V., S. K. Lyons, L. Olson, F. A. Smith, A. B. Wilson, and Y. Yom-Tov. 2006. Ecotypic variation in the context of global climate change: Revisiting the rules. Ecology Letters 9:853–869.

Minch, E., A. Ruiz-Linares, D. B. Goldstein, M. W. Feldman, and L. L. Cavalli-Sforza. 1995. MICROSAT software. Stanford University Press, Stanford, California.

Mitchell-Olds, T., J. H. Willis, and D. B. Goldstein. 2007. Which evolutionary processes influence natural genetic variation for phenotypic traits? Nature Reviews Genetics 8:845–856.

Mock, K. E., T. C. Theimer, O. E. Rhodes, Jr., D. L. Greenberg, and P. Keim. 2002. Genetic variation across the historical range of the Wild Turkey (*Meleagris gallopavo*). Molecular Ecology 11:643–657.

Møller, A. P., and J. J. Cuervo. 1998. Speciation and feather ornamentation in birds. Evolution 52:859–869.

Monge, C., and F. León-Velarde. 1991. Physiological adaptation to high altitude: Oxygen transport in mammals and birds. Physiological Reviews 71:1135–1172.

Monmonier, M. 1973. Maximum-difference barriers: An alternative numerical regionalization method. Geographical Analysis 3:245–261.

Morales-Pérez, J. E., M. A. A. González-Ortega, and P. G. Domínguez. 2000. Records of the Blackbanded Woodcreeper *Dendrocolaptes picumnus* in Chiapas, Mexico. Bulletin of the British Ornithologists' Club 120:133–136.

Moritz, C. 1994. Applications of mitochondrial DNA analysis in conservation: A critical review. Molecular Ecology 3:401–411.

Moritz, C. 1999. Conservation units and translocations: Strategies for conserving evolutionary processes. Hereditas 130:217–228.

Moritz, C. 2002. Strategies to protect biological diversity and the evolutionary processes that sustain it. Systematic Biology 51:238–254.

Moritz, C., and C. Cicero. 2004. DNA barcoding: Promise and pitfalls. PLoS Biology 2:1529–1531.

Morjan, C. L., and L. H. Rieseberg. 2004. How species evolve collectively: Implications of gene flow and selection for the spread of advantageous alleles. Molecular Ecology 13:1341–1356.

Moyle, R. G., C. E. Filardi, C. E. Smith, and J. Diamond. 2009. Explosive Pleistocene diversification and hemispheric expansion of a "great speciator." Proceedings of the National Academy of Sciences USA 106:1863–1868.

Mumme, R. L. 2002. Scare tactics in a Neotropical warbler: White tail feathers enhance flush-pursuit foraging performance in the Slate-throated Redstart (*Myioborus miniatus*). Auk 119:1024–1035.

Mumme, R. L., M. L. Galatowitsch, P. G. Jabłoński, T. M. Stawarczyk, and J. P. Cygan. 2006. Evolutionary significance of geographic variation in a plumage-based foraging adaptation: An experimental test in the Slate-throated Redstart (*Myioborus miniatus*). Evolution 60:1086–1097.

Myers, N., R. A. Mittermeier, C. G. Mittermeier, G. A. B. da Fonseca, and J. Kent. 2000. Biodiversity hotspots for conservation priorities. Nature 403:853–858.

Nagel, L., and D. Schluter. 1998. Body size, natural selection, and speciation in sticklebacks. Evolution 52:209–218.

National Marine Fisheries Service. 1991. Pacific Salmon, *Oncorhynchus* spp., and the definition of "species" under the Endangered Species Act. Federal Register: 56:10542.

NAVARRO-SIGÜENZA, A. G., AND A. T. PETERSON 2004. An alternative species taxonomy of Mexican birds. Biota Neotropica 4(2). [Online.] Available at www.biotaneotropica.org.br/v4n2/pt/.

NAVARRO-SIGÜENZA, A. G., A. T. PETERSON, A. NYARI, G. M. GARCÍA-DERAS, AND J. GARCÍA-MORENO. 2008. Phylogeography of the *Buarremon* brush-finch complex (Aves, Emberizidae) in Mesoamerica. Molecular Phylogenetics and Evolution 47:21–35.

NEE, S., E. C. HOLMES, R. M. MAY, AND P. H. HARVEY. 1994. Extinction rates can be estimated from molecular phylogenies. Philosophical Transactions of the Royal Society of London, Series B 344:77–82.

NEGRO, J. J., G. R. BORTOLOTTI, J. L. TELLA, K. J. FERNIE, AND D. M. BIRD. 1998. Regulation of integumentary colour and plasma carotenoids in American Kestrels consistent with sexual selection theory. Functional Ecology 12:307–312.

NEI, M. 1987. Molecular Evolutionary Genetics. Columbia University Press, New York.

NEIGEL, J. E., AND J. C. AVISE. 1986. Phylogenetic relationships of mitochondrial DNA under various demographic models of speciation. Pages 515–534 *in* Evolutionary Processes and Theory (S. Karlin and E. Nevo, Eds.). Academic Press, New York.

NELSON, M. P., J. A. VUCETICH, AND M. K. PHILLIPS. 2007. Normativity and the meaning of *endangered*: Response to Waples et al. 2007. Conservation Biology 21:1646–1648.

NEWTON, A., Ed. 1862. A List of the Birds of Europe: By Professor J. H. Blasius. Reprinted from the German, with the author's corrections. Matchett and Stevenson, Norwich, England.

NEWTON, I. 2003. Speciation and Biogeography of Birds. Academic Press, London.

NICHOLLS, J. A., J. J. AUSTIN, C. MORTIZ, AND A. W. GOLDIZEN. 2006. Genetic population structure and call variation in a passerine bird, the Satin Bowerbird, *Ptilonorhynchus violaceus*. Evolution 60:1279–1290.

NIXON, K. C., AND Q. D. WHEELER. 1990. An amplification of the phylogenetic species concept. Cladistics 6:211–223.

NOSIL, P. 2008. Speciation with gene flow could be common. Molecular Ecology 17:2103–2106.

OBERHOLSER, H. C. 1906. A description of a new *Querquedula*. Proceedings of the Biological Society of Washington 19:93–94.

O'BRIEN, S. J., AND E. MAYR. 1991. Bureaucratic mischief: Recognizing endangered species and subspecies. Science 251:1187–1188.

OFFICE OF THE SOLICITOR. 2007. The meaning of "in danger of extinction throughout all or a significant portion of its range." Memorandum M-37013 from the Solicitor to the Director of the U.S. Fish and Wildlife Service (16 March). U.S. Department of the Interior, Office of the Solicitor, Washington, D.C. [Online.] Available at www.doi.gov/solicitor/M37013.pdf.

OHTA, T. 2002. Near-neutrality in evolution of genes and gene regulation. Proceedings of the National Academy of Sciences USA 99:16134–16137.

OLESEN, J. M., J. BASCOMPTE, Y. L. DUPONT, AND P. JORDANO. 2007. The modularity of pollination networks. Proceedings of the National Academy of Sciences USA 104:19891–19896.

OLMSTEAD, R. G. 1995. Species concepts and plesiomorphic species. Systematic Botany 20:623–630.

OLSON, S. L., AND H. F. JAMES. 1982. Prodromus of the fossil avifauna of the Hawaiian Islands. Smithsonian Contributions to Zoology 365:1–59.

OLSON, S. L., AND H. F. JAMES. 1991. Descriptions of thirty-two new species of birds from the Hawaiian Islands: Part I. Non-Passeriformes (S. L. Olson and H. F. James, Eds.). Ornithological Monographs, no. 45.

OLSON, S. L., AND H. F. JAMES. 1995. Nomenclature of the Hawaiian Akialoas and Nukupuus (Aves: Drepanidini). Proceedings of the Biological Society of Washington 108:373–387.

ORME, C. D. L., R. G. DAVIES, M. BURGESS, F. EIGENBROD, N. PICKUP, V. A. OLSON, A. J. WEBSTER, T.-S. DING, P. C. RASMUSSEN, R. S. RIDGELY, AND OTHERS. 2005. Global hotspots of species richness are not congruent with endemism or threat. Nature 436:1016–1019.

ORME, C. D. L., R. G. DAVIES, V. A. OLSON, G. H. THOMAS, T.-S. DING, P. C. RASMUSSEN, R. S. RIDGELY, A. J. STATTERSFIELD, P. M. BENNETT, I. P. F. OWENS, AND OTHERS. 2006. Global patterns of geographic range size in birds. PLoS Biology 4:1276–1283.

OYLER-MCCANCE, S. J., N. W. KAHN, K. P. BURNHAM, C. E. BRAUN, AND T. W. QUINN. 1999. A population genetic comparison of large- and small-bodied sage grouse in Colorado using microsatellite and mitochondrial DNA markers. Molecular Ecology 8:1457–1465.

OYLER-MCCANCE, S. J., F. A. RANLSER, L. K. BERKMAN, AND T. W. QUINN. 2007. A rangewide population genetic study of Trumpeter Swans. Conservation Genetics 8:1339–1353.

OYLER-MCCANCE, S. J., J. ST. JOHN, AND T. W. QUINN. This volume. Rapid evolution in lekking grouse: Implications for taxonomic definitions. Pages 170–173 *in* Avian Subspecies (K. Winker and S. M. Haig, Eds.). Ornithological Monographs, no. 67.

PÄCKERT, M., C. DIETZEN, J. MARTENS, M. WINK, AND L. KVIST. 2006. Radiation of Atlantic goldcrests *Regulus regulus* spp.: Evidence of a new taxon from the Canary Islands. Journal of Avian Biology 37:364–380.

PADIAL, J. M., AND I. DE LA RIVA. 2006. Taxonomic inflation and the stability of species lists: The perils of the Ostrich's behavior. Systematic Biology 55:859–867.

PAETKAU, D., R. SLADE, M. BURDEN, AND A. ESTOUP. 2004. Genetic assignment methods for the direct, real-time estimation of migration rate: A simulation-based exploration of accuracy and power. Molecular Ecology 13:55–65.

PAGE, R. D. M. 1996. TREEVIEW: An application to display phylogenetic trees on personal computers. Computer Applications in the Biosciences 12:357–358.

PAMILO, P., AND M. NEI. 1988. Relationships between gene trees and species trees. Molecular Biology and Evolution 5:568–583.

PANHUIS, T. M., R. BUTLIN, M. ZUK, AND T. TREGENZA. 2001. Sexual selection and speciation. Trends in Ecology and Evolution 16:364–371.

PARADIS, E. 2004. Can extinction rates be estimated without fossils? Journal of Theoretical Biology 229:19–30.

PARCHMAN, T. L., AND C. W. BENKMAN. 2002. Diversifying coevolution between crossbills and black spruce on Newfoundland. Evolution 56:1663–1672.

PARKES, K. C. 1982. Subspecific taxonomy: Unfashionable does not mean irrelevant. Auk 99:596–598.

PARZUDAKI, É. 1856. Catalogue de Oiseaux d'Europe offerts, en 1856, aux Ornithologistes. Régidé d'aprés les dernières classifications de le Prince Bonaparte. Paris.

PATTEN, M. A. 2009. 'Subspecies' and 'race' should not be used as synonyms. Nature 457:147.

PATTEN, M. A. This volume. Null expectations in subspecies diagnosis. Pages 170–173 in Avian Subspecies (K. Winker and S. M. Haig, Eds.). Ornithological Monographs, no. 67.

PATTEN, M. A., G. MCCASKIE, AND P. UNITT. 2003. Birds of the Salton Sea: Status, Biogeography, and Ecology. University of California Press, Berkeley.

PATTEN, M. A., AND C. L. PRUETT. 2009. The Song Sparrow, Melospiza melodia, as a ring species: Patterns of geographic variation, a revision of subspecies, and implications for speciation. Systematics and Biodiversity 7:33–62.

PATTEN, M. A., J. T. ROTENBERRY, AND M. ZUK. 2004. Habitat selection, acoustic adaptation, and the evolution of reproductive isolation. Evolution 58:2144–2155.

PATTEN, M. A., AND B. D. SMITH-PATTEN. 2008. Biogeographical boundaries and Monmonier's algorithm: A case study in the northern Neotropics. Journal of Biogeography 35:407–416.

PATTEN, M. A., AND P. UNITT. 2002. Diagnosability versus mean differences of Sage Sparrow subspecies. Auk 119:26–35.

PAVLOVA, A., R. M. ZINK, S. V. DROVETSKI, Y. RED'KIN, AND S. ROHWER. 2003. Phylogeographic patterns in Motacilla flava and Motacilla citreola: Species limits and population history. Auk 120:744–758.

PAXINOS, E. E., H. F. JAMES, S. L. OLSON, M. D. SORENSON, J. JACKSON, AND R. C. FLEISCHER. 2002. MtDNA from fossils reveals a radiation of Hawaiian geese recently derived from the Canada Goose. Proceedings of the National Academy of Sciences USA 99:1399–1404.

PAYNTER, R. J., JR., Ed. 1968. Check-list of Birds of the World, vol. 14. Museum of Comparative Zoology, Cambridge, Massachusetts.

PEARSON, D. L., AND M. A. PLENGE. 1974. Puna bird species on the coast of Peru. Auk 91:626–631.

PENNOCK, D. S., AND W. W. DIMMICK. 1997. Critique of the evolutionarily significant unit as a definition for "distinct population segment" under the U.S. Endangered Species Act. Conservation Biology 11:611–619.

PÉREZ-EMÁN, J. L. 2005. Molecular phylogenetics and biogeography of the Neotropical redstarts (Myioborus; Aves, Parulinae). Molecular Phylogenetics and Evolution 37:511–528.

PETERS, J. L., ET AL. 1934–1987. Check-list of the Birds of the World, vols. I–XVI. Museum of Comparative Zoology, Cambridge, Massachusetts.

PETERSON, A. T. 2006. Taxonomy is important in conservation: A preliminary reassessment of Philippine species-level bird taxonomy. Bird Conservation International 16:155–173.

PETERSON, A. T., AND R. G. MOYLE. 2008. An appraisal of recent taxonomic reappraisals based on character scoring systems. Forktail 24:110–112.

PETERSON, A. T., AND A. G. NAVARRO-SIGÜENZA. 1999. Alternate species concepts as bases for determining priority conservation areas. Conservation Biology 13:427–431.

PETERSON, A. T., A. G. NAVARRO-SIGÜENZA, AND K. P. JOHNSON. 2006. Consistency of taxonomic treatments: A response to Remsen (2005). Auk 123:885–887.

PETIT, R. J., AND L. EXCOFFIER. 2009. Gene flow and species delimitation. Trends in Ecology and Evolution 24:386–393.

PHILLIMORE, A. B., C. D. L. ORME, R. G. DAVIES, J. D. HADFIELD, W. J. REED, K. J. GASTON, R. P. FRECKLETON, AND I. P. F. OWENS. 2007. Biogeographical basis of recent phenotypic divergence among birds: A global study of subspecies richness. Evolution 61:942–957.

PHILLIMORE, A. B., AND I. P. F. OWENS. 2006. Are subspecies useful in evolutionary and conservation biology? Proceedings of the Royal Society of London, Series B 273:1049–1053.

PHILLIMORE, A. B., I. P. F. OWENS, R. A. BLACK, J. CHITTOCK, T. BURKE, AND S. M. CLEGG. 2008. Complex patterns of genetic and phenotypic divergence in an island bird and the consequences for delimiting conservation units. Molecular Ecology 17:2839–2853.

PHILLIMORE, A. B., AND T. D. PRICE. 2008. Density-dependent cladogenesis in birds. PLoS Biology 6:e71.

PHILLIMORE, A. B., AND T. D. PRICE. 2009. Ecological influences on the temporal pattern of speciation. Pages 240–256 in Speciation and Patterns of Diversity (R. Butlin, J. Bridle, and D. Schluter, Eds.). Cambridge University Press, Cambridge, United Kingdom.

PHILLIPS, J. C. 1923. A Natural History of the Ducks, vol. 2. Houghton Mifflin, Boston.

PIELOU, E. C. 1991. After the Ice Age: The Return of Life to Glaciated North America. University of Chicago Press, Chicago, Illinois.

PIRY, S., A. ALAPETITE, J.-M. CORNUET, D. PAETKAU, L. BAUDOUIN, AND A. ESTOUP. 2004. GENECLASS2: A software for genetic assignment and first-generation migrant detection. Journal of Heredity 95:536–539.

POLLARD, D. A., V. N. IYER, A. M. MOSES, AND M. B. EISEN. 2006. Widespread discordance of gene trees with species tree in *Drosophila*: Evidence for incomplete lineage sorting. PLoS Genetics 2: 1634–1647.

POOLE, A., AND F. GILL, Eds. 1992–2002. The Birds of North America. Academy of Natural Sciences, Philadelphia, and American Ornithologists' Union, Washington, D.C.

POSADA, D., AND K. A. CRANDALL. 1998. MODELTEST: Testing the model of DNA substitution. Bioinformatics 14:817–818.

POSTMA, E., AND A. J. VAN NOORDWIJK. 2005. Gene flow maintains a large genetic difference in clutch size at a small spatial scale. Nature 433:65–68.

POWER, D. M. 1969. Evolutionary implications of wing and size variation in the Red-winged Blackbird in relation to geographic and climatic factors: A multiple regression analysis. Systematic Zoology 18:363–373.

PRATT, H. D. 1980. Intra-island variation in the 'Elepaio on the Island of Hawaii. Condor 82:449–458.

PRATT, H. D. 1982. Relationships and speciation of the Hawaiian thrushes. Living Bird 19:73–90.

PRATT, H. D. 1989. Species limits in Akepas (Drepanidinae: *Loxops*). Condor 91:933–940.

PRATT, H. D. 1990. Bird the world's islands now. Birding 22:10–15.

PRATT, H. D. 1992. Systematics of the Hawaiian "creepers" *Oreomystis* and *Paroreomyza*. Condor 94:836–846.

PRATT, H. D. 2005. The Hawaiian Honeycreepers: Drepanidinae. Oxford University Press, Oxford, United Kingdom.

PRATT, H. D. 2008. Penduline-tits to Shrikes [Plates 31–35]. *In* Handbook of Birds of the World, vol. 13 (J. del Hoyo, A. Elliott, and D. A. Christie, Eds.). Lynx Edicions, Barcelona, Spain.

PRATT, H. D. This volume. Revisiting species and subspecies of island birds for a better assessment of biodiversity. Pages 170–173 *in* Avian Subspecies (K. Winker and S. M. Haig, Eds.). Ornithological Monographs, no. 67.

PRATT, H. D., P. L. BRUNER, AND D. G. BERRETT. 1979. America's unknown avifauna: The birds of the Mariana Islands. American Birds 33:227–235.

PRATT, H. D., P. L. BRUNER, AND D. G. BERRETT. 1987. A Field Guide to the Birds of Hawaii and the Tropical Pacific. Princeton University Press, Princeton, New Jersey.

PRATT, H. D., AND T. K. PRATT. 2001. The interplay of species concepts, taxonomy, and conservation: Lessons from the Hawaiian avifauna. Pages 68–80 *in* Evolution, Ecology, Conservation, and Management of Hawaiian Birds: A Vanishing Avifauna (J. M. Scott, S. Conant, and C. Van Riper III, Eds.). Studies in Avian Biology, no. 22.

PRICE, T. 2008. Speciation in Birds. Roberts, Greenwood, Colorado.

PRICE, T. D., AND M. M. BOUVIER. 2002. The evolution of F_1 postzygotic incompatibilities in birds. Evolution 56:2083–2089.

PRICE, T. D., A. QVARNSTRÖM, AND D. E. IRWIN. 2003. The role of phenotypic plasticity in driving genetic evolution. Proceedings of the Royal Society of London, Series B 270:1433–1440.

PRITCHARD, J. K., M. STEPHENS, AND P. DONNELLY. 2000. Inference of population structure using multilocus genotype data. Genetics 155:945–959.

PRUETT, C. L. 2002. Phylogeography and population genetic structure of Beringian landbirds. Ph.D. dissertation, University of Alaska, Fairbanks.

PRUETT, C. L., P. ARCESE, Y. CHAN, A. WILSON, M. A. PATTEN, L. F. KELLER, AND K. WINKER. 2008a. Concordant and discordant signals between genetic data and described subspecies of Pacific coast Song Sparrows. Condor 110:359–364.

PRUETT, C. L., P. ARCESE, Y. CHAN, A. WILSON, M. A. PATTEN, L. F. KELLER, AND K. WINKER. 2008b. The effects of contemporary processes in maintaining the genetic structure of western Song Sparrows (*Melospiza melodia*). Heredity 101:67–74.

PRUETT, C. L., D. D. GIBSON, AND K. WINKER. 2004. Amak Island Song Sparrows (*Melospiza melodia amaka*) are not evolutionarily significant. Ornithological Science 3:133–138.

PRUETT, C. L., AND K. WINKER. 2005. Northwestern song sparrow populations show genetic effects of sequential colonization. Molecular Ecology 14:1421–1434.

PRUETT, C. L., AND K. WINKER. This volume. Alaska Song Sparrows (*Melospiza melodia*) demonstrate that genetic marker and method of analysis matter in subspecies assessments. Pages 170–173 *in* Avian Subspecies (K. Winker and S. M. Haig, Eds.). Ornithological Monographs, no. 67.

PRUM, R. O. 1997. Phylogenetic tests of alternative intersexual selection mechanisms: Trait macroevolution in a polygynous clade (Aves: Pipridae). American Naturalist 149:668–692.

PRUM, R. O. 2006. Anatomy, physics, and evolution of structural colors. Pages 245–353 *in* Bird Coloration, vol. 1: Mechanisms and Measurements (G. E. Hill and K. J. McGraw, Eds.). Harvard University Press, Cambridge, Massachusetts.

PRYKE, S. R., AND S. C. GRIFFITH. 2009. Postzygotic genetic incompatibility between sympatric color morphs. Evolution 63:793–798.

Pyle, P., and J. Engbring. 1985. Checklist of the birds of Micronesia. 'Elepaio 46:57–68.

Quinn, T. W. 1992. The genetic legacy of Mother Goose—Phylogeographic patterns of Lesser Snow Goose *Chen caerulescens caerulescens* maternal lineages. Molecular Ecology 1:105–117.

Quinn, T. W., and A. C. Wilson. 1993. Sequence evolution in and around the mitochondrial control region in birds. Journal of Molecular Evolution 37:417–425.

Rabosky, D. L. 2009a. Ecological limits and diversification rate: Alternative paradigms to explain the variation in species richness among clades and regions. Ecology Letters 12:735–743.

Rabosky, D. L. 2009b. Ecological limits on clade diversification in higher taxa. American Naturalist 173:662–674.

Rabosky, D. L. 2010. Extinction rates should not be estimated from molecular phylogenies. Evolution 64: in press.

Rabosky, D. L., and I. J. Lovette. 2008. Density-dependent diversification in North American wood warblers. Proceedings of the Royal Society of London, Series B 275:2363–2371.

Raikow, R. J. 1994. A phylogeny of the woodcreepers (Dendrocolaptinae). Auk 111:104–114.

Ramey, R. R., II, H.-P. Liu, C. W. Epps, L. M. Carpenter, and J. D. Wehausen. 2005. Genetic relatedness of the Preble's meadow jumping mouse (*Zapus hudsonius preblei*) to nearby subspecies of *Z. hudsonius* as inferred from variation in cranial morphology, mitochondrial DNA and microsatellite DNA: Implications for taxonomy and conservation. Animal Conservation 8:329–346.

Ramos-Onsins, S. E., and J. Rozas. 2002. Statistical properties of new neutrality tests against population growth. Molecular Biology and Evolution 19:2092–2100.

Rand, A. L., and M. A. Traylor. 1950. The amount of overlap allowable for subspecies. Auk 67:169–183.

Rand, D. M. 1996. Neutrality tests of molecular markers and the connection between DNA polymorphism, demography, and conservation biology. Conservation Biology 10:665–671.

Rapoport, E. H. 1982. Areography: Geographical Strategies of Species. Pergamon Press, Oxford, United Kingdom.

Rea, A. M., and K. L. Weaver. 1990. The taxonomy, distribution, and status of Coastal California Cactus Wrens. Western Birds 21:81–126.

R Development Core Team. 2008. R: A language and environment for statistical computing. R Foundation for Statistical Computing, Vienna. [Online.] Available at www.R-project.org.

Reddy, S. 2008. Systematics and biogeography of the shrike-babblers (*Pteruthius*): Species limits, molecular phylogenetics, and diversification patterns across southern Asia. Molecular Phylogenetics and Evolution 47:54–72.

Reding, D. M., J. T. Foster, H. F. James, H. D. Pratt, and R. C. Fleischer. 2008. Convergent evolution of "creepers" in the Hawaiian honeycreeper radiation. Biology Letters 5:221–224.

Remsen, J. V., Jr. 1984. High incidence of "leapfrog" pattern of geographic variation in Andean birds: Implications for the speciation process. Science 224:171–173.

Remsen, J. V., Jr. 2005. Pattern, process, and rigor meet classification. Auk 122:403–413.

Remsen, J. V., Jr. This volume. Subspecies as a meaningful taxonomic rank in avian classification. Pages 170–173 *in* Avian Subspecies (K. Winker and S. M. Haig, Eds.). Ornithological Monographs, no. 67.

Remsen, J. V., Jr., C. D. Cadena, A. Jaramillo, M. Nores, J. F. Pacheco, M. B. Robbins, T. S. Schulenberg, F. G. Stiles, D. F. Stotz, and K. J. Zimmer. 2009. A classification of the bird species of South America. [Online.] American Ornithologists' Union, Washington, D.C. Available at www.museum.lsu.edu/~Remsen/SACCBaseline.html.

Remsen, J. V., Jr., and S. K. Robinson. 1990. A classification scheme for foraging behavior of birds in terrestrial habitats. Pages 144–160 *in* Avian Foraging: Theory, Methodology, and Applications (M. L. Morrison, C. J. Ralph, J. Verner, and J. R. Jehl, Jr., Eds.). Studies in Avian Biology, no. 13.

Rensch, B. 1960. Evolution above the Species Level. Columbia University Press, New York.

Reznick, D., D. H. Rodd, and L. Nunney. 2004. Empirical evidence for rapid evolution. Pages 244–264 *in* Evolutionary Conservation Biology (R. Ferrière, U. Dieckmann, and D. Couvet, Eds.). Cambridge University Press, Cambridge, United Kingdom.

Rheindt, F. E., and R. O. Hutchinson. 2007. A photoshot odyssey through the confused avian taxonomy of Seram and Buru (southern Moluccas). BirdingASIA 7:18–38.

Rice, W. R., and E. E. Hostert. 1993. Laboratory experiments on speciation: What have we learned in 40 years? Evolution 47:1637–1653.

Ricklefs, R. E. 2006. Global variation in the diversification rate of passerine birds. Ecology 87:2468–2478.

Ricklefs, R. E. 2007. Estimating diversification rates from phylogenetic information. Trends in Ecology and Evolution 22:601–610.

Ricklefs, R. E. 2009. Speciation, extinction and diversity. Pages 257–277 *in* Speciation and Patterns of Diversity (R. Butlin, J. Bridle, and D. Schluter, Eds.). Cambridge University Press, Cambridge, United Kingdom.

Ride, W. D. L. 1999. International Code of Zoological Nomenclature, 4th ed. International Trust for Zoological Nomenclature, London.

Ridgely, R. S., T. F. Allnutt, T. Brooks, D. K. McNicol, D. W. Mehlman, B. E. Young, and J. R. Zook. 2003. Digital distribution maps of the birds of the Western Hemisphere, version 1.0. NatureServe, Arlington, Virginia.

RIDGELY, R. S., AND G. TUDOR. 1994. The Birds of South America, vol. 2: The Suboscine Passerines. University of Texas Press, Austin.

RIDGWAY, R. 1881. Nomenclature of North American birds, chiefly contained in the U.S. National Museum. Bulletin of the U.S. National Museum 21:3–94.

RIDGWAY, R. 1901. The birds of North and Middle America: Family Fringillidae. Bulletin of the U.S. National Museum, no. 50, part 1.

RIDGWAY, R. 1902. The birds of North and Middle America. Bulletin of the U.S. National Museum, no. 50, part 2.

RIDGWAY, R., AND H. FRIEDMANN. 1901–1950. The Birds of North and Middle America: A descriptive catalogue of the higher groups, genera, species, and subspecies of birds known to occur in North America, from the Arctic lands to the Isthmus of Panama, the West Indies and other islands of the Caribbean Sea, and the Galapagos Archipelago. Bulletin of the U.S. National Museum, no. 50, parts 1–11.

RIDLEY, M. 2004. Evolution, 3rd ed. Blackwell, Malden, Massachusetts.

RIESING, M. J., L. KRUCKENHAUSER, A. GAMAUF, AND E. HARING. 2003. Molecular phylogeny of the genus *Buteo* (Aves: Accipitridae) based on mitochondrial marker sequence. Molecular Phylogenetics and Evolution 27:328–342.

RIND, F. C., AND P. J. SIMMONS. 1992. Orthopteran DCMD neuron: A reevaluation of responses to moving objects. I. Selective responses to approaching objects. Journal of Neurophysiology 68:1654–1666.

RISING, J. D. 1996. A Guide to the Identification and Natural History of the Sparrows of the United States and Canada. Academic Press, San Diego.

RISING, J. D. 2007. Named subspecies and their significance in contemporary ornithology. Pages 45–54 *in* Festschrift for Ned K. Johnson: Geographic Variation and Evolution in Birds (C. Cicero and J. V. Remsen, Jr., Eds.). Ornithological Monographs, no. 63.

RISING, J. D., D. A. JACKSON, AND H. B. FOKIDIS. 2001. Geographic variation in size and shape of Savannah Sparrows. Wilson Journal of Ornithology 121:253–264.

RISING, J. D., D. A. JACKSON, AND H. B. FOKIDIS. 2009. Geographic variation in plumage pattern and coloration of Savannah Sparrows. Wilson Journal of Ornithology 121:253–264.

RISING, J. D., AND K. M. SOMERS. 1989. The measurement of overall body size in birds. Auk 106:666–674.

ROBERTS, D. W. 2008. Statistical analysis of multidimensional fuzzy set ordinations. Ecology 89:1246–1260.

ROGERS, A. R. 1995. Genetic evidence for a Pleistocene population explosion. Evolution 49:608–615.

ROGERS, A. R., AND H. HARPENDING. 1992. Population growth makes waves in the distribution of pairwise genetic differences. Molecular Biology and Evolution 9:552–569.

ROGERS, J. S. 1972. Measures of genetic similarity and genetic distance. University of Texas Studies in Genetics 7:145–153.

ROHWER, S., L. K. BUTLER, AND D. R. FROEHLICH. 2005. Ecology and demography of East–West differences in molt-scheduling of Neotropical migrant passerines. Pages 87–105 *in* Birds of Two Worlds: The Ecology and Evolution of Migration (R. Greenberg and P. P. Marra, Eds.). Johns Hopkins University Press, Baltimore, Maryland.

ROHWER, S., AND P. W. EWALD. 1981. The cost of dominance and advantage of subordination in a badge signaling system. Evolution 35:441–454.

ROHWER, V. G., S. ROHWER, AND J. H. BARRY. 2008. Molt scheduling of western Neotropical migrants and up-slope movement of Cassin's Vireo. Condor 110:365–370.

RONQUIST, F., AND J. P. HUELSENBECK. 2003. MRBAYES 3: Bayesian phylogenetic inference under mixed models. Bioinformatics 19:1572–1574.

RONQUIST, F., J. P. HUELSENBECK, AND P. V. D. MARK. 2005. MRBAYES 3.1 Manual. School of Computational Science, Florida State University, and Division of Biological Sciences, University of California at San Diego.

ROSENBERG, N. A. 2002. The probability of topological concordance of gene trees and species trees. Theoretical Population Biology 61:225–247.

ROSENBERG, N. A. 2007. Statistical tests for taxonomic distinctiveness from observations of monophyly. Evolution 61:317–323.

ROSENBERG, N. A., AND R. TAO. 2008. Discordance of species trees with their most likely gene trees: The case of five taxa. Systematic Biology 57:131–140.

ROSENZWEIG, M. L. 1978. Geographical speciation: On range size and the probability of isolate formation. Pages 172–194 *in* Proceedings of the Washington State University Conference on Biomathematics and Biostatistics (D. Wollkind, Ed.). Washington State University, Pullman.

ROTHSCHILD, M. 1983. Dear Lord Rothschild: Birds, Butterflies and History. Balaban Publishers, Glenside, Pennsylvania.

ROZAS, J., J. C. SÁNCHEZ-DELBARRIO, X. MESSEGUER, AND R. ROZAS. 2003. DnaSP, DNA polymorphism analyses by the coalescent and other methods. Bioinformatics 19:2496–2497.

RUBIN, D. B. 1987. Multiple Imputation for Nonresponse in Surveys. Wiley, New York.

RUEGG, K. 2007. Divergence between subspecies groups of Swainson's Thrush (*Catharus ustulatus ustulatus* and *C. u. swainsoni*). Pages 67–77 *in* Festschrift for Ned K. Johnson: Geographic Variation and Evolution in Birds (C. Cicero and J. V. Remsen, Jr., Eds.). Ornithological Monographs, no. 63.

RUEGG, K. C., AND T. B. SMITH. 2002. Not as the crow flies: A historical explanation for circuitous migration in Swainson's Thrush (*Catharus ustulatus*). Proceedings of the Royal Society of London, Series B 269:1375–1381.

RUNDLE, H. D., AND P. NOSIL. 2005. Ecological speciation. Ecology Letters 8:336–352.

SABROSKY, C. W. 1955. Postscript to a survey of infraspecific categories. Systematic Zoology 4:141–142.

SANGSTER, G. 2000. Taxonomic stability and avian extinctions. Conservation Biology 14:579–581.

SANTER, R. D., F. C. RIND, R. STAFFORD, AND P. J. SIMMONS. 2006. Role of an identified looming-sensitive neuron in triggering a flying locust's escape. Journal of Neurophysiology 95:3391–3400.

SAVIDGE, J. A. 1987. Extinction of an island forest avifauna by an introduced snake. Ecology 68:660–668.

SCHAFER, J. L. 1999. NORM: Multiple imputation of incomplete multivariate data under a normal model, version 2. Software for Windows 95/98/NT. [Online.] Available at www.stat.psu.edu/~jls/misoftwa.html.

SCHEMSKE, D. W., AND P. BIERZYCHUDEK. 2007. Spatial differentiation for flower color in the desert annual *Linanthus parryae*: Was Wright right? Evolution 61:2528–2543.

SCHLEGEL, H. 1844. Kritische Übersicht der europäischen Vögel. Arnz und Comp, Leiden.

SCHMIDT-NIELSEN, K., AND J. L. LARIMER. 1958. Oxygen dissociation curves of mammalian blood in relation to body size. American Journal of Physiology 195:424–428.

SCHODDE, R., AND I. J. MASON. 1999. The Directory of Australian Birds: Passerines. CSIRO Publishing, Collingwood, Australia.

SEDDON, N., R. M. MERRILL, AND J. A. TOBIAS. 2008. Sexually selected traits predict patterns of species richness in a diverse clade of suboscine birds. American Naturalist 171:620–631.

SEEBOHM, H. 1881. Catalogue of the Birds of the British Museum, vol. 5. British Museum, London.

SELANDER, R. K. 1971. Systematics and speciation in birds. Pages 57–147 *in* Avian Biology, vol. 1 (D. S. Farner and J. R. King, Eds.). Academic Press, New York

SÉLYS LONGCHAMPS, E. 1842. Faune belge. Première partie. Indication méthodique des mammifères, oiseaux, reptiles et poissons, observés jusqu'ici en Belgique. Pages i–xii, 1–310 *in* Faune Belge, Liège, Belgium.

SHARPE, R. B. 1909. A Hand-list of the Genera and Species of Birds, vol. 5. British Museum (Natural History), London.

SHARPE, R. S. 1968. The evolutionary relationships and comparative behavior of prairie chickens. Ph.D. dissertation, University of Nebraska, Lincoln.

SHIN, H. S., AND P. G. JABŁOŃSKI. 2008. Integration of optimality, neural networks, and physiology for field studies of the evolution of visually-elicited escape behaviors of Orthoptera: A minireview and

prospects. Journal of Ecology and Field Biology 31:89–95.

SHORT, L. L. 1967. A review of the genera of grouse (Aves, Tetraoninae). American Museum Novitates, no. 2289:1–39.

SHRIVER, M. D., E. J. PARRA, S. DIOS, C. BONILLA, H. NORTON, C. JOVEL, C. PFAFF, C. JONES, A. MASSAC, N. CAMERON, AND OTHERS. 2003. Skin pigmentation, biogeographical ancestry and admixture mapping. Human Genetics 112:387–399.

SIBLEY, C. G. 1954. The contribution of avian taxonomy. Systematic Zoology 3:105–110.

SIBLEY, C. G., AND B. L. MONROE, JR. 1990. Distribution and Taxonomy of Birds of the World. Yale University Press, New Haven, Connecticut.

SIBLEY, C. G., AND B. L. MONROE, JR. 1993. A supplement to *Distribution and Taxonomy of Birds of the World*. Yale University Press, New Haven, Connecticut.

SIMPSON, G. G. 1961. Principles of Animal Taxonomy. Columbia University Press, New York.

SITES, J. W., JR., AND K. A. CRANDALL. 1997. Testing species boundaries in biodiversity studies. Conservation Biology 11:1289–1297.

SKALSKI, J. R., R. L. TOWNSEND, L. L. McDONALD, J. W. KERN, AND J. J. MILLSPAUGH. 2008. Type I errors linked to faulty statistical analyses of endangered subspecies classifications. Journal of Agricultural, Biological, and Environmental Statistics 13:199–220.

SLABBEKOORN, H., AND T. B. SMITH. 2002. Bird song, ecology and speciation. Philosophical Transactions of the Royal Society of London, Series B 357:493–503.

SLIKAS, B., I. B. JONES, S. R. DERRICKSON, AND R. C. FLEISCHER. 2000. Phylogenetic relationships of Micronesian white-eyes based on mitochondrial sequence data. Auk 117:355–365

SLUD, P. 1964. The birds of Costa Rica: Distribution and ecology. Bulletin of the American Museum of Natural History, no. 128.

SMITH, H. M., AND F. N. WHITE. 1956. A case for the trinomen. Systematic Zoology 5:183–190.

SMITH, J. W., AND C. W. BENKMAN. 2007. A coevolutionary arms race causes ecological speciation in crossbills. American Naturalist 169:455–465.

SMITH, J. W., C. W. BENKMAN, AND K. COFFEY. 1999. The use and misuse of public information by foraging Red Crossbills. Behavioral Ecology 10:54–62.

SNOW, D. W. 1954. Trends in geographical variation in the Palearctic members of the genus *Parus*. Evolution 8:19–28.

SNOW, D. W. 1997. Should the biological be superseded by the phylogenetic species concept? Bulletin British Ornithologists' Club 117:110–120.

SNOWBERG, L. K., AND C. W. BENKMAN. 2007. The role of marker traits in the assortative mating within Red Crossbills, *Loxia curvirostra* complex. Journal of Evolutionary Biology 20:1924–1932.

SNYDER, L. L., AND H. G. LUMSDEN. 1951. Variation in *Anas cyanoptera*. Occasional Papers of the Royal Ontario Museum of Zoology 10:1–18.

SOKAL, R. R., AND F. J. ROHLF. 1995. Biometry: The Principles and Practice of Statistics in Biological Research, 3rd ed. W.H. Freeman, New York.

SOL, D., D. G. STIRLING, AND L. LEFEBVRE. 2005. Behavioral drive or behavioral inhibition in evolution: Subspecific diversification in Holarctic passerines. Evolution 59:2669–2677.

SOLEK, C., AND L. SZIJJ. 2004. Cactus Wren (*Campylorhynchus brunneicapillus*). *In* The Coastal Scrub and Chaparral Bird Conservation Plan: A Strategy for Protecting and Managing Coastal Scrub and Chaparral Habitats and Associated Birds in California. California Partners in Flight and PRBO Conservation Science, Stinson Beach, California.

SORENSON, M. D., A. COOPER, E. E. PAXINOS, T. W. QUINN, H. F. JAMES, S. L. OLSON, AND R. C. FLEISCHER. 1999. Relationships of the extinct moa-nalos, flightless Hawaiian waterfowl, based on ancient DNA. Proceedings of the Royal Society of London, Series B 266:2187–2193.

SPAULDING, A. 2007. Rapid courtship evolution in grouse (Tetraonidae): Contrasting patterns of acceleration between the Eurasian and North American polygynous clades. Proceedings of the Royal Society of London, Series B 274:1079–1086.

STANFORD, C. B. 2001. The subspecies concept in primatology: The case of Mountain Gorillas. Primates 42:309–318.

STARRETT, A. 1958. What *is* the subspecies problem? Systematic Zoology 7:111–115.

STATTERSFIELD, A. J., AND D. R. CAPPER, Eds. 2000. Threatened Birds of the World: The Official Source for Birds on the IUCN Red List. Lynx Edicions, Barcelona, Spain.

STEADMAN, D. W. 2006. Extinction and Biogeography of Tropical Pacific Birds. University of Chicago Press, Chicago, Illinois.

STEJNEGER, L. 1884. On the use of trinomials in American ornithology. Proceedings of the U.S. National Museum 7:70–81.

STERN, D. L., AND V. ORGOGOZO. 2008. The loci of evolution: How predictable is genetic evolution? Evolution 62:2155–2177.

STEULLET, A., AND E. DEAUTIER. 1950. Una nueva subespecie de *Dendrocolaptes pallescens* Pelzeln. Hornero 9:175–177.

STOECKLE, M., AND K. WINKER. 2009. A global snapshot of avian tissue collections: State of the enterprise. Auk 126:684–687.

STORER, R. W. 1982. Subspecies and the study of geographic variation. Auk 99:599–601.

STRESEMANN, E. 1921. Die Spechte der Insel Sumatra. Archiv für Naturgeschichte 87 (Abteilung A):64–120.

STRESEMANN, E. 1931. Die Zosteropiden der indoaustralischen Region. Mittheilungen aus dem Zoologischen Museum zu Berlin 17:201–238.

STRESEMANN, E. 1936. The Formenkreis-theory. Auk 53:150–158.

STRESEMANN, E. 1975. Ornithology from Aristotle to the Present. Harvard University Press, Cambridge, Massachusetts.

STRICKBERGER, M. W. 2000. Evolution, 3rd ed. Jones and Bartlett, Boston.

STUDENT [GOSSET, W. S.]. 1908. The probable error of a mean. Biometrika 6:1–25.

SUN, J. X., J. C. MULLIKIN, N. PATTERSON, AND D. E. REICH. 2009. Microsatellites are molecular clocks that support accurate inferences about history. Molecular Biology and Evolution 26:1017–1027.

SUZUKI, Y., AND N. F. NIJHOUT. 2007. Genetic basis of adaptive evolution of a polyphenism by genetic accommodation. Journal of Evolutionary Biology 21:57–66.

SWEI, A., P. V. BRYLSKI, W. D. SPENCER, S. C. DODD, AND J. L. PATTON. 2003. Hierarchical genetic structure in fragmented populations of the Little Pocket Mouse (*Perognathus longimembris*) in southern California. Conservation Genetics 4:501–514.

SWOFFORD, D. L. 2003. PAUP*: Phylogenetic Analysis Using Parsimony (*And Other Methods), version 4. Sinauer Associates, Sunderland, Massachusetts.

TACHA, T. C., P. A. VOHS, AND W. D. WARDE. 1985. Morphometric variation of Sandhill Cranes from midcontinental North America. Journal of Wildlife Management 49:246–250.

TAKAHATA, N., AND M. SLATKIN. 1990. Genealogy of neutral genes in two partially isolated populations. Theoretical Population Biology 38:331–350.

TARR, C. L., AND R. C. FLEISCHER. 1994. Mitochondrial DNA variation and evolutionary relationships in the Amakihi complex. Auk 110:825–831.

TARR, C. L., AND R. C. FLEISCHER. 1995. Evolutionary relationships of the Hawaiian Honeycreepers (Aves: Drepanidinae). Pages 147–159 *in* Hawaiian Biogeography: Evolution on a Hot Spot Archipelago (W. L. Wagner and V. A. Funk, Eds.). Smithsonian Institution Press, Washington, D.C.

TATENO, Y., M. NEI, AND F. TAJIMA. 1982. Accuracy of estimated phylogenetic trees from molecular data. I. Distantly related species. Journal of Molecular Evolution 18:387–404.

TIAN, C., R. KOSOY, A. LEE, M. RANSOM, J. W. BELMONT, P. K. GREGERSEN, AND M. F. SELDIN. 2008a. Analysis of East Asia genetic substructure using genomewide SNP arrays. PLoS ONE 3(12):e3862.

TIAN, C., R. M. PLENGE, M. RANSOM, A. LEE, P. VILLOSLADA, C. SELMI, L. KLARESKOG, A. E. PULVER, L. QI, P. K. GREGERSEN, AND M. F. SELDIN. 2008b. Analysis and application of European genetic substructure using 300 K SNP information. PLoS Genetics 4(1):e4.

Tickell, W. L. N. 2003. White plumage. Waterbirds 26:1–12.

Todd, W. E. C. 1950. The northern races of *Dendrocolaptes certhia*. Journal of the Washington Academy of Sciences 40:236–238.

Topp, C. M., and K. Winker. 2008. Genetic patterns of differentiation among five landbird species from the Queen Charlotte Islands, British Columbia. Auk 125:461–472.

Traylor, M. A., Jr., Ed. 1979. Check-list of Birds of the World, vol. 8. Museum of Comparative Zoology, Harvard University, Cambridge, Massachusetts.

Underwood, J. N., L. D. Smith, M. J. H. Van Oppen, and J. P. Gilmour. 2007. Multiple scales of genetic connectivity in a brooding coral on isolated reefs following catastrophic bleaching. Molecular Ecology 16:771–784.

Unitt, P. 2004. San Diego County Bird Atlas. Proceedings of the San Diego Society of Natural History, no. 39.

Unitt, P. 2008. San Diego Cactus Wren, *Campylorhynchus brunneicapillus sandiegensis*. Pages 300–305 *in* California Species of Special Concern (W. D. Shuford and T. Gardali, Eds.). Western Field Ornithologists, Camarillo, California, and California Department of Fish and Game, Sacramento.

U.S. Department of the Interior and U.S. Department of Commerce. 1996. Policy regarding the recognition of distinct vertebrate population segments under the Endangered Species Act. Federal Register 61:4722–4725.

U.S. Fish and Wildlife Service. 1983. Listing and Recovery Priority Guidelines. Federal Register 48:43098–43105.

U.S. Fish and Wildlife Service and National Marine Fisheries Service. 1980. Rules for listing endangered and threatened species, designating critical habitat, and maintaining the lists. Federal Register 45:13010–13026.

U.S. Fish and Wildlife Service and National Marine Fisheries Service. 1996. Policy regarding the recognition of distinct vertebrate population segments under the Endangered Species Act. Federal Register 61:4722–4725.

U.S. Fish and Wildlife Service and National Marine Fisheries Service. 2000. Controlled propagation of species listed under the Endangered Species Act. Federal Register 65:56916–56922.

Uy, J. A. C., and G. Borgia. 2000. Sexual selection drives rapid divergence in bowerbird display traits. Evolution 54:273–278.

Uy, J. A. C., R. G. Moyle, and C. E. Filardi. 2009. Plumage and song differnces mediate species recognition between incipient flycatcher species of the Solomon Islands. Evolution 63:153–164.

Valkiūnas, G., and T. A. Iezhova. 2001. A comparison of the blood parasites in three subspecies of the Yellow Wagtail *Motacilla flava*. Journal of Parasitology 87:930–934.

van Balen, S. 2008. Family Zosteropidae (White-eyes). Pages 402–485 *in* Handbook of Birds of the World, vol. 13: Penduline-tits to Shrikes (J. del Hoyo, A. Elliott, and D. A. Christie, Eds.). Lynx Edicions, Barcelona, Spain.

VanderWerf, E. A. 1998. 'Elepaio (*Chasiempis sandwichensis*). *In* The Birds of North America, no. 344 (A. Poole and F. Gill, Eds.). Academy of Natural Sciences, Philadelphia, and American Ornithologists' Union, Washington, D.C.

VanderWerf, E. A., A. Cowell, and J. L. Rohrer. 1997. Distribution, abundance, and conservation of O'ahu 'Elepaio in the southern leeward Ko'olau Range. 'Elepaio 57:99–105.

van Oosterhout, C., B. Hutchinson, D. Wills, and P. Shipley. 2004. MICROCHECKER: Software for identifying and correcting genotyping errors in microsatellite data. Molecular Ecology Notes 4:535–538.

van Rossem, A. J. 1936. The Orange-bellied Redstart of western Central America. Condor 38:117–118.

Vogler, A. P., and R. DeSalle. 1994. Diagnosing units of conservation management. Conservation Biology 8:354–363.

Vucetich, J. A., M. P. Nelson, and M. K. Phillips. 2006. The normative dimension and legal meaning of *endangered* and *recovery* in the U.S. Endangered Species Act. Conservation Biology 20:1383–1390.

Wakeley, J. 2006. Coalescent Theory: An Introduction. Roberts, Greenwood Village, Colorado.

Waples, R. S., P. B. Adams, J. Bohnsack, and B. L. Taylor. 2007a. A biological framework for evaluating whether a species is threatened or endangered in a significant portion of its range. Conservation Biology 21:964–974.

Waples, R. S., P. B. Adams, J. Bohnsack, and B. L. Taylor. 2007b. Normativity redux. Conservation Biology 21:1649–165.

Waples, R. S., and O. Gaggiotti. 2006. What is a population? An empirical evaluation of some genetic methods for identifying the number of gene pools and their degree of connectivity. Molecular Ecology 15:1419–1439.

Watling, D. 2001. A Guide to the Birds of Fiji and Western Polynesia. Environmental Consultants, Suva, Fiji.

Wayne, R. K., and P. A. Morin. 2004. Conservation genetics in the new molecular age. Frontiers in Ecology and the Environment 2:89–97.

Weir, J. T. 2006. Divergent timing and patterns of species accumulation in lowland and highland Neotropical birds. Evolution 60:842–855.

Weir, J. T., E. Bermingham, M. J. Miller, J. Klicka, and M. A. González. 2008. Phylogeography of a morphologically diverse Neotropical montane species, the Common Bush-Tanager (*Chlorospingus ophthalmicus*). Molecular Phylogenetics and Evolution 47:650–664.

WEIR, J. T., AND D. SCHLUTER. 2008. Calibrating the avian molecular clock. Molecular Ecology 17:2321–2328.

WEST-EBERHARD, M. J. 2003. Developmental Plasticity and Evolution. Oxford University Press, New York.

WETMORE, A. 1944. A collection of birds from northern Guanacaste Costa Rica. Proceedings of the U.S. National Museum 95:25–80.

WHEELER, Q. D., AND R. MEIER, Eds. 2000. Species Concepts and Phylogenetic Theory: A Debate. Columbia University Press, New York.

WHEELER, Q. D., AND K. C. NIXON. 1990. Another way of looking at the species problem: A reply to de Queiroz and Donoghue. Cladistics 6:77–81.

WHEELER, Q. D., AND N. I. PLATNICK. 2000. The phylogenetic species concept. Pages 55–69 *in* Species Concepts and Phylogenetic Theory: A Debate (Q. D. Wheeler and R. Meier, Eds.). Columbia University Press, New York.

WIENS, J. A. 1982. Forum: Avian subspecies in the 1980's. Auk 99:593.

WILES, G. J. 2005. A checklist of the birds and mammals of Micronesia. Micronesica 38:141–189.

WILEY, E. O. 1978. The evolutionary species concept reconsidered. Systematic Biology 27:17–26.

WILEY, E. O. 1981. Phylogenetics: The Theory and Practice of Phylogenetic Systematics. Wiley-Interscience, New York.

WILGENBUSCH, J. C., D. L. WARREN, AND D. L. SWOFFORD. 2004. AWTY: A system for graphical exploration of MCMC convergence in Bayesian phylogenetic inference. [Online.] Available at ceb.csit.fsu.edu/awty.

WILLIAMS, M. 1991. Introductory remarks: Ecological and behavioural adaptations of Southern Hemisphere waterfowl. Pages 841–842 *in* Acta XX Congressus Internationalis Ornithologici (M. N. Clout and D. C. Paton, Eds.). New Zealand Ornithological Congress Trust Board, Wellington.

WILLIS, E. O. 1972. The behavior of Plain-brown Woodcreepers, *Dendrocincla fuliginosa*. Wilson Bulletin 84:377–420.

WILLIS, E. O. 1979. Behavior and ecology of two forms of White-chinned Woodcreepers (*Dendrocincla merula*, Dendrocolaptidae) in Amazonia. Papéis Avulsos de Zoologia, São Paulo 33:27–66.

WILLIS, E. O. 1982. The behavior of Black-banded Woodcreepers (*Dendrocolaptes picumnus*). Condor 84:272–285.

WILLIS, E. O. 1992. Comportamento e ecologia do arapaçu-barrado *Dendrocolaptes certhia* (Aves, Dendrocolaptidae). Boletim do Museu Paraense Emílio Goeldi, série Zoologia 8:151–215.

WILLIS, E. O., AND Y. ONIKI. 1978. Birds and army ants. Annual Review of Ecology and Systematics 9:243–263.

WILLIS, E. O., AND Y. ONIKI. 2001. On a nest of the Planalto Woodcreeper, *Dendrocolaptes platyrostris*, with taxonomic and conservation notes. Wilson Bulletin 113:231–233.

WILLMANN, R., AND R. MEIER. 2000. A critique from the Hennigian species concept perspective. Pages 101–118 *in* Species Concepts and Phylogenetic Theory (Q. D. Wheeler and R. Meier, Eds.). Columbia University Press, New York.

WILSON, A., P. ARCESE, L. F. KELLER, C. L. PRUETT, K. WINKER, M. A. PATTEN, AND Y. CHAN. 2009. The contribution of island populations to in situ genetic conservation. Conservation Genetics 10:419–430.

WILSON, E. O. 1994. Naturalist. Island Press, Washington, D.C.

WILSON, E. O., AND W. L. BROWN, JR. 1953. The subspecies concept and its taxonomic application. Systematic Zoology 2:97–111.

WILSON, R. E., M. EATON, AND K. G. MCCRACKEN. 2008. Color divergence among Cinnamon Teal (*Anas cyanoptera*) subspecies from North America and South America. Ornitología Neotropical 19:307–314.

WILSON, R. E., AND K. G. MCCRACKEN. 2008. Specimen shrinkage in Cinnamon Teal. Wilson Journal of Ornithology 120:390–392.

WINKER, K. 1996. The crumbling infrastructure of biodiversity: The avian example. Conservation Biology 10:703–707.

WINKER, K. 1998. Suggestions for measuring external characters of birds. Ornitología Neotropical 9:23–30.

WINKER, K. 2009. Reuniting genotype and phenotype in biodiversity research. BioScience 59:657–665.

WINKER, K. This volume. Subspecies represent geographically partitioned variation, a gold mine of evolutionary biology, and a challenge for conservation. Pages 170–173 *in* Avian Subspecies (K. Winker and S. M. Haig, Eds.). Ornithological Monographs, no. 67.

WINKER, K., D. A. ROCQUE, T. M. BRAILE, AND C. L. PRUETT. 2007. Vainly beating the air: Species-concept debates need not impede progress in science or conservation. Pages 30–44 *in* Festschrift for Ned K. Johnson: Geographic Variation and Evolution in Birds (C. Cicero and J. V. Remsen, Jr., Eds.). Ornithological Monographs, no. 63.

WITTENBERGER, J. F. 1978. The evolution of mating systems in grouse. Condor 80:126–137.

WOOD, D. S. 1992. Color and size variation in eastern White-breasted Nuthatches. Wilson Bulletin 104:599–611.

WOODS, T., AND S. MOREY. 2008. Uncertainty and the Endangered Species Act. Indiana Law Journal 83:529–536.

WRIGHT, S. 1943. Isolation by distance. Genetics 28:114–138.

YOUNG, J. R. 1994. The influence of sexual selection on phenotypic and genetic divergence of Sage Grouse. Ph.D. dissertation, Purdue University, West Lafayette, Indiana.

YOUNG, J. R., C. E. BRAUN, S. J. OYLER-MCCANCE, J. W. HUPP, AND T. W. QUINN. 2000. A new species of sage-grouse (Phasianidae: *Centrocercus*) from southwestern Colorado. Wilson Bulletin 112:445–453.

Young, J. R., J. W. Hupp, J. W. Bradbury, and C. E. Braun. 1994. Phenotypic divergence of secondary sexual traits among Sage Grouse populations. Animal Behaviour 47:1353–1362.

Zimmer, K. J., and M. L. Isler. 2003. Family Thamnophilidae (typical antbirds). Pages 448–681 *in* Handbook of the Birds of the World, vol. 8: Broadbills to Tapaculos (J. del Hoyo, A. Elliott, and D. A. Christie, Eds.). Lynx Edicions, Barcelona, Spain.

Zink, R. M. 1988. Evolution of Brown Towhees: Allozymes, morphometrics and species limits. Condor 90:72–82.

Zink, R. M. 1989. The study of geographic variation. Auk 106:157–160.

Zink, R. M. 1994. The geography of mitochondrial DNA variation, population structure, hybridization, and species limits in the Fox Sparrow (*Passerella iliaca*). Evolution 48:96–111.

Zink, R. M. 1997. Phylogeographic studies of North American birds. Pages 301–324 *in* Avian Molecular Evolution and Systematics (D. P. Mindell, Ed.). Academic Press, London.

Zink, R. M. 2002. A new perspective on the evolutionary history of Darwin's finches. Auk 119:864–871.

Zink, R. M. 2004. The role of subspecies in obscuring avian biological diversity and misleading conservation policy. Proceedings of the Royal Society of London, Series B 271:561–564.

Zink, R. M. 2005. Natural selection on mitochondrial DNA in *Parus* and its relevance for phylogeographic studies. Proceedings of the Royal Society of London, Series B 272:71–78.

Zink, R. M. 2006. Rigor and species concepts. Auk 123:887–891.

Zink, R. M., and G. F. Barrowclough. 2008. Mitochondrial DNA under siege in avian phylogeography. Molecular Ecology 17:2107–2121.

Zink, R. M., G. F. Barrowclough, J. L. Atwood, and R. C. Blackwell-Rago. 2000. Genetics, taxonomy, and conservation of the threatened California Gnatcatcher. Conservation Biology 14:1394–1405.

Zink, R. M., and R. C. Blackwell-Rago. 2000. Species limits and recent population history in the Curve-billed Thrasher. Condor 102:881–886.

Zink, R. M., and D. L. Dittmann. 1992. Review of *The Known Birds of North and Middle America*, Part II, by A. R. Phillips. Wilson Bulletin 104:764–767.

Zink, R. M., and D. L. Dittmann. 1993a. Gene flow, refugia, and evolution of geographic variation in the Song Sparrow (*Melospiza melodia*). Evolution 47:717–729.

Zink, R. M., and D. L. Dittmann. 1993b. Population structure and gene flow in the Chipping Sparrow and a hypothesis for evolution in the genus *Spizella*. Wilson Bulletin 105:399–413.

Zink, R. M., S. V. Drovetski, S. Questiau, I. V. Fadeev, E. V. Nesterov, M. C. Westberg, and S. Rohwer. 2003. Recent evolutionary history of the Bluethroat (*Luscinia svecica*) across Eurasia. Molecular Ecology 12:3069–3075.

Zink, R. M., S. V. Drovetski, and S. Rohwer. 2002a. Phylogeographic patterns in the Great Spotted Woodpecker *Dendrocopos major* across Eurasia. Journal of Avian Biology 33:175–178.

Zink, R. M., A. E. Kessen, T. V. Line, and R. C. Blackwell-Rago. 2001. Comparative phylogeography of some aridland bird species. Condor 103:1–10.

Zink, R. M., and J. T. Klicka. 1990. Genetic variation in the Common Yellowthroat and some allies. Wilson Bulletin 102:514–520.

Zink, R. M., D. F. Lott, and D. W. Anderson. 1987. Genetic variation, population structure, and evolution of California Quail. Condor 89:395–405.

Zink, R. M., and M. C. McKitrick. 1995. The debate over species concepts and its implications for ornithology. Auk 112:701–719.

Zink, R. M., and J. V. Remsen, Jr. 1986. Evolutionary processes and patterns of geographic variation in birds. Pages 1–69 *in* Current Ornithology, vol. 4 (R. F. Johnston, Ed.). Plenum Press, New York.

Zink, R. M., J. D. Rising, S. Mockford, A. G. Horn, J. M. Wright, M. Leonard, and M. C. Westberg. 2005. Mitochondrial DNA variation, species limits, and rapid evolution of plumage coloration and size in the Savannah Sparrow. Condor 107:21–28.

Zink, R. M., S. Rohwer, S. Drovetski, R. C. Blackwell-Rago, and S. L. Farrell. 2002b. Holarctic phylogeography and species limits of Three-toed Woodpeckers. Condor 104:167–170.

Zink, R. M., W. L. Rootes, and D. L. Dittmann. 1991. Mitochondrial DNA variation, population structure, and evolution of the Common Grackle (*Quiscalus quiscula*). Condor 93:318–329.

ORNITHOLOGICAL MONOGRAPHS

(Continued from back cover)

No. 41. *Hindlimb Myology and Evolution of Old World Suboscine Passerine Birds (Acanthisittidae, Pittidae, Philepittidae, Eurylaimidae).* R. J. Raikow. viii + 81 pp. 1987. $15.00.

No. 42. *Speciation and Geographic Variation in Black-tailed Gnatcatchers.* J. L. Atwood. vii + 74 pp. 1988. $10.00.

No. 43. *A Distributional Survey of the Birds of the Mexican State of Oaxaca.* L. C. Binford. viii + 418 pp. 1989. $20.00.

No. 44. *Recent Advances in the Study of Neogene Fossil Birds: I. The Birds of the Late Miocene-Early Pliocene Big Sandy Formation, Mohave County, Arizona* (K. J. Bichart); *II. Fossil Birds of the San Diego Formation, Late Pliocene, Blancan, San Diego County, California* (R. M. Chandler). vi + 161 pp. 1990. $20.00.

Nos. 45 & 46. *Descriptions of Thirty-two New Species of Birds from the Hawaiian Islands: Part I. Non-Passeriformes* (S. L. Olson and H. F. James), 88 pp.; *Part II. Passeriformes* (H. F. James and S. L. Olson), 88 pp. 1991. Bound together (not available separately). $25.00 ($22.50).

No. 47. *Parent-Offspring Conflict and Its Resolution in the European Starling.* E. Litovich and H. W. Power. 71 pp. 1992. $15.00 ($12.00).

No. 48. *Studies in Neotropical Ornithology Honoring Ted Parker.* J. V. Remsen Jr., Ed. xiv + 918 pp. 1997. $49.95 ($39.95).

No. 49. *Avian Reproductive Tactics: Female and Male Perspectives.* P. G. Parker and N. T. Burley, Eds. v + 195 pp. 1998. $20.00 ($16.00).

No. 50. *Avian Community, Climate, and Sea-Level Changes in the Plio-Pleistocene of the Florida Peninsula.* S. D. Emslie. iii + 113 pp. 1998. $20.00 ($16.00).

No. 51. *A Descriptive and Phylogenetic Analysis of Plumulaceous Feather Characters in Charadriiformes.* C. J. Dove. iii + 163 pp. 2000. $19.95 ($15.96).

No. 52. *Ornithology of Sabah: History, Gazetteer, Annotated Checklist, and Bibliography.* F. H. Sheldon, R. G. Moyle, and J. Kennard. vi + 285 pp. 2001. $25.00 ($22.50).

No. 53. *Evolution of Flightlessness in Rails (Gruiformes: Rallidae): Phylogenetic, Ecomorphological, and Ontogenetic Perspectives.* B. C. Livezey. x + 654 pp. 2003. $10.00 ($9.00).

No. 54. *Population Dynamics of the California Spotted Owl (Strix occidentalis occidentalis): A Meta-Analysis.* A. B. Franklin, R. J. Gutiérrez, J. D. Nichols, M. E. Seamans, G. C. White, G. S. Zimmerman, J. E. Hines, T. E. Munton, W. S. LaHaye, J. A. Blakesley, G. N. Steger, B. R. Noon, D. W. H. Shaw, J. J. Keane, T. L. Mcdonald, and S. Britting. viii + 54 pp. 2004. $10.00 ($9.00).

No. 55. *Obligate Army-ant-following Birds: A Study of Ecology, Spatial Movement Patterns, and Behavior in Amazonian Peru.* S. K. Willson. x + 67 pp. 2004. $10.00 ($9.00).

No. 56. *Prehistoric Human Impacts on California Birds: Evidence from the Emeryville Shellmound Avifauna.* J. M. Broughton. xii + 90 pp. 2004. $10.00 ($9.00).

No. 57. *Management of Cowbirds and Their Hosts: Balancing Science, Ethics, and Mandates.* C. P. Ortega, J. F. Chace, and B. D. Peer, Eds. viii + 114 pp. 2005. $10.00 ($9.00).

No. 58. *Ernst Mayr at 100: Ornithologist and Naturalist.* W. J. Bock and M. R. Lein, Eds. viii + 109 pp. 2005. $20.00 ($18.00).

No. 59. *Modeling Approaches in Avian Conservation and the Role of Field Biologists.* S. R. Beissinger, J. R. Walters, D. G. Catanzaro, K. G. Smith, J. B. Dunning, Jr., S. M. Haig, B. R. Noon, and B. M. Stith. viii + 56 pp. 2006. $10.00 ($9.00).

No. 60. *Current Topics in Avian Disease Research: Understanding Endemic and Invasive Diseases.* R. K. Barraclough, Ed. viii + 111 pp. 2006. $10.00 ($9.00).

No. 61. *Patterns of Migratory Connectivity in Two Nearctic–Neotropical Songbirds: New Insights from Intrinsic Markers.* M. Boulet and D. R. Norris, Eds. viii + 88 pp. 2006. $10.00 ($9.00).

No. 62. *Storm-petrels of the Eastern Pacific Ocean: Species Assembly and Diversity along Marine Habitat Gradients.* L. B. Spear and D. G. Ainley. viii + 77 pp. 2007. $10.00 ($9.00).

No. 63. *Festschrift for Ned K. Johnson: Geographic Variation and Evolution in Birds.* C. Cicero and J. V. Remsen, Jr., Eds. viii + 114 pp. 2007. $10.00 ($9.00).

No. 64. *Conservation of Grassland Birds in North America: Understanding Ecological Processes in Different Regions.* R. A. Askins, F. Chavez-Ramirez, B. C. Dale, C. A. Haas, J. R. Herkert, F. L. Knopf, and P. D. Vickery. viii + 46 pp. 2007. $10.00 ($9.00).

Order from: Buteo Books, 3130 Laurel Road, Shipman, VA 22971, 1-800-722-2460; e-mail allen@buteobooks.com; or www.buteobooks.com. Prices in parentheses are for AOU members. For a complete list of Ornithological Monographs including both in-print and out-of-print books, please visit the American Ornithologists' union website at www.aou.org.

No. 65. *Reproduction and Immune Homeostasis in a Long-lived Seabird, the Nazca Booby (Sula granti).* V. Apanius, M. A. Westbrock, and D. J. Anderson. viii + 46 pp. 2008. $40.00 ($20.00).

No. 66. *Cladistics and the Origin of Birds: A Review and Two New Analyses.* Frances C. James and John A. Pourtless IV. viii + 78 pp. 2009. $40.00 ($20.00).

Order from: University of California Press, Journals and Digital Publishing, www.ucpressjournals.com, or e-mail customerservice@ucpressjournals.com.

LaVergne, TN USA
09 March 2011
219345LV00002B